Life on the edge

For Irena

Life on the edge

Peter Danckwerts GC MBE FRS
brave, shy, briiliant

by Peter Varey

First published by PFV Publications 2012

Copyright © Peter Varey 2012

Peter Varey has asserted his right under the Copyright, Designs and
Patents Act 1988 to be identified as the author of this work.

First published in Great Britain in 2012 by

PFV Publications

2 Petersfield

Cambridge CB1 1BB

A CIP catalogue record for this book
is available from the British Library

ISBN 978-0-9538440-2-9

Front cover: A pastel on paper portrait of
Lieutenant Peter Danckwerts GC RNVR by William Dring, 1941
© National Maritime Museum, Greenwich, UK,
image number PT0799
Back cover: Pembroke Street, Cambridge: Peter Danckwerts' base from 1959
Courtesy Cambridge University Department of
Chemical Engineering and Biotechnology

Text typeset in Sabon and Formata with Berthold Block headings
Book design by John Carrod, Cambridge, UK

Foreword

Peter Danckwerts was born in 1916 to a naval family in 'genteel poverty' and enjoyed a golden youth at Winchester College and Oxford. His ambition was to be a chemist but war caught up with him. He volunteered in July 1940 and, two weeks' training later, Temporary Sub-lieutenant Danckwerts found himself responsible for dealing with unexploded bombs and mines in the Thames estuary. With the Blitz came huge parachute mines and many of them failed to explode. The fuses were very unstable but Peter kept a cool head and won a George Cross. Sent to Gibraltar in 1942, he faced Italian frogmen riding human torpedoes and bearing limpet mines. Called to invade Sicily, he absent-mindedly walked into a minefield and was shipped home. Medal stripes, plaster and crutches enhanced his natural attractiveness to women; in fact, people generally liked him and he seemed to have contacts wherever he went. Demobbed at 29 after two years in Combined Operations, he journeyed to Massachusetts Institute of Technology to find out how Americans made chemicals in useful quantities. On his return, no British company would admit to needing a chemical engineer, so he joined the new Shell department in Cambridge and, during six years of 'academic indolence', applied science to industrial practice and made himself an international reputation. He briefly joined Christopher Hinton in atomic energy production but was soon lured back to academia—first in London and then in Cambridge as Shell professor. He finally met the right woman—a Junoesque aristocrat related to Lord Lucan—on a London dance floor. An imposing and laconic public presence hid a razor-sharp and wide-ranging intellect, a fine sense of irony and a gift with words. His time in charge in Cambridge was a golden age.

About the author

Peter Varey was born in Watford, brought up in Brighton and graduated in chemistry at Cambridge University. After a spell at ICI Manchester, he taught industrial processes – first in Kingston, Surrey, and then in Caracas, Venezuela. Back in the UK, he edited *The Chemical Engineer* magazine, published books and organized events for the Institution of Chemical Engineers. Now he is a freelance writer. Married to Polish jewellery designer Irena Maria Varey and now based in Cambridge, he lives half a mile from the house where Peter Danckwerts spent the last 20 years of his life.

Contents

Volunteering for special duties

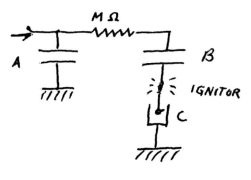

IT WAS THE MORNING of Wednesday 24 July 1940. Peter Danckwerts—23 years old, just under six foot tall, dark and handsome in a Teutonic kind of way—had arrived for an interview in the London recruiting office of the Royal Naval Volunteer Reserve (RNVR). He found himself sitting opposite a seasoned interrogator: Winchcombe van Grutten, known to his friends as 'Winch'. Van Grutten, a WW1 veteran and a peacetime assistant secretary at the Cambridge University Appointments Board, was skilled at assessing raw young graduates and fitting them into holes of the right shape. Peter was still quite raw; anxious to go to war, he had just resigned after a year in his first job as a 'research chemist' with Fullers' Earth Union. Van Grutten told Peter that the RNVR was currently recruiting officers for 'special duties'. Peter recalled:

Chemists as such were not required, but 'scientists' would do. He drew for me a simple circuit diagram and asked 'What would happen if capacitor A were charged?' Naturally, my Oxford chemistry course had not prepared me for this. He did his best for me but eventually said, 'Well, actually it is the basis of the whole thing. But I suppose you're keen, aren't you?' 'Dead keen', I replied. He then told me that the special duties were bomb disposal.

Winch van Grutten's circuit diagram

The following Sunday, 28 July, found Peter on a train heading north for Lincolnshire. His commission had been fast-tracked; he had, after all, the advantage of a strong naval background on both sides of his family. His father, Rear-admiral Victor Danckwerts, was director of plans at the Admiralty and although he was by then 'more or less unemployed as a result of his negative attitude towards some of the First Sea Lord

Churchill's more bizarre plans', he put in a good word for his son. A visit to the tailors Moss Bros then provided Peter with the uniform of a commissioned officer: temporary sub-lieutenant RNVR (special branch). The phrase 'special branch' may create a feeling of unease these days, but there was nothing sinister about it then: the category included such exotic specialists as meteorologists and interpreters. But Peter was in for something more nerve-racking than weather forecasting.

His destination on day one was the Joint Services bomb disposal school at RAF Manby, near Louth. There he got his first glimpse of what he would be up against:

I was surprised by the sheer size of the German 250kg SC bomb [SC = *Spreng Cylindrisch*, a thin-walled general purpose bomb] on display—I had expected something about the size of a 6-inch artillery shell, not a dustbin. The Germans had launched a raid [in October 1939 at Scapa Flow] on the British fleet (in the same spirit of over-optimism which led the RAF to bomb the German fleet at Kiel—naval damage in both cases zero) and some bombs had fallen ashore and failed to explode. These provided the exhibits for our classes.

Peter had always pictured the fuse as sticking out of the nose of the bomb, but he found that it was actually set in the side, in a tube which extended in some cases right across the body of the bomb. The fuse, with a head which looked rather like the end of a domestic light bulb, was held

---

BACKGROUNDER:

TITLES AND UNIFORM

'Temporary' or 'acting' was a very common appendage to WW2 service titles. Young officers in the lowest grade were on trial. Those who had earned promotion to higher levels were kept temporary for quite a time; for them there was also an element of trial, but as far as top brass was concerned 'temporary' also meant paying at the rate of the rank below. It diverted less money from fuel or weaponry. Still, the uniform itself gave its wearer a thrill. An RNVR officer sported wavy gold lines on his jacket cuffs (hence the everyday name for the RNVR: 'Wavy Navy') in contrast to the straight line and simple curl used by the professional navy. As a sub-lieutenant Peter wore a single wavy gold line. A green stripe below the gold one showed his attachment to the special branch. Its shade differed subtly from that of the electrical branch, and was a source of some pride.

The Royal Navy sub-lieutenant's cuff badge is a simple curl; the corresponding RNVR badge is a wavy one

in place by a locking ring. It was detonated electrically: the 'basis of the whole thing' that van Grutten had shown him at interview. Now he got to understand it:

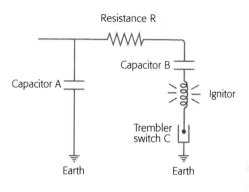

The bomb fuse circuit diagram

As the bomb left the aircraft, a high tension battery flicked a charge into condenser A [which temporarily stored the energy]. During the few seconds in which the charge then leaked through the high resistance R into the firing condenser B, the bomb had fallen safely clear of the aircraft. It was then ready to explode on impact, or after a slight pyrotechnic delay, by operation of the 'trembler switch' C.

The design of these German 'electrical condenser resistance' or ECR fuses (condenser is another word for capacitor) was brilliantly simple and quite robust. It required none of the high standards of machining involved in the production of mechanical fuses which the British RAF favoured. The German Luftwaffe used ECR fuses in all their bombs throughout WW2. The components of the electrical circuit were enclosed, with the fuse det-onator, within the electrical unit. The simplest impact fuse was a Type 15. Peter commented:

Like the end of a domestic light bulb: the head of a Type 15 bomb fuse
Courtesy Steve Venus aka The Fuzeman

It was extremely safe from the air-crew's point of view, an important con-sideration when some types of bomb detonated at the moment of release from the aircraft. Morale amongst aircrews depended very much on the appre-hension that whatever they carried or dropped was not going to kill them.

The explosion of a bomb, or a mine, is not a single event but a very rapid sequence of events. The main charge is a powerful explosive but not too unstable, otherwise it might blow up the factory which manufactures it or the lorry drivers, seamen and aviators who transport it. The main charge explosive in German WW2 mines was hexanite—a mixture of TNT (trinitrotoluene, 60%), HNDA (the polysyllabic mouthful hexanitrodiphenylamine, 24%) and aluminium powder (16%). Hexanite melted easily for casting into the mine casing; when lit by a match it burned harmlessly, a bit like a hyperactive sparkler. Once cast inside the mine casing, it had to be set off by the explosion of a rather-more-unstable explosive: the primer or booster, usually picric acid (trinitrophenol). This in turn was set off by the even-less-stable 'gaine', a mixture of wax and PETN (pentaerythritol tetranitrate). Finally, there was the initiating element, like a finger to topple a row of dominoes: a detonator. The electrical detonator (in bomb fuses) fired a flame into the gaine to set it off; a mechanical detonator (in mine fuses) achieved the same result by spring-driven impact. In explosive terms, the tiniest domino—the detonator—toppled dominoes of increasing size which, within a split-second, knocked over the largest—the main charge. To ensure safety during transport of bombs and mines it was usual practice to hold two of the dominoes in the row well apart; they would be brought together immediately before launch.

Amongst the sparse literature at the Manby school there was a 1930s catalogue produced by the German firm of Rheinmetall-Börsig. A patent taken out by Rheinmetall covered the principle of the ECR bomb fuse; the Germans were trying to sell almost identical fuses to the British air ministry just before the war. So there was very little reason for surprise when the first fuses were recovered from unexploded German bombs.

A bomb fuse with gaine attached (l)... and unscrewed (r)
Courtesy Steve Venus aka The Fuzeman

Defusing a Type 15 was relatively simple and Peter learned about several techniques. One was to reverse the charging step: press a two-pin 'Crabtree' discharger against the two plungers at the end of the fuse (which made it look like the end of a light bulb). Pressure connected a circuit to the bomb's two condensers (A and B) and, on earthing, the two condensers discharged harmlessly. A second technique was to unscrew the locking ring so that the fuse and other bits and pieces could be removed from the fuse pocket. A third was to physically remove the main charge: powders could be extracted through the filling hole and cast fillings dealt with by cutting a hole in the body of the bomb well away from the fuse pocket, injecting steam and letting an emulsified mixture of water and explosive drain away.

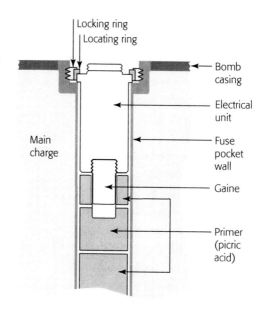

Components of a Type 15 fuse

The main matter of interest in the Manby course, Peter discovered, was in the possibility of the Germans starting to use delay-action fuses. At that

---

BACKGROUNDER:
IDENTIFYING FUSES
German manufacturer Rheinmetall was meticulous in placing manufacturing information, including an identification number, on the head of each fuse. This allowed them, as manufacturers, to trace any fault back to its origin. It also was rather helpful for British defusers because it meant that they could identify undamaged fuses at a glance. The identification number convention was very helpful; the all-important

code number appeared surrounded by a circle:
• electrical impact fuses for general use with a number ending in '5', like Type 15;
• fuses with a time-delay ended in '7', like Type 17;
• if there was an anti-handling or anti-disturbance element the fuse number ended in '0', as in Type 50 and ZUS 40 fuses.
• the ending '4' applied to special fuses, of which more in Chapter 2.

time no-one knew what
form such fuses might
take. In fact the Germans
already had such a fuse in
production. It was labelled
Type 17 and by the end of
August 1940, scarcely a
month after Peter's course,
they had begun using it.
Initiation was by the same
electrical mechanism as for
the Type 15 fuse, but when
the bomb hit the ground a
clockwork mechanism took
over and ran for a pre-set
interval before detonating
the bomb. Peter realised
that there would be subtle
philosophy involved here:

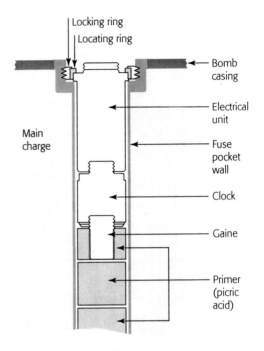

Type 17 delay-action fuse

A proportion of delay-
action bombs among the load, exploding at unpredictable intervals after the raid,
inhibits and demoralises rescuers and repair teams, and denies use of railway lines,
tunnels and runways as long as their presence is suspected. Also dud bombs,
inevitably a substantial proportion of those dropped, may become even more
effective weapons than those which go off, so long as their presence is known and
there is no way of identifying them as duds.

Unexploded bombs were not usually found just lying about. This was
a point, Peter occasionally complained, that somehow or other did not
emerge from his course at Manby. Falling at near-sonic velocity on to a
road or field, bombs could penetrate up to 9 metres (30 feet) into the
ground, and usually not in a straight line. The public perception of a bomb
disposal expert, Peter knew, was of someone who sauntered in a debonair
manner up to a bomb with a stethoscope and a spanner, and rendered it
safe without taking the cigarette from his mouth. It was a myth; in due
course Peter found he needed shovels, timber and pumps. Then, protected
by wooden shuttering, he could delve for long periods in stinking, water-
logged mud.

After a week with the RAF on bombs, Peter travelled to Portsmouth to

The main gateway to *HMS Vernon*
Courtesy Australian War Memorial, negative number PO5468.001

learn about mines. They were the Navy's responsibility. The Naval Torpedo School, *HMS Vernon*, sat on the quayside. Today little remains: a few of the original buildings and the wartime gateway. The Gunwharf Quays shopping centre has taken its place. The mines course began with the conventional horned and moored type—'contact' mines. They required a ship to bump into them before they performed. These were the mines with which Britain had blockaded German ports and sealed the English Channel so successfully during WW1. Moored to the sea floor, they floated a metre or two below the surface. To his surprise Peter found that the existence and position of minefields were supposed to be disclosed by international protocol. Furthermore the design of such mines was covered by an ancient international convention. If the mooring wire broke, or was cut by a minesweeper, tension on the mooring spindle was relaxed; the spindle was supposed to be spring-loaded, so that it moved up into the body of the mine and broke the firing circuit. The mine would float to the surface but did not pose any danger to passing ships. Peter was sceptical:

Apart from the improbability of such a safety device working in practice—due, for instance, to incrustation by barnacles—it emerged that the fiendish enemy had left his options open. German mines were fitted externally with a switch which could be turned to *Ein* or *Aus*. If set to *Ein* it remained armed even if it broke adrift. When called upon to deal with a German mine cast up on a beach, the first thing to do was to look at the switch and, if necessary and physically possible, use a screwdriver to change *Ein* to *Aus*.

Approximate positions of ships sunk by mines off the UK east coast from September 1939 to December 1940
Data courtesy Chatham Historic Dockyard

But contact mines were already rather old hat. During the mid-1930s, the Germans—conscious of the battering they took from British conventional mines in WW1—had developed an 'influence' mine. Contact with a ship was not necessary; the mine could sense a steel ship passing nearby, and this was enough to trigger detonation. Magnetic mines were designed to be placed on the seabed by ship or submarine; without a mooring cable, they were unsweepable by traditional methods. The Germans were delighted and somewhat surprised to find that the British had not developed countermeasures to magnetic mines between the wars, and began laying them in the autumn of 1939 in an attempt to close North Sea ports. Because magnetic mines sat on the seabed, they were ideally suited to the job of closing harbours and estuaries. In the first three months of the war magnetic mines sank British vessels totalling some 180,000 tons off East

Coast ports. It could not be allowed to go on. The British had little idea how these magnetic mines worked and made feverish efforts, with the encouragement of the First Sea Lord Winston Churchill, to find out so that ways could be devised to disable them.

First find your mine! Recovery of an intact magnetic mine from a ship or submarine engaged in laying them looked a very long shot. But fortunately the Luftwaffe, jealous of the German Navy's success with magnetic mines, had lobbied successfully to lay them from the air by parachute. This was good news for the British, because sooner or later magnetic mines were bound to be dropped somewhere where recovery and defusing was feasible. Sure enough, in late November 1939 a disorientated German aircraft dropped two in shallow water in the Thames estuary at Shoeburyness. Both failed to explode. They were recovered and, in a heroic action, John Ouvry and Charles Baldwin from *HMS Vernon* defused one of them. Ouvry and Baldwin's success allowed *HMS Vernon* to lay bare all the secrets of the magnetic mine. The Shoeburyness trophy attracted

BACKGROUNDER:
DEFUSING THE FIRST UNEXPLODED
MAGNETIC MINE

The night of 22 November 1939 was a lucky one for the British Navy. A German plane dropped two magnetic mines in shallow water on a military testing range off Shoeburyness. They failed to explode. A guard spotted 'two sailors kitbags suspended from parachutes' descending. He alerted *HMS Vernon* in Portsmouth and soon after dawn next morning two brave men—Lieutenant Commander John Ouvry and Chief Petty Officer Charles Baldwin—used a set of non-magnetic tools in an attempt to defuse one of the magnetic mines. They were in unknown territory. A back-up team of two— Roger Lewis and Archie Vearncombe— watched their every move. Following standard practice, Ouvry agreed a plan of action with Lewis who then retired

to a safe distance. The second mine was waiting nearby, in case Ouvry and Baldwin were unlucky. They were dealing with an unknown design and any action of many—removing a screw, snipping a wire—could have been fatal. In the event they successfully removed the main fuse and disarmed other bits and pieces; Lewis and Vearncombe helped in manipulating and stripping the mine before it was taken by lorry to Portsmouth for detailed examination. King George VI decorated the two officers, Ouvry and Lewis, with a Distinguished Service Order each. Their bravery did not qualify for a Victoria Cross because it was not 'in the face of the enemy'; the George Cross and Medal to reward bravery in these circumstances were only created in September 1940. Baldwin and Vearncombe each received a Distinguished Service Medal.

many distinguished visitors to Portsmouth, Churchill amongst them.

The Navy's newly-acquired understanding of the workings of the magnetic mine triggered an intense effort to devise counter-measures. A technique called 'degaussing' quickly emerged; the Germans measured the strength of magnetic fields in gauss, after a famous German investigator of magnetism, so the name for the British counter-measure seemed appropriate. The process of building a ship—all that bashing and crashing on metal plates—tends to line up magnetic dipoles in the metal and turns the ship into a huge floating magnet with its own magnetic field all around it. As the ship moves through the water, the field moves with it and adds to, or subtracts from, the Earth's magnetic field. The effect is tiny but a magnetic mine can sense it and responds by detonating. Degaussing 'cancelled' the effect the ship had on the Earth's magnetic field by applying an equal effect in the opposite direction. This made the ship invisible, or sufficiently close to invisible, to the mine.

First attempts to achieve degaussing were rather basic—wrapping a coil of very thick wire round the ship and running a heavy current continuously through it, in one direction for ships built in the northern hemisphere and in the other for those from the southern hemisphere. Simpler and more practical variations quickly emerged. A streamlined version called 'wiping' proved useful for smaller ships. The effect of wiping gradually wore off, so wiped ships had to be wiped again at regular intervals throughout WW2. But degaussing and wiping had the drawback that they only protected individual ships. An alternative—triggering magnetic mines safely—would help all ships, and the 'double L' sweeping technique was devised to do this. In one variation of it, two degaussed minesweepers travelled alongside each other a few hundred yards apart. Each vessel towed two cables, three inches in diameter—one quite short and the other up to 600 yards long. Each of the four cables had a copper electrode on the end. On pulsing a current of some 2,000 amps down the lines— first positive and then negative—seawater completed the circuit in a loop between the four electrodes. This set up a magnetic field sufficiently strong to trigger any magnetic mines on the seabed between the two cables. Electrodes dissolved like sugar and had to be wrapped in sacking; this kept the concentrated copper solution formed during one pulse near the electrode so that it would re-plate the electrode with copper on the reverse pulse.

During his few days in Portsmouth, Peter learned nothing about the secrets of magnetic mines. That may seem strange now, but at the time an unexploded magnetic mine was a rare and precious treasure; the team at *Vernon* did not want enthusiastic amateurs playing about with them, either blowing them (or themselves) up or dismantling them so comprehensively that they could never be reconstructed. The *Vernon* team was, after all, responsible for devising defusing methods and keeping abreast of new developments: so it reserved these mines for its own people. But the experts themselves were still learning.

On the morning of Tuesday 6 August while Peter was attending a class given by a craggy Chief Petty Officer (technical) there was a sharp explosion nearby. 'Ah', exclaimed the CPO, 'they're testing that double depth-charge thrower.' The reality was rather more sinister. *Vernon* personnel had been dismantling a newly-recovered and apparently fully-disarmed German magnetic mine. It was the ninth or tenth such mine to be recovered and sent to *Vernon*. In the process of dismantling all the bits and pieces, the magnetic unit at the rear of the mine was the last step. The *Vernon* team undid the bolts which secured it and started to ease the unit away from the body.

There was a booby-trap: the withdrawal of a stud attached to the rear door of the mine completed an electrical circuit which fired a charge. It was only a small one but had sufficient power to destroy the magnetic unit and damage things close at hand; three men were killed instantly and three more died later of their wounds.

Peter mused:

There must have been some technical reason why this small charge—probably a grim professional joke rather than a regular design feature—had not been positioned where it would have set off the main charge of 750kg of explosive. If it had, our course would have come to a premature conclusion. As it was, we learned a salutary lesson.

Reconstruction of the pieces revealed the reason for the booby-trap: the magnetic unit had been modified to make sweeping more difficult. A delay-action 'clicker' caused the mine to ignore the first five vessels which passed overhead. The sixth detonated the mine. In the interests of greater safety, the dismantling of mines by *Vernon* staff was moved inland to a disused quarry set in the South Downs and christened *HMS Mirtle*.

A few days later Peter left *HMS Vernon*. After training for a fortnight he was about to be thrown in at the deep end. Posted for operational duties to the Thames estuary and based in London, he would get there in good time for the start of the Blitz. Quite soon he got a close look at magnetic mines arriving by parachute and, unexpectedly, not into the water of the Thames estuary but on to dry land in London and its suburbs. They were the Navy's responsibility and required Peter's personal attention.

BACKGROUNDER:
BOOBY-TRAP BOMBS

Hard on the heels of the delay-action bomb fuse Type 17 came anti-handling devices (or booby-traps) to discourage defusers. The nastiest was the ZUS 40, a very simple device slipped in below the fuse. Under the detonator, the gaine fitted into a recess in the ZUS 40 and held back a striker with a spring pushing on it. If the main fuse were to be withdrawn more than a small way from the fuse pocket, the striker would slip past the gaine and hit a cap which detonated a second gaine and exploded the bomb. The introduction of the ZUS 40 in August 1940 led to a general instruction that— until other means became available—the fuses on

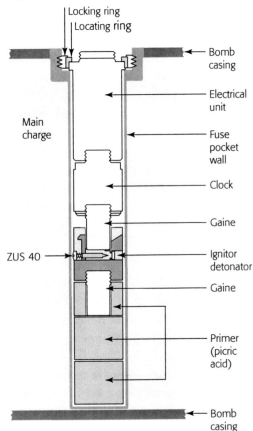

Type 17 fuse with a ZUS 40 fitted

parachute mines should be removed remotely, by attaching a string to the fuse and retiring a suitable, rather long, distance in case a ZUS 40 were fitted behind the fuse. In fact, this booby-trap was rarely found, if at all, on parachute mines. The Type 50 fuse was another kind of booby-trap. The Germans placed extra resistances within the electrical fuse of a bomb to delay

the charging of the condensers for several minutes, with the idea of allowing the bomb to hit its target and come to rest within it. Only then would the fuse be primed and await any disturbance. A very sensitive device used instead of a trembler switch meant that the disturbance needed was very slight indeed. The Type 50 fuse was first encountered in early September 1940.

Meeting a parachute mine 2

AS NAVAL BOMB DISPOSAL OFFICER for the Thames estuary, Peter found himself responsible for an area of water stretching from Teddington to the west as far as Southend to the east. He was assigned to the stone frigate *HMS President*, and his area coincided with the territory of the flag officer in charge, Admiral Martin Dunbar-Naismith, a retired submariner with a WW1 Victoria Cross won in the Dardanelles. Nasmith was one of many retired but highly experienced naval officers hauled back into service life to help organise the war effort. Their headquarters was the Port of London Authority building overlooking Trinity Square and the Tower of London. Peter found himself surprisingly well-equipped:

My predecessor had scared the living daylights out of [Nasmith] on the subject of delay-action bombs, and I inherited a motor-boat on the river, a CPO, or Chief Petty Officer (technical)—one of the salt of the earth, as they could take on any job from seamanship to electrical wiring—a three-ton truck with red mudguards and an obstinate old driver, recalled to the colours after many years as a London bus driver. I also had a crew of six gormless [dumb] teenagers dressed as matlows [sailors].

The precise meaning of 'bomb disposal' depended on where the unexploded bomb fell. Those landing in open country could be blown up where they lay. But those which fell close to sites vital to the war effort or to everyday life couldn't simply be detonated where they lay but had to be made safe and taken away. The work was done in pairs. The tradition was that an officer took the lead in defusing bombs and mines; his CPO was there to help, advise and generally do anything the officer asked him to do, but not to take initiatives.

Any bombs which fell within the area of the river Thames and its surrounding docks and failed to explode were Peter's responsibility. He was supposed to deal with mines too, but as these were usually placed in estuaries and approaches to ports rather than in the ports themselves, he didn't see anything of them—at least at first. One of his initial tasks was to impose some kind of authority on his team.

I was notably lacking in officer-like qualities. Nobody ever taught me about saluting—how and whom—or about not smoking on ships' boats, or drinking the Loyal Toast sitting down. Of course I had my CPO to keep the crew shaved and non-mutinous, but my driver was not about to take suggestions from me as definite orders. He used to drive our truck off to wherever he thought the crew might be able to get fish and chips.

The Blitz—Hitler's attempt to force Britain into submission by terrorising first London and then other major cities—began on 7 September 1940. Although Peter's area of responsibility was right in the thick of it, at first he found that his services were rarely called upon. The Army dealt with bombs falling outside the docks, and most of those coming down inside the docks fell into water or into piles of rubble, to which much of the dock buildings had been reduced by earlier bombs. It was unlikely that anyone would see or report unexploded bombs falling in these areas. Meanwhile Army bomb disposal units were being worked to death, in some cases literally so.

One major challenge with bombs was identification. When one dropped on land it left little more than an insignificant hole, frequently masked by fallen masonry. The question then was had it exploded or not, for a bomb going deep into soft ground could explode underground and leave little sign at the entry point. Careful examination would often reveal incontestable signs of explosion, Peter had discovered, which made it unnecessary to do any digging.

After a time one became very quick at making a diagnosis, but (I blush to say it) over-confidence and lack of time would occasionally lead one astray. This did not, fortunately, lead to any disasters but one such misapprehension recoiled on my head with unusual promptness and, I fear, did my reputation for infallibility no good.

The 'misapprehension' happened one morning after a raid, when Peter received a report saying 'Delay-action bomb, believed 1,000kg, in roadway near east gate of X dock'. He located the dock policeman who had made the report and they went together to inspect the hole. After taking a good look Peter decided that, judging by the cracking and subsidence of the road surface plus the blackening and erosion around the entry hole, the bomb had exploded underground. Turning to the luckless policeman he said:

You can write this one off right away. Well, I must be getting along—got a lot of other reports to look into, all bogus I expect.

The following afternoon a sewerman, walking along the sewer running beneath the subsided roadway of X dock, stumbled over a 250kg bomb.

The odd misdiagnosis like this did not seem to affect the admiration which Peter's crew showed him. Some of that admiration was due to his sense of humour in repartee when they were out on a job, but it was his readiness to tackle any situation that really impressed them. To investigate the unexploded bomb in this case he dressed up appropriately.

The London sewer-men were a dedicated band of brothers, like the miners. They kitted me out in their ancient working clothes, consisting of something like the bottom half of a diver's suit, a duffle coat and a slouch hat. Then they sent me down the manhole with fraternal wishes backed up by folk tales about rats as big as cats and flash floods. I actually found the bomb, lying in a gentle stream of sewage, and my crew was genuinely proud of me—not because of the explosive but because of the shit.

Outside the dock areas, the hard-working Army disposal teams had competition, usually short-lived, from some unexpected quarters. The Earl of Suffolk, who had excellent contacts in the Army, not to mention society in general, ran his own bomb disposal gang. Peter had heard stories about it:

They took orders from no-one. The gang included his chauffeur [Fred Hards]—whom he used to summon, so it was said, by firing his revolver into the ground—and his secretary [Eileen Morden] who used to stand on the edge of the excavation and take shorthand notes [which later served as advice to bomb disposal trainees]. The first I heard of Suffolk was in connection with a very expensive house in Park Lane. A bomb had gone through the house and penetrated a parquet floor at ground level. The Royal Engineers had done this excavation with respectful care, laying tarpaulins to protect the parquet and so on. When the bomb was exposed, Suffolk took over. He used a gadget designed to deactivate the fuse by firing a bullet through it, by remote control. Of course, the bomb went off and the house came down. Later on [in 1941] Suffolk was working on a bomb with his gang standing around the top of the excavation and up went the lot—Suffolk, secretary, chauffeur and all. *Not* good practice.

In spite of this somewhat dismissive account, Suffolk had some technical skill and commanded considerable esteem in high places. Prior to his bomb disposal activities, the Earl and his private secretary had played an exotic-sounding part in smuggling goods from Paris to London as the Nazis moved into France: the goods included some French scientists, $10million worth of industrial diamonds and a consignment of heavy water. Suffolk's approach to these activities had earned him the nickname 'Mad Jack'. By mid-1941 he had tackled 34 bombs successfully before the 35th turned out to be booby-trapped and undid him; he got a posthumous George Cross.

The morning-after-the-night-before experience of Londoners when the Blitz began in earnest was of streets, buildings and railways blocked off by signs announcing 'Danger—unexploded bomb'. Barriers were erected based on reports either by an eye-witness or by someone who had heard a whistle and no subsequent explosion, or perhaps had simply discovered a suspicious small hole in the ground. These phantom bombs caused great dislocation but in Peter's judgement most of them didn't exist at all, and simply interfered with everyday life.

Air raid wardens were likely to get very officious if people ignored the signs, and yet 'there was a war on' and it might have been expected that any civilian who accepted a small risk in order to get to work might merit congratulation. I bet that the Russians took a more brutal and realistic line about unexploded bombs.

Suddenly, in mid-September 1940, the scope of Peter's responsibilities changed dramatically. The Luftwaffe, warming to its task, reached for the largest ordnance it could find: the magnetic mine. Its aircraft were already laying magnetic mines at sea, and the aircrews had not found it easy. Accurate placement by day meant flying at low speed and altitude, making aircraft vulnerable; dropping mines accurately at night was a real challenge, as events at Shoeburyness in November 1939 had shown. But launch from air to sea had also required some modifications to the design of the mine. In the first place, it was not as sturdy as a bomb. While the total weight was around a tonne, the mine casing was of weak, non-magnetic

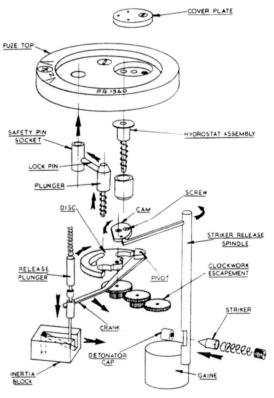

'A miracle of ingenuity': an exploded view of a Type 34A bomb fuse

aluminium. Dropped from the air, the casing would not withstand impact with water; a large parachute had to be added. This brought another challenge: the detonator and the delicate magnetic detector unit needed protection from the shock of a huge parachute opening behind the mine, not to mention the later impact of the mine on water. And the Germans were particularly concerned that if a magnetic mine were to be dropped by mistake on dry land rather than in water

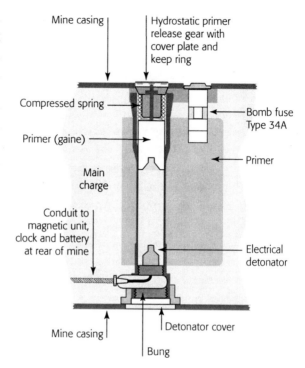

The parachute mine's fuse pockets

and failed to explode, British defusers might discover how it worked. The solution was to add a 'bomb fuse', able to detonate the main charge and destroy the evidence. The bomb fuse in use in September 1940 was Type 34A, a miracle of ingenuity.

The fuse pocket of the unmodified magnetic mine ran right across the body of the mine. An electrical detonator was secured at one end of the pocket, and it was activated by the mine's magnetic unit. At the other end of the pocket was the primer, held in position with a large spring compressed behind it. The separation of primer and detonator allowed safe transport of the mine before use. Once the mine was released into the sea and sank beneath the waves, a hydrostatic control within the primer release gear allowed the spring to force the primer along the pocket and into contact with the detonator. The mine was then armed.

When the Type 34A fuse was added, it fitted into a separate and shorter fuse pocket alongside the other and at the same end as the primer release gear. It had its own detonator, gaine and primer, and the mechanism controlling it was mechanical, not electrical—a timer mechanism not unlike the delay-action of an old camera shutter although with several functions.

Here is the sequence of events envisaged by the designers for launch of the magnetic mine from the air. Before launch, and as a safety precaution, a peg sticking out of a tear-off disc locked the timer mechanism of the bomb fuse. As the mine was released, a lanyard attached to the aircraft stripped the disc off, yanking the peg out of the timer mechanism and setting it in motion. (The same lanyard pulled a safety fork out of the primer release gear in the adjacent mine fuse pocket so that in due course the mine could be primed.) Another peg and lanyard arrangement controlled the opening of the heavy parachute. There was a powerful jerk as the canopy deployed; it all happened within seven seconds of launch, and during this period the timer kept the fuse impact mechanism locked. After that the timer stopped and remained so until the mine hit something, either water or land. Impact threw a weight forward within the fuse and that restarted the timer. It had 17 seconds to run.

If the mine fell into water more than 4.5 metres (15 feet) deep, water pressure would force a plunger into the fuse to stop the timer before 17 seconds were reached. From then on the bomb fuse was irrelevant; the magnetic firing mechanism controlled detonation. The magnetic firing unit at the rear of the mine was a relatively large device spanning the whole diameter of the casing. Located in a water-tight compartment, it was mounted on gimbals to keep it level and cushioned by a rubber disc to give the unit some protection. A separate pressure switch started a clock which armed this unit and, once on the seabed, the mine awaited a passing ship. But if the mine came down on land, or in water less than 15 feet deep, the hydrostatic plunger did not intervene. The Type 34A fuse, after buzzing away for 17 seconds, was supposed to detonate the main charge.

In September 1940 both the Luftwaffe and the German Navy were still using magnetic mines in the war at sea but because the British had learned how to counter them—especially with the double L sweeping technique— they preferred a more recent influence mine, the acoustic mine, instead. Acoustic mines responded to sound waves in water: a passing propeller, for instance. HMS Vernon soon countered the acoustic mine too, by adding to their degaussed double L minesweepers a device to make a loud banging noise under water. This triggered acoustic mines before the minesweeper got too close to them. All the same, by mid-1940 magnetic

The Port of London Authority building now coverted to flats

mines were becoming obsolescent. When the Luftwaffe pressed to use magnetic mines in the Blitz, the German Navy protested that they were unsuitable for use on land but, as was customary in disputes between the two, the Luftwaffe got its way. Magnetic mines packed several hundred kilograms of explosive in their aluminium casing, a very high ratio of explosive to total weight, and were capable of enormous blast damage. The Germans had already got a ringside view of just how enormous. In May 1940 British newsreels and press releases revealed in graphic detail what happened when a Heinkel HE-111 trying to lay magnetic mines at sea crashed in Clacton-on-Sea, Essex, and its cargo exploded.

On the night of 16 September 1940 Peter was relaxing in the basement of the Port of London Authority building, which he judged to be a very safe, comfortable place during the Blitz. Outside there was a lot of noise and the floor of the building heaved from time to time. Then his underground bunker received a report from an air raid precautions (ARP) controller in a North London borough. A large dark cylindrical object had fallen in a local shopping street; the Army bomb disposal officer who was responsible for the area had a feeling, the report continued, that the cylinder might be a magnetic mine. He had never seen one before, so he couldn't be sure.

He wanted to know whether, if he put it on a lorry and took it away, it was likely to explode. I rang him up and said I thought there was every chance of it, but by that time, it turned out, he had already removed the mine. The officer had had a look at the various fittings and decided one of them looked like a fuse. He had unscrewed it and it *was* a fuse. The next thing he started to unscrew appeared to have a spring behind it, forcing it outwards, so he hastily screwed it up again. Deciding to leave well

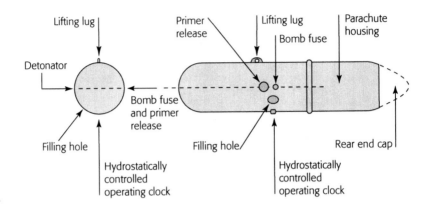

End and side views of a Type C parachute mine

alone, he loaded the mine on a lorry and dumped it at an isolated spot. He was probably the luckiest man in London that night, but his luck did not last the year, poor chap.

The next morning Peter drove to the site with Lieutenant Commander Dick Ryan, a 37-year-old Admiralty officer. Ryan was attached to *HMS Vernon*, had experience of defusing magnetic mines and was authorised to tackle them. They arranged to meet the lucky Army bomb disposal

BACKGROUNDER:

TYPES OF MAGNETIC MINE

The British used a simple code to identify German WW2 magnetic mines. Type A (or GA, where G = German), which the Germans called LMA (where LM = *LuftMine*) weighed 500kg and carried 305kg (670 pounds) of explosive. Its big brother, Type B, used in the same period, packed 700kg. By the time Peter met his first magnetic mine, Types A and B had been superseded by a large Type C and its smaller equivalent Type D. Type D differed from A and Type C from B by nothing more than a change in the polarity of the magnetic triggering mechanism. In contrast the later Type G magnetic mine, first recovered in May 1941, was a much-improved affair. The fusing system included a booby-trap to impede recovery from water, a bomb fuse and a photo-electric cell triggered on exposure to light. The first step in rendering it safe was to remove the rear cover in total darkness and mask the photo-electric cells. The pale blue casing of a Type G mine was shaped like a bomb. It had a pointed nose and was strong enough not to need a parachute. On land it could bury itself up to 10 metres deep; that took a lot of digging out, with the constant risk that vibration might set off the timer and with no certainty that the magnetic sensor had not been activated. Only fragments of its Bakelite tail near the entry point might give a clue.

officer and wanted to see the cylinder for themselves, as well as two
others which had dropped close by. These mines, they surmised, might
have been intended for the Thames. There were plenty of reservoirs
and streams in the Lea Valley area and it was possible that the German
pilot had mistaken one of them for the river. They found the mine where
their host had left it, lying on open ground in the middle of a large
common. It was painted dark green and measured more than two and a
half metres (a little over eight feet) long by two-thirds of a metre (about
two feet) in diameter. It was a Type C, weighing a tonne and containing
700kg (1,535lbs) of high explosive plus a lot of gadgets. A very large and
heavy parachute was spread out behind it. Some eight metres in diameter,
the canopy was attached by 28 artificial silk cords which were soon to
become much in demand amongst society ladies as a nightwear accessory;
alternatively, when unravelled each cord
contained enough yarn to knit a sweater.

Ryan began a lecture-demonstration.
Turning to the Army bomb disposal
officer, he confirmed that the pocket in the
side of the cylinder did indeed house the
fuse. It turned out that the officer still had
the fuse, stuffed into his trouser pocket.
Ryan suggested that he should take it out
right away, as it was quite capable of exploding. The cap at the inside end of the
fuse, the gaine, contained enough explosive
to remove a limb. Peter was paying full
attention:

Head of a Type 34A fuse with the
cover plate removed
Courtesy Steve Venus aka The Fuzeman

'You see,' [Ryan] continued, 'these fuses are sometimes very sensitive if they don't
go off at the right time. When the mine hits the ground the fuse buzzes for about 25
seconds [17, according to the manuals] and then sets the mine off. If the mine falls
on water, it is sinking during those 25 [17] seconds, and if the water is deep enough
the pressure pushes in a pin behind this rubber diaphragm here and stops the fuse
firing, so leaving the magnetic mechanism in charge of the mine.'

The identifying number on the top of the fuse was 34A. Ryan explained
that there were two important things about the fuse from the defuser's
point of view. The first was that if the mine hit the ground and failed to
explode, the fuse would be in a very dangerous condition. Even a very
slight disturbance might trigger it and set off the mine. There were several

reasons why the fuse was unreliable. The Germans kept their high standards of precision but, asked to design in a hurry in response to the Luftwaffe's insistence on launching from the air, there were unforeseen weaknesses. Even with a parachute to soften the landing, the G-forces to which the mine would be subjected on impact were considerable and might jolt one moving part out of line with another so that the timer stopped. The inertia switch also turned out to be fallible. On impact a little weight was supposed to slide smoothly along a groove in the inertia block until it dropped into a hole, releasing a plunger and setting off the 17-second pre-detonation sequence. In practice it sometimes jammed in

---

BACKGROUNDER:

HOW TO GAG A MAGNETIC MINE

The intrepid *HMS Vernon* defusers worked out several ingenious techniques to disable the Type 34A bomb fuse in an unexploded parachute mine before dismantling it. Lieutenant Commander Roger Lewis had dealt with the first magnetic mine carrying a Type 34A fuse in May 1940 (the Shoeburyness fuse was the earlier 24A). One way of disabling the fuse was to bring pressure to bear on it, imitating water pressure below 15 feet. The key item to use was a rubber motor horn—allegedly confiscated from London taxi-cabs—attached to a length of brass tubing including a tap. The brass tube was threaded at the other end so that it could be screwed on to matching thread on the fuse head: a weakness in the German design which the British found most convenient. The defuser first inflated the rubber horn with a bicycle pump and closed the tap. He then screwed the brass tubing onto the fuse and opened the tap. The pressure of air in the horn imitated water pressure and a slight click told

the defuser that the fuse was gagged. After this the fuse could be freed and removed by hand. The enthusiastic inventors of this method referred to themselves as 'hornblowers'. Another weakness in the early design of a Type 34A fuse was the 'thrupenny bit', the metal cover plate on the fuse head secured by a screw. The plate was the size and thickness of a silver three pence coin of the day, with three holes drilled in it. The holes allowed water to access the hydrostatic valve within, and one of these holes was in a direct line with the valve itself. So by inserting a thin rod—a dart—into the hole, defusers could push down the spring-loaded plunger and stop the timer. Peter didn't mention using either of these techniques in September 1940, although he may have used them later. German designers in due course eliminated the weaknesses with the Type 34B fuse: it frustrated the hornblowers with a small hole drilled in the wall of the fuse pocket to release air pressure, and foiled the dart-users by re-machining the access point for water with two right-angle turns in it. Measure and counter-measure.

its groove. Then, of course, there was grit, dust or sand; the tiniest fragment was enough to halt proceedings, at least temporarily. War Cabinet minutes suggest that some 300 parachute mines fell in the two-and-a-half weeks following 17 September 1940. Of these 60% failed to explode and 125 of the remainder were made safe.

Ryan's second point was that if the fuse were to jam before it had buzzed for the full 17 seconds and was then un-jammed by some kind of disturbance—the vibration caused by a bus going past in a road nearby or by the feet of an approaching mine disposal expert—it would only have the rest of the 17 seconds to run before it fired. The message to his listeners was: if they heard a faint buzzing when they were working on a mine to run like hell, and preferably to a pre-identified 'funk hole', in the hope that they might get away with it. The fuse might still have a full 17 seconds to run or perhaps only one or two; there was no way of knowing. But in this respect the mine defuser was relatively fortunate; bombs detonated immediately they were disturbed.

Ryan picked up the fuse the Army bomb disposal officer had given him. The gaine had been unscrewed but Ryan knew that the fuse itself would probably still be in a sensitive state. He wanted to show just how sensitive. He rolled it across the grass, and after ten seconds or so it fired with a sharp crack. 'Hmmm, just as well you happened to take that out before you moved the mine,' he told the Army bomb disposal man. 'Well, let's get the other oddments out.' On the opposite side of the mine body he showed them a small cover held in place by four screws. Removing them, he revealed a bung underneath. It was very tight, and Ryan needed a special tool to move it. Under that, and right at the bottom of the pocket, was the electric detonator for magnetic firing. Another special tool dealt with that. 'By gum, these Jerries screw things up solid,' he commentated. 'Here it comes, treat it gently, cut the wires and short them.' Peter was watching carefully as Ryan continued:

'Now, roll the mine over. Opposite the detonator [and in a large fuse pocket alongside that of the bomb fuse] is the primer release gear and the primer.' He unscrewed the keep-ring with a pin-spanner, and at the last turn there was a hollow whoof and a spring, three foot long, shot out of the pocket. 'No, it's all right, you can come back, that always happens. Well, that's got rid of the primer.'

Finally Ryan pointed out to his audience the operating clock set into the body of the mine casing at right angles to the fuse pockets. When the mine was placed at sea, the mechanism would be started by pressure of water when the mine reached sufficient depth and, after a delay of some 30 minutes to allow the mine to settle, it switched on the magnetic unit at the rear of the mine. After that the mine was ready to be triggered by any magnetic disturbance in the water above. It took a big pin-spanner and a lot of brute strength to get the retaining ring to turn. Peter watched as Ryan eventually managed it and extracted what Peter described as...

Magnetic mine operating clock:
'a beautiful gadget'

...a beautiful gadget, trailing coloured wires and with a case made of Perspex, so that one could look at the works.

Ryan continued:

'That's all we're going to touch. The magnetic unit is in the tail-end here, but Jerry has a habit of putting a small charge there, just big enough to demolish the unit—and the investigator—when the cover is taken off.'

That occasional addition of a booby-trap was something *HMS Vernon* had learned about by painful experience during Peter's training course. Ryan took Peter on to investigate the other two mines in the area; they found that *Vernon* personnel had travelled up by car from Portsmouth, dealt with them both and returned to the south coast. When Peter and Ryan returned to the Port of London Authority building in the late afternoon Peter had one of the Type 34A fuses with him. He took it to bits that evening, with the enthusiastic help of his CPO John Beadle and of an ex-torpedo officer who was working in the naval control office and had a passion for gadgets and explosives. By the time they had finished they reckoned they understood it pretty thoroughly. It was just as well: they were about to see a lot more of them.

Out on the loose 3

DURING THE NIGHT OF 17 SEPTEMBER, after Peter's daytime introduction to a magnetic mine delivered by parachute, German activity in the skies over London continued with vigour. The Port of London Authority building, where most of the naval team had by now taken up residence, was very sturdy but even so the team was acutely aware of the bombs dropping around them. The sound of gunfire rose to a crescendo as enemy aircraft passed overhead. Peter's telephone rang. It was the duty officer:

'Something in your line here.'

'Not a bomb?'

'Chap seems to think it's a mine. He's in a bit of a flap—
better have a word with him.'

Peter heard the harassed voice of an ARP (air raid precautions) controller somewhere in South London. Three large objects on parachutes had dropped in his area. Peter recorded what happened next:

The local Army bomb disposal person declined to deal with them (sensible chap) as he'd never seen one before and, anyway, mines belonged to the Navy. Meanwhile some astronomical number of people had been evacuated and could the Navy do something about it as soon as possible, please?

Peter explained how much he would like to help, but pointed out that he was not really supposed to deal with parachute mines. *HMS Vernon* had issued firm instructions about this. The logic didn't seem to impress the ARP controller.

'My God, who *does* deal with mines?'

'I'm afraid that the nearest people are down in Portsmouth. It would take them some time to get here.'

Now the British public traditionally had great trust in the Senior Service in those days, and Peter was acutely aware of it. He found it humiliating to consider breaking that trust. On the one hand he knew very well the reason why *HMS Vernon* restricted access to parachute mines. On the other hand it was clear to Peter that, for some reason, the enemy was now intentionally dropping mines inland, using them as bombs. He did not want to miss the chance of a genuine job—and one that, unlike his work with bombs dropped without parachutes, required no digging. So he got on the phone to *HMS Vernon*. After some time and effort he got them to understand that something ought to be done about all those evacuated people. They agreed that it would take their own team far too long to get there. And anyway, Peter explained, he knew all about these mines after his outing that same day with Dick Ryan. The *Vernon* duty officer, with great reluctance, finally agreed that Peter could tackle the South London parachute mines.

Immediately Peter woke the torpedo officer who had joined him in taking apart the mine fuse during the evening and his CPO John Beadle, another torpedo-man who would have never forgiven him if he had been left behind. Both were very good with gadgets and at getting things unscrewed. Peter knew that he was supposed to use non-magnetic tools…

…but none of our mines had been in water, and there was no reason to suppose that their magnetic units had been activated. All the same, most of the things were very hard to unscrew unless you had tools of the right shape. But we got a lot of screwdrivers, hammers and cold- chisels and, most important of all, we took a ball of string which is the essential thing for bomb disposal.

The three men set off in an Admiralty car through the Blitz. Ten minutes after their departure, *HMS Vernon* rang back to the Port of London Authority building, withdrawing the permission granted earlier. But it was too late: the latest recruits to parachute mine disposal were on their way.

It was a horrible night, full of craters in the roads, wrecked trams, blazing gas mains and anti-aircraft batteries banging away, largely

---

BACKGROUNDER:
BRIEF LIVES

Dick Ryan, who had shown Peter how to deal with a parachute mine the night before, was killed by one a week later. Earlier he had been one of two officers who had stripped the first unexploded Type C magnetic mine, found in a German aircraft which had crashed at Clacton. During the Blitz Ryan tackled six of these mines. One was in a canal; he had to work waist deep in mud and water, groping for the fuse below the surface. If the timer had resumed ticking, escape would have been impossible. Then, on 21 September 1940, he and his CPO, Reginald Ellingworth, tackled a mine hanging by its parachute in a Dagenham warehouse. It was a situation recognized as the most dangerous of all as there was nowhere to hide. This one exploded as the two men entered the building and killed them both.

ineffectually but quite boosting for public morale. Peter's driver turned out to be imperturbable in negotiating all these hazards, but even he didn't know quite where they were, and none of them knew where the three mines were.

We swept masterfully through the Blackwall tunnel, which was closed to traffic, and on into the echoing streets of South London. It wasn't until we saw a man in a dressing gown carrying a suitcase and running hard that we knew we were getting warm. We detained him firmly, and he pointed out where the mine was—practically in his back garden, he complained. We assured him that the experts had now arrived and that he would be able to return home in no time. Then we went to inspect our first mine.

A parachute mine in south London, 1940, showing on the mine casing (l to r) bomb fuse and primer release, filling hole and operating clock
Courtesy Australian War Memorial
PO5468.010

It lay in a passageway between two houses—black and sinister. This was a solemn moment for the three of them, but Peter got the impression that none of them was frightened. A touch of cold sweat on the palms of the hands, perhaps, but their predominant sensation was one of excitement at being about to 'pull something off'. The excitement was allied, Peter surmised...

...with that technical interest which so often leaves no room for fear in the somewhat one-track mind of the specialist.

They probably knew that a mine explodes faster than the human nervous system can react, so that should anything go wrong they would know nothing about it. And there was a philosophical aspect too: Peter found it difficult to convince himself that anything so solid and stable could really disintegrate under his hand or, indeed, that there was any imminent possibility of his own sudden disintegration. For the possibility of such a fate to be taken seriously, it needed a narrow escape or a ghastly example. They came later. For the moment:

The dangerous part of a job such as this did not usually last very long, and moderate danger in small doses taken at will is an exhilarating stimulant—not a depressant, like the interminable agony of being shelled. Nor was it to be compared with that hateful business of waiting for a flying bomb to stop. Essentially the situation is under one's own control, and there is no-one else to blame if anything goes wrong.

The first thing the three men noticed as a result of Dick Ryan's impromptu training session with Peter was that the all-important bomb fuse Type 34A was underneath. That meant rolling the whole mine over, before 'that ingenious but untrustworthy gadget' could be extracted. With infinite gentleness, Peter and John Beadle rolled the mine casing, while the torpedo-man kept his ear as near to the top of the fuse as he could, listening for any indication that it had begun to run.

We all tensed ourselves for an Olympic sprint should that tell-tale buzzing start. But all was well, and as soon as the fuse was accessible we got the pin-spanner on to it; unfortunately some strong-armed German armourer had made that keep-ring superhumanly tight, and it needed several sharp taps with a hammer to start it. By the time we had it off the fuse was still quiescent. It now lay free in its pocket, and ready to be merely pulled out.

Tempting as it seemed, they didn't pull the fuse out immediately. Peter knew that it could be in a very delicate condition and that the Germans sometimes used a ZUS 40 to booby-trap bomb fuses. They might be doing the same thing with mines. This was the moment when the ball of string came into its own. It had to be hitched delicately to the outside of the fuse head where a few turns of screw thread were handily exposed. (Later on defusers used a specially-produced collet to screw onto the thread; the collet had a ring on it for attaching the string.) Peter and his colleagues retreated, paying the string out as they went. They clambered over several garden walls until finally taking shelter behind the corner of a house. Peter yanked the string. Nothing happened. When he let go of the string, it sprang back as if it were elastic; somewhere the line was caught in a bush. It seemed hopeless to try and untangle it in the dark, so all three men scrambled back to the mine, cut the string, tied on a new piece and retreated again. This time a pull on

Training on a parachute mine 1: removing a Type 34A fuse. Note the string for yanking the fuse out from a distance

Training on a parachute mine 2:
removing the large spring from
behind the primer

the string felt right, and they returned to find the fuse lying on the ground along side the mine. The 'twitchy' part was over.

The three of them sat down on the mine and had a cigarette. It was then that Peter noticed that the mine was covered in a thick layer of greasy soot. It must have been camouflage to make the mine, slung under the aircraft before launch, less obvious under the scrutiny of anti-aircraft searchlights. Black grease was all over their faces, hands and trousers.

I was about to comment on this when there was a sharp 'crack' from beneath our feet, and we jumped up as one. But it was only the fuse, which I had thrown on the ground, firing harmlessly. It was at least a convincing demonstration that caution was not out of place in this job, and we congratulated ourselves on the delicacy of our rolling operation. All that remained was to remove the other oddments which, if the form was according to the book, should be perfectly safe. It was now that I realised the value of having in the party two torpedo-men of ripe vintage, from whom the unscrewing of well-nigh immovable fittings in inaccessible positions and semi-darkness called forth a determination magnificent to watch.

The final task to be completed before moving on to the next job was to unshackle the huge parachute at the back end of the mine. This was a potential souvenir, however unwieldy, and Peter was determined to have it. It took three of them to carry and almost filled up the back of their car.

I have only vague recollections of the rest of that night. I had been fired, I must admit, by the competitive spirit, for I knew that *Vernon* would be hard on our heels come day-light. I wanted to show them that they weren't the only people who could deal with mines, and also to have a run for our money before we were stopped. Both my companions wanted parachutes too now that I had got one, so that we were all in a state of exasperation mixed with nervous tension as we groped our slow and frustrated way about the southern counties.

In this state Peter retained only fleeting memories. He remembered calling at a police station and finished half a bottle of flat beer while trying to persuade an unwilling sergeant to come and show them where a mine was. He recalled, perhaps predictably, a debonair and feminine warden who promised blithely to lead them to the exact spot and confessed, half a fruitless hour later, that she hadn't actually seen the mine herself. Eventually they found a mine balanced weirdly and precariously on its nose in an open field, put their tools down in the grass nearby and couldn't find them again. An age-long fingertip search of a vast sweep of recreation-ground ensued. Once the cursing was over they broke open an emergency toolkit in a nearby bus depot. Peter was challenged at one stage as a possible enemy parachutist and blessed elsewhere as the saviour of evacuated old ladies. Finally, as dawn broke, he recalled a half-hour search in a housing estate which yielded nothing but a story of 'a soldier who picked the mine up and carried it away'. The sun was now up and as they dealt with a mine in a field beside a Kentish gas-works, they found they had to fight back an interested audience.

We relied on hammer and cold-chisel to start the keep-rings. We were sustained by ARP wardens, police and the general populace who had been evacuated from their homes and were anxious to get back. They occasionally took up collections, gave us boxes of chocolates or bottles of whisky. There were cries of 'good old Navy'. I am sure the women would have offered themselves had circumstances been propitious—at least there was no shortage of *them*. Finally, how gratefully I remember fried eggs and bacon at a police canteen. We were many miles, and twelve hours, from home, sooty, raw-nerved and exhausted, but we had left three mines as safe as dustbins. We had our three parachutes...and we had not blown ourselves up.

When they got back to base there was a good deal of explaining to do. The torpedo-man, who had played truant, was hauled back to his office. The Admiralty was concerned about whether Peter's team had really made their mines safe, and wanted to know where they had got their non-magnetic tools from. They learned that a *Vernon* party of four had eventually arrived on a second trip from Portsmouth, had split into two pairs and scoured the home counties, but the mines they found had apparently been rendered safe already—by team Danckwerts. This put a few distinguished noses out of joint. One of the *Vernon* four may have recognised the Danckwerts' style: Lieutenant Geoff Hodges had taught physical education at Peter's old school—Winchester College—during Peter's time there in the early 1930s. The leader of the *Vernon* team muttered, as they moved from one defused mine to another, 'I think I'd better catch up with this chap and warn him in case he kills himself in his hurry.' On the other hand Douglas Ritchie, the outstanding wartime chief executive of the London port emergency committee who was knighted at the end of the war for his service, told his friends what a 'wonderful fellow' young Danckwerts was for his dynamism in tackling parachute mines.

After lunch Peter was summoned to the Admiralty and observed a considerable uproar in progress in the torpedoes and mining department. It seemed that his three mines were a small fraction of the total, disclosed by the light of day; Peter estimated at the time that one in five parachute mines failed to explode. And each was surrounded by evacuated houses, barricaded streets and great disruption of everyday life. The Navy's view was that every mine should be looked at, to search for new refinements in the mechanisms. But it didn't have enough specialists, and requested the RNVR training school *HMS King Alfred* on the sea-front in Hove to invite newly-fledged officers to volunteer. So this small and rather odd branch of the Senior Service was eventually staffed entirely by RNVR officers. Meanwhile Peter and his team still had to face the music:

I was relieved to be kindly received, but it was pointed out that our methods had been unorthodox and undesirable; working at night was likely to cause more explosions than it prevented and, as I should still be needed, my position should be regularised. I received a proper set of tools and the Chief and I were told to regard ourselves as attached to the [*Vernon*] department.

Now that Peter and CPO Beadle's freelance period was over, the job came down to earth, but there was still what Peter described as 'a good deal of amusement and excitement' about it. They became part of *Vernon's* London branch. Every morning there would be a fresh crop of mines, often in improbable and inaccessible spots, and Peter, CPO Beadle and their team would be assigned to two or three:

We would dash off, crashing the red lights for the hell of it (there was an indescribable satisfaction in this, when one had a policeman-convincing excuse) and run our quarry

Peter's three-ton truck with driver and red mudguards for crashing red lights

Happy hunters: Peter's team return to base with souvenirs

down in some hitherto unguessed-at corner of London. Having done the job we would retire with a lordly 'You can return to your homes now, good people' and report to headquarters that they could cross another one off their lists. Nothing very exciting, for some reason, happened to us; none of our mines ever buzzed or blew up (although we had one false alarm which gave me as unpleasant a thirty seconds as I ever hope to experience).

The 'false alarm' may have been an incident which was subsequently officially reported, although it doesn't seem to fit exactly with Peter's comment that 'none of our mines ever buzzed'. According to the report, Peter and his CPO found two mines hanging from their parachutes with their noses on a warehouse floor. They approached, but their footsteps started up the clock in one of them. They retreated, although in a ware-house situation there could be no sanctuary. Mercifully the clock stopped. They approached again and dealt with the mine that had stopped buzzing, knowing all the time that its clock was highly sensitive and could only have a few seconds left to run. Then they defused the second mine. Peter wrote modestly, citing some situations his colleagues had faced:

[Neither] did we encounter then, as others did, mines inside gasometers, welded to live rails on the Tube or hanging from the roof of the Palladium (with free tickets *ad lib* in prospect). On one occasion, after a job close to a famous Anglo-American firm, our working day ended decisively at 3pm with a whisky party in the managing director's office.

There were plenty of stories of his own. One bomb, in Peter's view 'the queerest I had anything to do with', was in a stick of four dropped across the Thames near Hungerford railway bridge and Charing Cross. Someone claimed to have seen the bombs fall and observed that one of them, in the middle of the river, had not exploded. Peter interviewed the steward of the Seven Seas club which occupied a vessel tied up on the Embankment nearby. The steward claimed that at least one bomb had gone off; the shock had left the ship's parrot hanging upside-down from its perch and swearing horribly. The Northern line of the Tube system ran under the river at this point; was there a delay-action bomb down there somewhere? Peter and a couple of Tube engineers went down into the tunnel and, avoiding the live rail just in case, stumbled along to a point about mid-stream and looked for evidence of an explosion. Some concrete had shed from joints in the steel sections of the tunnel but it wasn't much to go on. So they arranged for a lighter to be moored above the suspect point and a naval diver went down to grope about. By some miracle he located the tail fins of a German bomb. A conversation Peter overheard in the course of the diver's efforts appealed to his sense of humour:

Diver's mate in lighter, speaking into microphone connected to helmet of diver, then at the bottom of the Thames: 'Are you there, diver?'

They did not know at the time what was the longest delay possible for a German delay-action bomb, but eventually Charing Cross station was re-opened and commuters poured back and forth as usual on the Northern line. As Peter summed it up: 'The bomb may still be there'.

On another occasion Peter and his team were working in a timber-yard in Bermondsey when a very old man sauntered in—this in the centre of what should have been a rigidly evacuated area. Peter asked him what he was doing there. Much to CPO John Beadle's scorn, he claimed that he had been a torpedo-man in the Navy, having started his career in sail as a boy and retired long before WW1. Consequently he reckoned he knew all about

Digging continues, but technology moves on: Peter's fuse-extractor on a 250kg SC bomb

explosive devices, and had come along to see if he could lend a hand. Peter thanked him, pointing out that they had all but finished. But the ancient continued to ramble on until Peter was electrified to hear him say, 'Ah, I done one of them things me-self, not long since'.

As we slowly got the story out of him, I remembered that, weeks before, we had found a 50kg bomb on the foreshore of the Thames, and that the fuse had been missing. Someone—some poacher—had taken it out and walked off with it. Enquiries had failed to find a trace of it.

Now the old man produced the fuse from his trouser pocket. He had been carrying it ever since, complete with gaine. He was so cut up at Peter's reaction, and so much attached to his fuse, that Peter let him keep it—after removing the gaine.

Only once more did I do a job at night (and heard all about it the next day). I weak-mindedly answered an agonised call from Islington and went to work by the light of spluttering incendiaries. Having with the usual caution laid out a long string across a street and round a corner, I was astonished to feel first a twitch and then a spasmodic and forceful pull on the line. After a cautious pause, I went round the corner and saw an air-raid warden, all unawares, striding down the street with the bight of my string round his boots.

Any team which survived a long campaign defusing parachute mines had had all the luck it could expect; certainly team Danckwerts enjoyed a lot of it. Survival beyond ten weeks was considered remarkable, and Peter would have been well aware of it. 1940 was the heroic age of bomb and parachute mine disposal; urgency and a lack of knowledge led to fantastic risks being taken and to many deaths. Occasionally there was light relief. One day Peter went with the 'Sage of Science', Desmond Bernal, to witness a civil defence trial. It had been designed, as Peter cynically described it, to show that the air-raid shelters built to government specifications were death traps.

The explosive charge used, for reasons of economy or verisimilitude, was an un-exploded German bomb. It was detonated and the shelter duly collapsed. What impressed me was that there was a perceptible interval between the explosion and the time at which Bernal ducked down into our shelter trench. He explained that he knew the velocity of the bomb fragments and could calculate their time of arrival in his head.

The brave men who dealt with unexploded devices during the Blitz were, quite reasonably, frequent candidates for awards for their bravery. The *London Gazette* is the traditional place for all announcements of promotions and medals, and on 20 December 1940 the paper reported a George Cross for Peter and a George Medal for his CPO John Beadle. The George Cross had been established on 24 September 1940 and Peter's was the ninth awarded. The official definition refers to 'acts of the greatest heroism or of the most conspicuous courage in circumstances of extreme danger'. Although it was primarily a civilian award, it could be given to military personnel for gallant conduct which was not in the face of the enemy and so not eligible for a Victoria Cross.

The investiture by King George VI took place much later, on 27 July 1941. The citation read: 'Sub-lieutenant Danckwerts had been under six weeks in the service. He set out without orders and with incomplete equipment, to deal with mines endangering his district. He had never before touched a mine, except under instruction, but he worked almost without rest for 48 hours and dealt successfully with 16 enemy mines. [Peter's own account of that first outing describes dealing with only three, and some condensation of the text may have occurred under the pressures of the time. *New Scientist* recorded in 1960 that his award was for 'one especially hazardous week's work']... He has worked through air raids without thought of self, and shown fine courage and exemplary devotion.' The Navy keeps the memory of those days alive: as recently as December 2009 the Ministry of Defence newspaper *Navy News* ran a précis of his achievements in the Blitz of 1940 as number 68 of its regular column Heroes of the Royal Navy.

The Blitz continued until May 1941, when the Germans became distracted by Hitler's decision to attack the Soviet Union. Peter was in the capital for much of this time; he found himself near St Paul's cathedral shortly after the major raid of 29 December 1940 which left the dome standing but most of its surroundings reduced to rubble. Overcome by emotion, he sat down on the cathedral steps and wept. As the Luftwaffe turned its attention to other British cities, Peter was posted to the provinces. In October 1941, newly promoted to temporary lieutenant, he was posted to Tyneside where he spent a hard winter through New Year 1942. Briefly he saw service in Londonderry in Northern Ireland. Throughout this time the Germans were modifying the triggering devices of their mines and bombs to make them more difficult to deal with if they failed to explode; in turn the British naval disposal teams, always at risk, recovered examples and sent them back to *HMS Vernon* where safe defusing methods could be worked out and procedures circulated to all those involved. By then most departments of *HMS Vernon* had been moved out of Portsmouth, which was suffering heavy air raids, some to the relative calm of Roedean Girls' School along the south coast east of Brighton. Officers recorded their delight at finding bells above their beds labelled 'Ring for a mistress'.

Once the crisis of the Blitz was over, top brass decided that making distinctions between mine disposal and bomb disposal was counter-productive. There was a realisation that it would be more efficient if all officers could disarm explosive devices of any class. So, in March 1942, Peter was sent back to a training school—this time *HMS Volcano* in Cumberland—for a month to get up to date. *HMS Volcano* occupied Holmrook Hall at Ravenglass on the Cumbrian coast just south of Sellafield, and had only specialised in bomb disposal training since the beginning of February that year. The Admiralty had let it be known as a cover story which would not alarm the locals and, much to the amusement of participants, that the house was being used as a home for shipwrecked and distressed sailors. This rambling pile, once the property of a relative of Lewis Carroll, was demolished soon after the war ended. In WW2, though, it was a key Allied establishment; following training there, specialists like Peter assumed the title of bomb and mine disposal officers.

Equipped with this new title, refreshed information and skills, Peter was posted in April 1942 to a new location with a quite different challenge. To prepare Peter's new superiors, his ultimate boss Llewellyn Llewellyn, the director of the naval unexploded bomb department (or DUBD) throughout WW2, wrote a report. Llewellyn, known affectionately to his men as 'Lulu', was an ex-chief inspector of naval ordnance brought back from retirement with the rank of captain. The pressure of more urgent matters meant that Lulu's report on Peter's conduct didn't get filed until August 1942. It read: 'A highly intelligent officer, zealous and inventive. Most courageous and awarded George Cross for outstanding bravery. Has the temperament of a scientist rather than that of an officer and does not mix well with other officers of a lower standard of education. Shows great initiative and can be relied on to act on his own.' Lulu had Peter down to specialise in any scientific capacity and, while marking 'intellectual ability' at nine out of ten, he awarded a generous eight out of ten for 'administrative ability'. Notoriously unattracted by paperwork, Peter must have been really interested in what he was doing.

Peter's George Cross had brought with it an extra duty: that of sitting

The Dring portrait
© National Maritime Museum, Greenwich, UK
Image number PT0799

for his portrait. The government had commissioned several official war artists whose job it was to capture the likeness of the many servicemen and officers who had distinguished themselves through bravery. One of the artists was William Dring, a product of the Slade School of Art in the 1920s but best remembered for this wartime work. For his portrait of Peter, Dring used pastels on paper, a combination he often employed to make his subjects look unposed, almost dreamy. By the time of the sitting Peter had been promoted to temporary lieutenant and sported two wavy stripes on his cuffs, with the exclusive green 'special branch' stripe between them. After the war the portrait was buried deep in the vaults of the National Maritime Museum in Greenwich. Now photographed and viewable on-line, it can be bought as a print. It reveals an almost shockingly youthful Peter standing before a map of his area of responsibility, apparently unscarred by his new life as national hero.

But the scars were there. Peter rarely spoke about his defusing activities; he was temperamentally unwilling to blow his own trumpet. When he did mention this period of his life, it was to emphasise that—as he saw it—he was exposed to danger for very short periods and nothing like to the same extent as those on the front line. His comments on wartime service echoed those of his contemporary at Balliol College Oxford,

Denis Healey: 'Long periods of boredom broken by short bursts of excitement'. But Peter's bursts were, to the average citizen, very dangerous indeed. Peter Pritchard, a medical student in London in 1939-40 and friend of Peter's younger brother Dick, noticed the effect. From mid-1939 and through the Blitz the two Peters met socially. Pritchard was impressed by Peter's insight into explosives; working from no more than the knowledge he had picked up at college and his boyhood experience of home-made bombs, he could outline on the back of an envelope how a nuclear bomb would work and be detonated. Once Peter got involved in bomb disposal he didn't say much about his new responsibilities, but Pritchard could see the change: he had become tense. In WW2 pink gin was cheap, accessible and a traditional way for naval officers to relax. Both men drank it; Peter Danckwerts never really threw off the habit.

BACKGROUNDER:

THE PRITCHARD CONNECTION

In the summer of 1939 Peter's brother Dick was studying at St Thomas's Hospital on the south bank of the Thames to be a doctor. On the same course was Peter Pritchard; the two had read medicine together at Gonville and Caius College, Cambridge. As his brother had a job not far away in Redhill, Dick introduced the two Peters. They got on well and subsequently met, says Pritchard, for an occasional 'night on the tiles' before and during the Blitz. Shortly afterwards the Danckwerts and Pritchards forged another link. One night when Dick was short of a dancing partner, Pritchard suggested that his sister Ann, a nurse, could fill in. The couple hit it off and before long made arrangements to get married. It was a secret affair at Caxton Hall, because Dick was a penniless student and knew that his father Victor would be furious. Soon Ann got pregnant and the two of them sheepishly turned up at the Danckwerts' family home in Emsworth, Hampshire, to confess. Dick's mother Joyce spent the whole night trying to persuade her husband to continue supporting them, but he wouldn't. So, without telling anyone, Joyce's sister Audrey Morse helped them out. Eventually Dick and Ann took £10 fares to Australia and brought up four children there.

Limpet mines and human torpedoes 4

ETER'S NEW POSTING in April 1942 was to Gibraltar. The journey by Sunderland flying boat took many tedious hours and delivered him to a new stone frigate: *HMS Cormorant*, the floating hulk of an old warship tied up in Gibraltar harbour. *Cormorant* was very short on accommodation, so a brick block on the quay nearby served as an extension, its dormitories provided with bars for the slinging of hammocks.

Peter's main task in Gibraltar, elaborating on his usual bomb safety role, was to counter the threat of Italian activities below the surface of the bay.

Photomontage of an Italian *maiale*;
the propeller is protected to avoid snagging in defensive netting
Courtesy *After the Battle*

The Italians had a flair for underwater technology and warfare; by the time Peter arrived in Gibraltar Italian frogmen had already made several attempts to attack Allied shipping using midget submarines. Each of these cylindrical craft, nicknamed 'human torpedo', was driven by a battery-driven propeller and equipped with flotation equipment. A crew of two frogmen sat astride the cylinder; their weapon was a delay-action mine packed into the torpedo's detachable nose cone. After several false starts aimed at Gibraltar Bay, one Italian attack—on 20 September 1941—had been successful and had shocked the Gibraltar establishment. Six frogmen with three human torpedoes arrived in the Bay in a mother submarine, the *Scirè*. Two of the torpedoes attacked ships at anchor in the bay, sinking the storage hulk *Fiona Shell* and damaging the motor vessel *Durham*. The third managed to get inside the harbour and immobilized the tanker

Denbydale. All three human torpedo crews escaped to the Spanish shore of the bay, sank and blew up their craft and made their way back to Italy. Intelligence reports from Spain recorded men dressed in diving suits in the area the day before the attacks, and two Germans in a smart touring car visiting briefly to observe. That first successful raid had put the wind up the Gibraltar authorities; another, in Alexandria harbour three months later, crippled two British battleships and made an even bigger impact. It seriously disturbed the balance of naval power in the Mediterranean. There were crisis meetings in London. The cumulative concern was probably behind Peter's posting.

His new surroundings were a cultural shock. By April 1942 Gibraltar (or 'The Rock' from its Arab name *Jabal-al-Tariq* where Jabal = Rock) was an isolated outpost of the Allied war effort. The German-Italian Axis dominated most of the Mediterranean. Malta was under heavy bombardment and in a parlous state; convoys passed through Gibraltar on their way to try and provide relief. Under severe attack from submarines throughout the Mediterranean and from the air as they neared Malta, many of the vessels failed to get through. It seemed only a matter of time before Malta succumbed. Conditions there, Peter heard tell, made Gibraltar feel like a holiday resort. All the same he noted soon after his arrival that those stationed in Gibraltar viewed the threat of Germans arriving at any moment from Africa by air, by sea or by land as 'quite plausible'.

Security was a serious matter on the Rock. When the alarms sounded, anything or anyone important took refuge in a labyrinth of tunnels honeycombing the Rock. The tunnels were being continuously extended, the tunnelers dumping debris from their efforts in the shallow waters of Gibraltar Bay, gradually extending the length of the single east-west runway. Peter had arrived in Gibraltar just as its governor, Lord Gort, met an Air Council deadline to extend the airstrip to 1,150 yards (1,050 metres). This was enough for Wellington bombers to fly in and out on their way to Egypt, so avoiding Malta where bombing by Axis forces was so fierce. But if Gibraltar were to serve as a staging post for the invasion of North Africa and then southern Europe, the length of runway required was 1,550 yards (1,400 metres). Work continued frenetically to achieve that target.

Gibraltar dockyard was a major centre of Allied shipping operations and, as bomb safety officer, Peter was responsible for dealing with any unexploded devices recovered in the course of defending the dockyard and the bay. In 1945, aged 28, he looked back:

Gib was no bad place for a run-ashore, compared with the average British dock-yard town in those days. Women, it is true, were as scarce as snakes in Ireland, and there was a strict eleven o'clock curfew. On the other hand there were lights and drink and food. In England whisky had just gone up to twenty-five shillings and nine pence a bottle, cigarettes to two shillings for twenty, and both were almost unobtainable; food was strictly rationed. In Gibraltar, whisky was ten shillings a bottle and cigarettes (as far as I remember) sixpence for twenty, both *ad lib*; there was no perceptible short-age or rationing of most sorts of food, except a few items such as eggs (eight pence each) and potatoes; sweets and clothes were both unrationed, and there was an unconditional petrol allowance.

These benefits were some compensation for other, less attractive aspects of life on the Rock. The total garrison of some 15,000 could be swollen by a further 4,000 when the fleet—battleships *Rodney* and *Nelson* plus their protective swarm of destroyers—was in port. With a tiny amount of habitable ground available, overcrowding was severe. Privacy was a luxury, solitude unknown. There was no leave and, with so many stuck in the same place with the same companions, claustrophobia and boredom were endemic. Peter found the almost total absence of women deeply depressing:

However irritating they may be at times, women by their mere presence exert a certain beneficial influence, and keep alive a type of social activity which provides variety and a touch of the unpredictable, even in a tightly closed social circle.

Local wives and daughters had been evacuated to the UK, Madeira or Jamaica. Even the population of the red light district had been repatriated to Spain. Although Spanish women crossed the frontier from La Línea on the northern boundary of Gibraltar in numbers each day, all except a few entertainers had to return to Spain by evening. Such was the demand for female company that a squad of Royal Marines used to moonlight as bodyguards for the girl entertainers as they moved between dressing room and stage. The handful of Wrens and nursing sisters who lived on the Rock in segregated quarters known as 'The Wrennery' was quite in-sufficient to compensate. Every minute of their spare time was fiercely con-tended for; their commanding officer and the matron vetted requests for dates and kept a blacklist of 'undesirables'. A beach on the east side of the Rock set aside for use by women—Sandy Bay—was secured behind wire and concrete defences.

Until now Peter had been dealing mainly with explosive devices descending from the air; in contrast, during his six months in Gibraltar there was only a handful of 'languid Italianate air raids'. Few of the bombers arrived over their target, because once the alarms were sounded announcing their imminent arrival all available armaments on the Rock let fly, causing what Peter described as 'a collective failure of Italian nerve'. As a result many of the bombs intended for Gibraltar fell on the luckless villagers of La Línea. Peter had a fairly cynical view of Italian ordnance. He felt that safety precautions often got in the way of performance. He recalled one example: a lanyard secured to the invading bomber pulled out a pin when the bomb was launched and this allowed a small propeller to turn. After a time delay, the propeller adjusted the bomb fuse configuration from 'safe' to 'explode on impact':

In the case of the Italians—ingenious but perverse—the propeller might unscrew the whole fuse from the bomb so that the two landed separately, with a great saving of life.

After one rare Italian raid Peter had to deal with a second example, a 'typically Italian super-bomb' which also featured a propeller. As most of the components had failed to explode, he was able to piece it together with great appreciation. What he found tended to reinforce his prejudices.

It was a large bomb-shaped canister and had a small propeller in the nose; this spun in the airstream when the bomb was falling and drove a small dynamo and also a counter which looked like a car mileometer. When the requisite number of metres had been ticked off, a small charge separated the nose from the rest of the canister and out fell a dozen or so anti-personnel bombs, each of which, in turn, armed itself as it fell and burst on striking the ground.

There was yet another twist which, Peter observed, exposed the fiendish 'cunning of the Latin mind':

The nose itself was full of explosive and shrapnel, and when it was blown off a fuse was lit causing it, in its turn, to burst as it approached the ground. But this was not all, for in the remainder of the canister was fixed yet another anti-personnel bomb— a large one this time—which itself burst on impact. Such, at any rate, was the sequence of events, no doubt described with dramatic force in the specification submitted to the Regia Aeronautica [the Italian airforce]; in practice the effect was less gratifying, except to the local bomb-disposal officers who reaped an exceptional harvest.

The underwater threat to Gibraltar's security was far more significant. Shortly after the declaration of war in June 1940, the Italians had set about modifying two of their submarines—the *Scirè* and the *Gondar*—to carry human torpedoes. From their base in La Spezia the *Gondar* went east to Alexandria and the spectacularly successful attack in December 1940. The

A 1940s Italian frogman kitted out to go swimming
Courtesy The National Archives UK

Scirè travelled west to Gibraltar.

Italian frogmen were equipped with wet suits, flippers (which made it practicable to swim long distances), breathing gear and a camouflage helmet with face net. They nicknamed the human torpedo *maiale*, or pig, because it was such a challenge to control. And they were brave men: the wet suits were very uncomfortable and the breathing equipment inherently dangerous. The torpedo body was 6.0 metres (20 feet) long and 0.5 metre (21 inches) in diameter. It travelled slowly—at a mere three knots—to avoid its human cargo being swept off and, because its propellers were battery-driven, the craft had a limited range of ten miles. A partially flooded tank, from which air could be vented or water driven out by compressed air, regulated the depth.

The plan of attack was as follows. The crew would approach its target under cover of darkness with heads and shoulders above the water, like periscopes, ready to submerge if there was a chance of being spotted. On arrival at the target vessel the two frogmen submerged the craft. They attached a clamp to each of the vessel's two bilge keels; these keels run either side along the hull of a vessel at the level of the lowest internal compartment and reduce roll, but they also proved very convenient for Italian mine-layers. Securing a rope between the two clamps, the frogmen detached the torpedo's nose cone with its 300kg of high explosive and slung it on the rope. After setting the clockwork fuses the crew beat a retreat on the torpedo minus the nose cone.

That was the theory; practice often proved frustrating. The *Scirè*'s first operation, in September 1940, was called off halfway through when spies on the Spanish mainland reported Gibraltar harbour empty of shipping. A second, in October 1940, got closer and found the battleship *Barham* in harbour. The *Scirè* edged into Gibraltar Bay to an area near the northern shore and released three human torpedoes into the bay. All three malfunctioned. Aborting, two of the three crews abandoned their torpedoes and met an agent on the Spanish coast who saw them back to

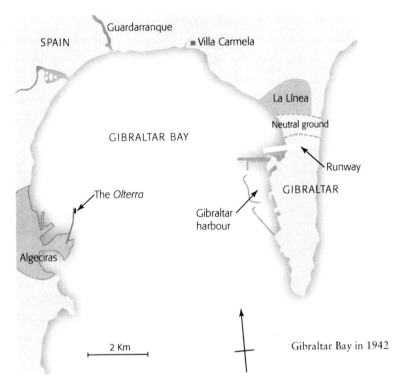

Gibraltar Bay in 1942

Italy. The third crew got into the harbour and, although hampered by reduced speed and manœuverability when water leaked into their batteries, got to within 70 metres of the *Barham*. Then their torpedo stuck fast on the bottom. Setting the timer on their nose cone, they abandoned ship and swam off towards Spain, only to be intercepted by the British. Later that same morning their interrogation was interrupted by a large but harmless explosion in the harbour. One of the abandoned torpedoes ran ashore on the Spanish coastline at La Línea and the British recovered it. Now they knew what they were up against.

The Italians, regrouping as the Tenth Light Flotilla under the determined command of Prince Junio Valerio Borghese, made a third attempt in May 1941. To lessen stress on the human torpedo crews—in earlier attempts they had become exhausted during the week-long trip in the *Scirè* from La Spezia even before mounting their torpedoes—the men travelled independently to Cádiz. Meanwhile the *Scirè*, with three *maiale* aboard, slipped silently into Cádiz harbour alongside the 6,500-ton Italian tanker *Fulgor* which had been interned by the Spanish. The crews boarded the *Scirè* from the *Fulgor*, and the submarine took men and torpedoes on the relatively short trip south to Gibraltar. The Italians' main targets were war-ships within the harbour, but they again found it empty and decided to attack smaller commercial vessels in the bay. There were more technical difficulties, this time in securing the nose cones to bilge keels. The mission was again called off and all six crewmen returned safely to Italy via Spain. Although these three attempts had all ended in failure, the Italians had gained valuable experience which contributed to their success at the fourth attempt in September 1941 which again made use of the *Fulgor*.

Before Peter's arrival in Gibraltar the Italians had begun to streamline their operations. Access to the waters of the bay through Spain had turned out to be quick and relatively simple, so why not launch attacks direct from Spanish coastline within the bay? The leading Italian frogman Lieutenant Licio Visintini (who had piloted the September 1941 human torpedo which disabled the *Denbydale*) identified an even better base than the *Fulgor*. Two years earlier, when Italy entered WW2, the 5,000-ton Italian tanker *Olterra* was standing off Algeciras and her captain, following standard practice, had run her on to a shoal in order to disable her. By the summer of 1942, following an arrangement with its Italian owner, the vessel was refloated and moored at the end of the outer pier of Algeciras harbour, a little more than 3 miles (6km) from Gibraltar harbour. The Italians explained to the Spanish authorities that they were going to repair the boilers and engine; under this cover and working with Pepo Martini, a young, swash-buckling Italian secret agent with a reputation for getting things done, they brought in human torpedo parts (hidden in consignments of boiler tubes), warheads (hidden in barrels of petrol) and workmen from Italy.

Under Visintini's directions, the Italians surreptitiously cut a four-foot square escape hatch in the forepeak (the lowest compartment right at the front of the vessel) below the waterline and flooded it with water. Aerial photographs told the British that something was afoot, but they weren't sure what and anyway could do nothing about it. By the end of 1942 human torpedoes could move freely from the *Olterra* out into the waters of Gibraltar Bay without detection. In fact crews undertook only three further operations. In December 1942, just after Peter had left Gibraltar, Visintini with his torpedo and companion managed to penetrate the harbour in spite of enhanced defences but they were killed by depth charges on the way out. Two more attacks in May and August 1943 were less ambitious and aimed at targets in the bay. The total yield of disabled shipping during two years of operation was some 68,000 tons.

Of course, Peter did not know what was going on across the water in Algeciras. He simply noted that from late 1941 until he left Gibraltar in November 1942, human torpedoes failed to damage any ship beyond repair. The September 1941 attack had ensured that the more important or valuable ships were berthed within the harbour with its protective moles and anti-submarine netting. The bomb safety team already knew that crews of human torpedoes carried wire cutters and lifting gear. So they added more layers of protection: floodlights near the harbour entrance plus the frequent dropping of depth charges at irregular intervals. It seemed to the British that although the explosions killed plenty of fish they probably also scared the human torpedo crews away. Frogmen captured later on confirmed that the Italians had a particular fear of depth charges. Peter judged at the time that, on his watch, the Italians never actually got into the harbour.

During most of 1942 the main success of the Italian underwater campaign was in the war of nerves. Everyone in Gibraltar worried about the possibility of attack. Worry and sleeplessness had an impact on productivity and diverted resources from more useful tasks. After the September 1941 attack an Underwater Working Party (UWP) had been formed, dedicated to countering the threat and checking shipping for attached mines, quaintly referred to in many of the contemporary reports as 'infernal machines'. Peter and the UWP imagined the German consul based in Algeciras, the largest town on the bay, would have a house on the seafront with a bow window and a powerful telescope. There he would sit, they imagined, to watch Allied shipping coming and going, able to assess the effectiveness of Italian efforts to slow it down.

Peter worked alongside Bill Bailey, a famous WW2 underwater mine disposal specialist. Bailey originally trained as a helmeted diver and had been leader of the UWP since late 1941. When compared with what the Italians had available, the British equipment for underwater work was primitive: swimming trunks, any old top to protect their bodies against barnacles and other underwater obstacles, a nose clip, goggles and plimsolls weighted with lead which kept divers upright while working under water. For breathing purposes Bailey had got hold of some sets of Davis submarine escape apparatus (DSEA). They were designed for rescuing crew from sunken submarines, and in the freezing waters of the Atlantic the DSEA set was reputed to have killed more people than it saved;

emerging submarine crew members often froze to death on the surface awaiting rescue. But the warmer waters of Gibraltar Bay were kinder. Bailey's team of divers breathed in pure oxygen as they searched for mines attached to the hulls of vessels in harbour or at anchor. They breathed out carbon dioxide which was absorbed by a canister of soda lime. It was a tricky business keeping the flow of oxygen high enough to avoid carbon dioxide poisoning. Peter retained a vivid picture of Bailey:

[He] spent so much time under water that he became virtually amphibious. I can see him now, his lean frame shivering with cold after twenty minutes' immersion, stopping only to change his oxygen bottle and answer a few questions between chattering teeth, before he disappeared again into the depths of the bay. His great ambition was to come across an Italian at work under a ship and have it out with him at five fathoms.

Later he did just that, struck first and slit the Italian's air pipe. His luckless victim tried to escape to the surface and was discovered later, floating and unconscious. He died shortly afterwards. Peter considered that with this incident:

[Bailey] had inaugurated an entirely new branch of hand-to-hand combat.

In spite of the efforts of the UWP, during the summer and autumn of

---

BACKGROUNDER:

BILL BAILEY'S WAR

Graduating in electrical engineering in the summer of 1940, Bill Bailey applied to the RNVR. He didn't have Peter's naval connections, so he had to face a boardroom full of 'scrambled egg'—a collection of captains and commanders with a lot of gold braid on their uniforms. Accepted as a temporary electrical sub-lieutenant, he trained briefly at *HMS Vernon* in Portsmouth and then took charge of an armed trawler looking for enemy mines off the north-east coast of Scotland. When the Germans began to drop magnetic mines on land, he was diverted and found himself on the same team as John Ouvry of Shoeburyness fame. Bailey dealt with his first unexploded magnetic mine in mid-September 1940 near Spurn Head at the mouth of the river Humber. In September 1941 he trained as a helmeted diver, was promoted to lieutenant and immediately sent to Gibraltar to help counter the Italian human torpedo threat. In July 1943 Bailey was one of a team who recovered bodies and documents when the Liberator aircraft carrying the Polish leader General Sikorski fell into the bay on taking off from Gibraltar. Later he took part in preparations for the D-Day landings and in the invasion itself, dealing with unexploded devices in France and the Low Countries. He ended the war a heavily decorated lieutenant-commander.

1942 Peter and the team would hear occasional explosions in the bay, and then a ship or two began to sink and had to be beached. Other ships would leave, bound for elsewhere, only to be mysteriously holed by explosions *en route*. It was clear that in none of these cases were the explosions in the same league as those caused by human torpedo nose cones, whose 300kg charges could break a ship's back. What the UWP didn't know was that, while the Italians were establishing a new base for human torpedoes at the end of Algeciras harbour arm, they had switched tactics. Frogmen were laying limpet mines.

In January 1942 one of the Italians who had volunteered for service with the Tenth Light Flotilla, Antonio Ramognino, had married a Spanish girl called Conchita. With the couple's agreement, the Italian consulate in Algeciras arranged for them to rent the Villa Carmela which stood behind a private beach on Mayorga Point near La Línea. The pretence was that Conchita had been ill and needed sea air to recover. The villa was a little more than 2 miles (3km) from Gibraltar dockyard. Using Villa Carmela as a base, frogmen could unobtrusively slip into the water, swim out—either to the harbour or to merchant ships at anchor, some as close as 600 metres from the shore—and plant their mines. Intelligence feedback from the Spanish mainland during the summer of 1942 had alerted the British to some kind of activity centred on Villa Carmela although the UWP could only speculate about exactly what was going on there. Spanish sources noted Ramognino coming and going daily in a small green Peugeot car, either east to La Línea or west towards Guadarranque. Sometimes his wife went with him: 'a damn fine-looking piece' wrote the British translator of information received.

Then in July 1942 a party of 12 Italians—known as the Gamma group—made its way to Madrid, the journey orchestrated by Pepo Martini and his agents. Three Italians entered Spain hidden in a truck, three crossed the Pyrenees on foot and six arrived in Barcelona as deckhands on an Italian freighter. Once in Madrid, cars drove them to Cádiz as 'crew replacements' for the *Fulgor*. From there they continued on to Algeciras and the Villa Carmela. Their underwater suits and mines, smuggled in earlier by Martini and company, awaited them. On the morning of 13 July the frogmen looked out across Gibraltar Bay to select their targets. That night they put on their wet suits and fins, blackened each other's faces and donned hairnets interlaced with seaweed as camouflage. Carrying mines they waded into the water and swam towards their targets. After laying the mines all 12 frogmen returned safely to Villa Carmela where Conchita was waiting with a hot breakfast, coffee and brandy. Next morning four merchant ships at anchor in the bay—the *Meta*, the *Shuna*, the *Empire Snipe* and the *Baron Douglas*—were crippled by explosions and had to be run ashore in a hurry.

In spite of this Italian success, the strong currents of Gibraltar Bay were kind to the UWP. They swept a mine or two from below targeted vessels; eye-witnesses of the first explosion to hit the *Shuna* said it took place on the surface. Immediately after the raid the UWP found a sinister but unidentified object floating in the bay. Peter's team took it away to Western Beach—a strip of sand on the isthmus beyond the airstrip and abutting Spanish territory which, in Gibraltar, passed as a secluded spot— to make sure it was not live.

What exactly was it? The UWP was not sure. At first it seemed that the main charge was thermite (a mixture of a metal—usually aluminium— and a metal oxide—frequently ferric oxide—which can react together to create, briefly, a high temperature but no explosion). Peter put on a small show in the admiral's ashtray where a sample burned brightly with a shower of sparks. So they assumed that what they had found was an incendiary device intended to set fire to spilled oil. But then there was a surprise. One of Peter's colleagues took a lump of the main charge to a remote corner of the Rock, inserted a detonator and retired to a respectful distance. There was a very big explosion indeed: the charge was hexanite and the object was a limpet mine. Peter described how it worked:

A swimmer could carry about half a dozen of them in a harness. The mine consisted of a 5lb [2.3kg] charge, big enough to make a sizeable hole in a ship's bottom, and had round it a small ring-shaped float which could be inflated by a tiny cylinder of compressed air. All one had to do was to allow it to float up under the flat part of a ship's bottom. There it would lodge until, in the course of time, the water dissolved a pellet of salt and up she went!

There were various kinds of limpet mine; the Italians called this one a 'leech' and, in spite of its name it was not held in position, as some

An Italian buoyancy limpet mine
seen from below
Courtesy The National Archives UK

others were, by magnets or clamps but simply by the buoyancy of its float. Detonation could be achieved by either of two firing switches: one held open by a salt pellet to delay action below the target vessel, the other controlled hydrostatically to destroy the mine should it reach the surface. Peter prepared a report for *HMS Vernon:*

> The mine was dented so that solder around the mechanism plate had cracked, and sea-water had entered. As a result, most of the wiring connected to the positive pole of the battery had dissolved electrolytically. The insulating sheaths remained in place and showed the original position of the wiring.

A little later another specimen was discovered in the boiler-room of the crippled *Empire Snipe*, into which it had been either blown by an earlier explosion or sucked by an inrush of water. It was badly damaged but appeared to be identical to the earlier find. It gave Peter the opportunity to weigh the explosive filling precisely and examine the detonator fully.

In September 1942 the Italian frogmen made a reprise attack from Villa Carmela, but they were less successful. The Gibraltar authorities had by then moved the convoy anchorage pattern away from the Spanish shore and close to Gibraltar's heavy guns. There was increased patrol boat activity and

BACKGROUNDER:
A LIMPET MINE EXPOSED

Peter's limpet mine firing circuit
Courtesy The National Archives UK

Peter sent his report—together with the bits and pieces retrieved—back to the mine design department in Leigh Park House, Havant, not far from Portsmouth. They circulated widely a description of the limpet mine in late August 1942. The mine body was a brass canister 24cm in diameter and 11cm deep. A rubber belt 8cm wide and 38cm outside diameter surrounded the mine, inflated as required from an air bottle with stopcock. Within the mine body were the main charge (5lbs, 2.3kg) and a detonator wired to a 4.5volt battery through two switches in parallel. The two switches were almost identical. Viewed from above, an Ebonite (hard rubber) block (d) carrying two firing contacts fitted inside a cylinder. Inside the cylinder a movable rubber diaphragm (a) made a water-tight seal. An insulated plunger (c) with a brass head attached and carrying a spring (b) passed through the Ebonite block to make contact with the

diaphragm. On the other side
of the diaphragm was a spindle
(e) and keep ring. Above
the diaphragm the sealed
mechanism was dry; below it,
wet. The two switches differed
only slightly. One controlled
firing under a vessel—say, 35
feet (10m) down. Its spindle
ended with an extension which
passed through the diaphragm
(making a water-tight seal) and
screwed into the end of the
plunger. A salt plug sitting
on the spindle compressed
the spring to lift the brass head
of the plunger into the 'open'
position. Once immersed the plug
slowly dissolved and the strong spring
pushed the plunger down against water
pressure until its brass head 'made' the
firing circuit. In contrast, the second
switch triggered detonation should the
mine reach the surface. Its spindle ended
with an Ebonite disc. Below a depth of
10 feet (3m) water pressure on the
diaphragm was enough—the spring on

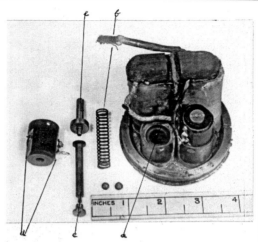

The firing mechanism partly stripped
Courtesy The National Archives UK

this plunger was weaker—to hold
the switch open. Above that the spring
closed the circuit. Neither Peter nor the
mine design team found any indication
of where a safety key might fit; they
presumed it must hold the second
switch open until the mine was deep
under the hull of a target vessel. Once
there the frogman would remove it,
fully inflate the rubber belt, position
the mine and retreat.

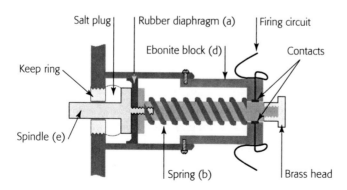

A reconstruction of the limpet mine deep-water firing switch (after Geoff Mason)

depth charge release, and that made life really difficult for the frogmen. They only succeeded in damaging the steamship *Ravenspoint*—albeit with two separate leech mines attached by two rather disorientated frogmen.

By October, the atmosphere on the Rock—already pretty hectic—was ratcheting up. This flap, Peter recalled looking back in 1945, 'for wild-eyed despair and confusion on a gigantic scale beat any of its predecessors hollow'. By that time he had already experienced the build-up to the Normandy landings; they gave only a faint idea, he felt, of 'the kind of hurly-burly' which was concentrated within the confines of Gibraltar. Speculation was rife.

Getting on with the paperwork: Peter as temporary lieutenant in late 1942, aged 25

For me the answer was authoritatively vouchsafed when I was allowed, early in November, to look inside a bulky paper-covered book entitled 'Torch Naval Operation Orders'. After so much secrecy it came as a bit of a shock to read in the first paragraph the blatant statement, 'The object of the operation is the occupation of Algeria and French Morocco...'

Operation Torch took off in early November and on 24 November Peter flew back to London from Gibraltar to prepare for a new posting. He handed over to Lionel 'Buster' Crabb, who had arrived in Gibraltar for bomb safety duties in late 1942. Crabb was as much an enthusiast for the underwater life as Bill Bailey, if not more so. In his next postings Peter found that colleagues passing through Gibraltar would keep him informed about Crabb's exploits. 'If you ever happen to be at the bottom of the sea,' they said, 'and you meet a short bloke with a red beard, you can bet your life it's Crabb'. Crabb became a household name in the UK 14 years later; when sent to investigate the hull of the Soviet cruiser *Ordzhonikidze*, which had arrived in Portsmouth harbour to deliver Soviet leaders Khrushchev and Bulganin, he disappeared without trace. The newspapers got hold of the story and an international incident ensued.

Once Italy stopped fighting in September 1943, the British were able to cross to Algeciras to investigate what exactly had been going on in the *Olterra*. They found the Italian human torpedo base: one serviceable human torpedo remained. Crabb commandeered it, christened it *Emily* and used it for a while as a taxi. Peter recorded that:

Subsequently several nervous people were seriously upset by the sight of the head and shoulders of a red-bearded man, wearing a DSEA, protruding in an upright pos-

---

BACKGROUNDER:
BUSTER CRABB'S WAR

Temporary Lieutenant Lionel Crabb RNVR arrived in Gibraltar in October 1942 after the usual rapid training course. He took over from Peter as Gibraltar's mine and bomb disposal officer but was quite a different character: he had no trust in anything scientific. Although his job was to dispose of explosive devices found by the underwater team, Crabb wanted to dive too. He was a poor swimmer but Bill Bailey taught him the ropes and he was under water from dawn till dusk checking vessel hulls for mines. Sociable and fearless, he was able to endure the great discomfort of the primitive diving equipment available. When, in the summer of 1943, Bailey broke an ankle and returned to the UK, Crabb took over. Crabb also ran a team of thirty or so young volunteers who travelled from Gibraltar to Spanish ports where merchant vessels were due to sail for Britain. 19-year-old Maurice Featherstone volunteered for the job in 1942 after his merchant ship had been torpedoed in the Atlantic. Classified as civilians, the volunteers were able to travel freely around neutral Spain and went as far afield as Barcelona. The youngsters—mostly ex-merchant navy—earned seven shillings and sixpence (37.5p) per dive, 'a small fortune in those days' says Featherstone. Some checked cargos for explosive devices; the rest examined bilge keels for limpet mines. The Italians avoided fixed-time fuses in Spanish ports, because if a ship were delayed it might sink in harbour and cause a diplomatic incident. They used an arming propeller instead, which set a clock ticking once the ship had left port and reached five knots. Members of Crabb's team could detach them safely and take them back to Gibraltar by train for Crabb to render safe. When Italy signed an armistice with the Allies in September 1943 and attacks by Italian frogmen stopped, Crabb got a well-deserved George Medal for his all-round efforts. 1944 found him clearing North African ports of mines and in May 1945, as principal diving officer for northern Italy, he earned an OBE for helping clear Venice and other Italian ports. Later in 1945 he was in Palestine dealing with Zionist mines. The 1958 British film *The Silent Enemy*, starring Laurence Harvey as Crabb, presents a majestically exaggerated version of Crabb's activities in Gibraltar.

ture from the water of the harbour, and moving at some speed without visible effort.

Emily, like all good Italian devices, was temperamental. She eventually dived deep, almost drowning Crabb and a companion, and was never recovered.

The announcement of the 1943 New Year honours included an MBE (Member of the Order of the British Empire) for Peter in recognition of his 'gallantry and undaunted devotion to duty' during 'mine disposal at Gibraltar in July 1942'. The full text of the citation seems to have gone astray, but it undoubtedly concerned Peter's work on the Italian limpet mine and how it performed. But by the time the award was announced Peter himself had already moved on towards the front line.

BACKGROUNDER:
BUSTER CRABB POST-WAR
Demobilised in early 1948 as temporary acting lieutenant-commander, Crabb's life became one of more-or-less secret diving work. Loosely attached to HMS *Vernon* in the early 1950s he had become a local character in Portsmouth, parading in a tweed suit and pork-pie hat and carrying a sword-stick with a big silver knob engraved with a crab. In October 1955 while the Soviet cruiser *Sverdlov* was visiting Portsmouth, both the British and the Americans noticed how remarkably manoeuvrable it was. To find out why, Crabb examined its hull by night: there was a retractable propeller in the bow. The following April several Soviet vessels, with leaders Khrushchev and Bulganin aboard the *Sverdlov*-class cruiser *Ordzhonikidze*, arrived in Portsmouth on an official visit. Crabb was commissioned to make a similar reconnaissance. No longer young, and always a heavy drinker, he was by now also rather unfit. He failed to return. One of the Soviet ships had spotted him, and Khrushchev remarked to that effect at a dinner party with British prime minister Anthony Eden. The British press caught on, the Soviets protested and Crabb became a legend. Soviet sources later suggested that he was shot in the water. A year later what was apparently Crabb's body—minus head and hands—surfaced in nearby Chichester harbour, along with a lot of conspiracy theories.

Chaos in North Africa 5

WITH OPERATION TORCH UNDERWAY, Peter spent December 1942 on a course on Whale Island, Portsmouth, getting acquainted with the latest advances in German and Italian bomb and mine technology. Then, in early January 1943, he made another long trip by flying boat to join the stone frigate *HMS Hannibal* in newly-liberated Algiers. The largest city in French North Africa, Algiers was one of the three areas where the invasion force of Operation Torch had concentrated. Nominally under the control of the Vichy French, Algiers switched sides with only token resistance. This was a lucky consequence, not of any unified support for the invading forces, but of political chaos, intrigue and chance. And Peter found when he got there that all three continued.

Relations between the Algerian French and the British in January 1943 were very difficult because of an incident two and a half years earlier. During the first year of WW2 the French in North Africa, like those in the homeland, had been unequivocally on the British side. When, on 22 June 1940, the residue of the French government under Marshal Pétain signed an armistice with the Germans, Winston Churchill pondered whether the new Vichy government might hand over the French fleet to Hitler. He judged they might and acted decisively. A British armed force immediately took French vessels moored at Devonport. The main French fleet was sheltering in Mers-el-Kébir, adjacent to Oman on the North African coast. There, on 3 July 1940, British ships standing offshore and commanded by Vice-admiral James Somerville on *HMS Hood* delivered an ultimatum. When it expired Somerville reluctantly opened fire and, in the course of ten minutes, destroyed several large vessels, killed 1,300 and wounded another 400. Not surprisingly, French attitudes hardened. To avoid the French imagining Operation Torch to be a British invasion, American forces had taken the lead with the British keeping a low profile.

In the run-up to Operation Torch, the Americans under General Eisenhower sought a French figurehead with whom they could come to an

The port of Algiers in 1943
Courtesy DeGolyer Library, Southern Methodist University, Dallas, Texas, USA, Ag2002.1454

agreement and who would then be able, on the strength of his reputation and popularity, to get the French in Algiers to back the Allied forces. De Gaulle was out of the question: he had sided with the British and the French armed forces would not cooperate. Eisenhower's men first flirted with General Maxime Weygand, a WW1 hero and commander of French forces driven back to Dunkirk by the Germans in 1940. He had a fine reputation with the people but proved too old and dispirited for the job. So the Allies turned to General Henri Giraud. Captured by the Germans in 1940, Giraud had been locked up in a castle near Dresden but had fired French imaginations in April 1942 by planning his own escape and finding his way safely to Switzerland. But Giraud wanted only military responsibility for North African activities, and strung out negotiations about details. He was still in Gibraltar negotiating when Operation Torch began. Meanwhile in Algiers itself Admiral François Darlan, who earlier in the war had been Pétain's deputy and a senior minister of the Vichy government, was the man on the spot. In spite of his reputation for having reached the rank of admiral without ever commanding a fighting ship, he was respected by the French armed forces. In the event it was he rather than Giraud who, following the Allied invasion and at the request of the Americans, ordered French troops to cease fire and join the Allies. Giraud arrived in Algiers shortly after the invasion but was only able to play second fiddle to Darlan. Then, on Christmas Eve 1942, a young monarchist assassinated Darlan. Thereafter there were fragile alliances, but the political atmosphere remained turbulent.

Into this environment Peter Danckwerts was precipitated. Nationalist attempts at disruption continued and bomb scares were not infrequent. Much later, when Peter scribbled a list of his wartime experiences which might interest a publisher, the title: 'A bomb in the casbah; evacuating veiled ladies' was amongst them. The title reflected both local tensions and his service responsibilities. If it was written it has not survived.

Allied armies had encountered more resistance at their other two landing sites: Tunis and Casablanca. Anti-personnel mines of German manufacture were everywhere. A French army captain, at the time *en route* from London to Tunisia, bought up large stocks of walking sticks and bales of string, to the astonishment of his supplier. The captain explained that the

---

BACKGROUNDER:
RECENT HISTORY IN ALGIERS
Viewed from the Mediterranean in 1943, Algiers looked like a seaside town on the south coast of France: white houses on the hill behind the port plus a heat haze. In 1943 there was a small indigenous sector, the medina, sometimes interchangeably called the casbah, just inland from the port. Algeria, in common with other French colonial outposts, had officially become part of France in 1871. But the local population was disenchanted with the French imperialists, who treated them poorly. After the Algerians had fought alongside the French in WW1, they expected better treatment but received no favours. So, during the 1930s, a nationalist movement grew up calling for separation from France. The French in Algiers had been posing, in quintessentially French style, as a great world power; when the Germans invaded France in 1940, native Algerians— some seven million out a total of eight million—regarded the Nazis as their liberators. But the new situation did not

deliver them any power. The French continued to run the country and were thoroughly divided amongst themselves. The Vichy government was nominally in control, and some saw opportunities in backing it. Their enthusiasm was tested as the war progressed: the natural abundance of cheap clothing, meat, fruit and vegetables, still evident in early 1941, declined as it was diverted to the Axis powers' war effort in Europe. Availability slumped and prices rose steeply. Other Frenchmen religiously supported General de Gaulle; they kept their allegiance secret but anticipated the General's appearance in Algiers at any time. The communists were another faction. And there was quite strong support amongst the reactionary colonialists for the reinstatement of the monarchy: the Orleanist dynasty in the shape of Henri, Count of Paris. Monarchists were motivated in part by nostalgia and partly by disenchantment with duplicitous politicians. The young activist who assassinated Admiral Darlan was a monarchist, but apparently acted alone.

Germans were using mines attached by a wire to a sort of mantrap which gripped when a man stepped into it. Tension in the wire primed the mine. When the wire was loosened, the mine exploded. His men had no mine detectors but had learned not to try and free themselves when they got caught. They dug a stick into the earth to replace their leg and maintain the tension—a walking stick would do. Then they attached a string to this stick, got clear and, when a little way off, yanked the string to pull the stick away. The mine then exploded harmlessly.

Besides the tensions on shore in Algiers, special operations groups were busy offshore and all along the southern Mediterranean coast. Amongst other things, they were collecting escaped prisoners of war (POWs), airmen who had been shot down and agents escaping from the south of France. A contemporary account records how a British vessel on one operation in early 1943 met up with a *feluca* (small fishing boat) manned by special operations sailors. From the *feluca* the sailors transferred to the vessel some POWs, airmen and a woman agent. The fugitives stayed in the officers' quarters; the officers' steward reported that the woman didn't trust anyone and slept with a pistol under her pillow. She had been pulled out of France for some reason; perhaps a German officer was paying too much attention to her, or maybe she was suspected of being an agent. Once the vessel returned to Algiers at dusk, it headed for a vacant berth and tied up there. A tug approached with a naval officer in the bows. He hailed: 'Ahoy there' but was ignored. So he amplified his greeting: 'Ahoy there, I've come to tow you away because you are sitting over an unexploded bomb'. This attracted some attention. It turned out he was the bomb and mine disposal officer: speculative stuff, but it could have been Peter.

Munitions were not the only hazards in Algiers. The NAAFI (all services canteen) established itself in an old warehouse just along the wharf from *HMS Hannibal* and passing naval traffic kept it well supplied. But it attracted rats. They were very big—about four times the size of a British rat—and would scamper down the mooring lines onto vessels berthed nearby. Rat guards on the mooring ropes were ineffective: like squirrels, rats are very resourceful and they still found ways to get on board. When the generators shut down for the night, squeaking and scurrying could be heard all around and the smell of rat urine lay heavy on the warm Mediterranean air. Traps brought limited relief: thirteen in one night was considered a good haul. Peter was no doubt a keen observer of all this, but had other interests in the local social scene.

A second title on Peter's list of memories for a publisher was entitled 'Scenes from an Algerian brothel'. The only surviving sign of this is a letter he sent to *New Scientist* in 1982. He had spotted an article by Richard Wilson, an information scientist based in Amsterdam, soliloquising on beards and their power as a declaration of male sexuality. Wilson then moved on to ask why it was so fashionable to shave them off. After all, he wrote, with the exception of the mythical Amazons, women never deal so carelessly with prominent secondary sexual characteristics. Peter pounced:

Richard Wilson's piece ... does not go to the root of the matter. Women throughout the ages have intermittently shown an inclination to rid themselves of all hair save that on the tops of their heads. Greek ladies of antiquity plucked their pubic hairs. The men didn't go for it (see *Lysistrata*). I remember sitting-in on *une exhibition* by two jolly French whores in Algiers. One of our party intemperately shouted, 'Too much hair!' The reply was, '*Je ne me rase pas—je ne suis pas Arabe.*'

As in Gibraltar, social life for the forces was, brothels apart, unremittingly male. But Peter's experience of the opposite sex in Algiers was a little wider than his *New Scientist* letter suggests. As usual, he made contacts: he may not have known many people when he arrived but he got to know some in record time. In Algiers he made the acquaintance of Marie-Louise Perreux, a French government journalist before WW2. When Peter, then aged 26, met her she was an attractive and experienced woman of 43.

With a diplomatic passport, Perreux had journeyed extensively in the Far East before the war. In 1940 she removed herself from Paris via Vichy France to Morocco and contributed now and then to the North African weekly *Tunisie-Algérie-Maroc* (TAM). By 1942 she had made her way to Algiers where local publishers Heintz printed her mem-

SALON DES ARTISTES FRANÇAIS 1927

PORTRAIT DE MADAME MARIE-LOUISE PERREUX
PAR FERDINAND HUMBERT
MEMBRE DE L'INSTITUT

Marie-Louise Perreux by Ferdinand Humbert

oire of those pre-war Asian journeys—*Croquis d'Asie*. Perreux was well-connected. She was, for instance, a confidante of Abbas Hilmi II, the last khedive, or viceroy, of Egypt and Sudan, with whom she kept up a flirtatious correspondence. (Abbas Hilmi II, the great-great-grandson of opportunist Albanian bandit and first khedive Muhammad Ali, got his marching orders from the British in 1914 when he backed the wrong side in WW1. Thereafter he established a base in Switzerland.) Peter kept a postcard bearing a portrait of Perreux by the French painter Ferdinand Humbert which hung in the Paris Salon in 1927. He annotated the reverse: '*Mon ancienne amie de la guerre*'.

Notwithstanding such blandishments, Peter's stay in Algiers was quite short. In preparation for Operation Husky, the invasion of Sicily, he was transported to Malta at the end of June 1943, ready to do his bit with the invading forces.

Sicily and the aftermath 6

AFTER HIS RECENT ARRIVAL IN MALTA, the call to invade Sicily with Operation Husky on 10 July 1943 found Peter in characteristically matter-of-fact mood:

My role, as usual, was entirely passive—I had been told [by the Navy's bomb safety headquarters in Britain] to go over and collect as many specimens as I could find of German mines, torpedoes, bombs and any unidentified objects. My superior officer told me that no-one would give me any help and I was wasting my time.

This quality of support was nothing out of the ordinary. A bomb safety officer was used to working alone, taking his own decisions and trusting his own judgment. Peter had been doing it now for three years, dealing with unexploded devices according to the nature of the threat they posed, collecting specimens of anything in that line that looked new or interesting and despatching them to *HMS Vernon* and the bomb design people back in Britain.

The attack on Sicily had been agreed at the insistence of Winston Churchill when the British prime minister met US president Franklin Roosevelt in Casablanca in mid-January 1943. The invasion of North Africa—Operation Torch—had been a success and now the leaders aimed to hammer out Allied strategy for going on into Europe. Because the crucial struggle over Stalingrad was in its final stages, Soviet leader Joseph Stalin absented himself but asked for the creation of a second front to help distract the Nazis from their Russian campaign. The Americans argued for a direct attack on Nazi Germany. Churchill persuaded them that it made more sense to go for Sicily; he called it the 'soft underbelly' of the Axis, although it turned out to be extremely tough in places. So after the Casablanca meeting, the Allies rapidly assembled their forces.

On D-day of Operation Husky Peter arranged to hitch a lift on a LCI(L)—landing craft, infantry, large—which was carrying 200 men of the 51st division of the British Eighth Army on the short trip north from Malta to Sicily. In the final stretch of the assault it was already the afternoon; the vessel headed at some speed—it was capable of 16 knots—towards the south-east tip of the island and the beaches on the south-west side of Cape Passero near Portopalo. Other Eighth Army units made landfall further up the coast towards Syracuse; there the slope of beaches was gentle and

---

BACKGROUNDER:
OPERATION HUSKY

The Axis forces defending Sicily on 10 July 1943 were the Italian Sixth Army under General Alfredo Guzzoni and some anxious Germans. Guzzoni commanded up to 300,000 men. About 30,000 German soldiers were deployed to back them up: two German divisions—the 15th and 90th panzer grenadiers—under the command of Field Marshal Albert Kesselring. Both Italians and Germans were keen to defend Sicily; they knew that giving the Allies a foothold there provided access to mainland Italy and southern Europe. For the Allies, a British general—Sir Harold Alexander with his Fifteenth Army group—was supposed to be in practical charge of the whole invasion. He had at his disposal Lieutenant General George Patton's American Seventh Army and General Bernard Montgomery's seasoned Eighth Army, mainly British with some Canadians. The supreme commander of Operation Husky was General Dwight Eisenhower. The total invasion force, which arrived on 10 July or—delayed by the bad weather—shortly after, numbered almost half a million. All but a relative handful of them came by sea in 2,600 vessels, one of the largest combined operations of WW2. A modest US airborne attack was reduced to disaster by American troops, facing strong resistance on the southern beaches of the Gulf of Gela west of Cape Passero, who mistook their own gliders for German aircraft and gave them lethal doses of friendly fire. But overall the invasion was a success, and within six weeks the Allies were in command of the entire island.

The invasion of Sicily, 10 July 1943

many of the landing craft ran ashore prematurely in shallow water, causing chaos. In contrast, the slope of beaches near Cape Passero was relatively abrupt. Peter and his group of invaders got to the shore quite quickly, barely getting their feet wet.

As we approached Sicily I was up on the bridge with the commanding officer, a callow lieutenant RNVR. We were about a mile off the Sicilian coast and the vessel was at full speed when the crew, with no orders, decided to let go the stockless anchor, designed for kedging the vessel off the beach were it to run ashore. Of course, the anchor and cable were carried away and never seen again. When we eventually hit the beach we were sideways on and in total confusion. I slung my boots round my neck, took a suitcase in each hand and tiptoed ashore.

There was no resistance; a south-easterly sirocco had reached gale force the day before, leaving a heavy swell and a churning sea for the landings and persuading Italian defenders that an invasion was unlikely. Peter had no immediate job, so he joined a 'picnic party' on the beach. The men he found there were 'beach commandos' who were supposed to be in command of the landing area. Their job, newly created for Operation Husky in an effort to bring order to the landings, was to communicate with incoming landing-craft by shouting or signalling. Thus they were supposed to coordinate arrivals and alert incoming vessels to hazards. Peter's beach commandos spent most of their time shouting at LSTs (landing ship, tanks) to 'float your ducks'. LSTs were larger two-deck vessels, each with a capacity of 60 amphibious DUKW craft, familiarly known as ducks. The LSTs were unable to get very close to the beaches so each duck would load up with troops, run down the LST ramp into the water, float ashore and then function as a six-wheeled land vehicle. Meanwhile, on the beach, refreshments were held up:

I was disappointed with my beach commandos because they didn't know how to brew a cup of tea. I had to show them how to soak a pile of sand with petrol and set light to it.

The headquarters team, to which Peter was loosely attached, established itself on the beach and, with still no sign of opposition, settled down for the night. The next day dawned invitingly. The wind had eased back and the warmth of an early Sicilian summer morning was more bearable than the searing and shadeless 40°C heat of the afternoon. That was yet to come. There was still no sign of the enemy and Peter had little to do. Beach defences seemed minimal—the Germans had requested on the previous day that the Italian troops lay mines, but even the local Sicilians found the

A beach commando communicates (by semaphore) with two landing craft—(l) a LCI(L) carrying infantry and (r) a LCT carrying tanks—off a Sicilian beach during Operation Husky
Courtesy Imperial War Museum, A 17921

July weather too hot and most of their explosives were later discovered in a heap behind Cape Passero lighthouse. So Peter wandered off along the beach, lost in thought. Only thirty-six hours earlier at a farewell party on Malta he had danced with a pretty Maltese girl; he always appreciated a good dancing partner and was probably pondering the experience.

It was to be his last dance for a while. Absent-mindedly he strayed into an area of the beach that had been mined, his foot snagged a trip-wire and he triggered a small anti-personnel mine. Shrapnel from the blast tore into his legs and feet. This was a *faux pas*, but there were extenuating circumstances. According to intelligence of the time, the forward edge of a main minefield was often unmarked. And whether it was marked or un-marked, the area in front of it might have been scattered with mines laid at random and unmarked. The mine Peter triggered was probably an Italian B4, a cylinder 13cm high and 8cm in diameter. It weighed only a kilogram, with a main charge of a mere 100 grams or so of TNT packed into an internal cylinder. Between that and the outer cover was scrap metal, or shrapnel. The mine could have been tethered to a small stake or even buried in the sand. Unlike some nastier Italian anti-personnel mines, this one didn't jump in the air after triggering and before detonating; that would have been lethal at close quarters. But it certainly rendered Peter a non-combatant.

When he came round there seemed to be a lot of blood about. He

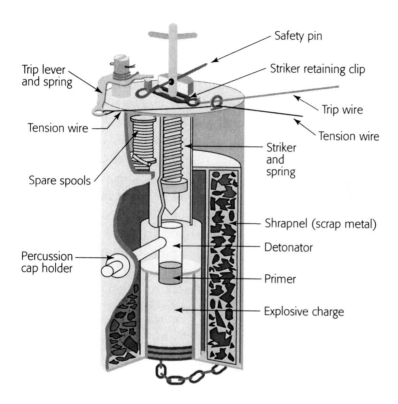

Safety pin

Trip lever and spring

Striker retaining clip

Trip wire

Tension wire

Tension wire

Striker and spring

Spare spools

Shrapnel (scrap metal)

Detonator

Percussion cap holder

Primer

Explosive charge

An Italian B4 mine, probably the type which injured Peter

gradually became aware that his left foot and right leg were both badly damaged. Shrapnel had broken a lot of bones. Realising that it was impossible to get up, let alone to walk back to headquarters, he shouted for a while. Eventually some soldiers appeared. He explained to them that he had strayed into a minefield and triggered a mine. 'For some reason,' Peter recorded dryly, 'they seemed reluctant to come closer'. But eventually a stretcher party turned up, treading carefully, and carried him off to the relative safety of a field hospital. A few days later he was evacuated to Tripoli with *HS Tairea*, the one remaining hospital ship in the area.

Before putting him in plaster, medical staff made a formal list of wounds. They found:

- right calf: 'through and through' wound;
- lower right leg: compound fracture;
- left foot: compound fractures of first, second and third metatarsals;
- right finger, arm, and face: minor wounds.

Conditions were such that infection was almost inevitable. For a while doctors worried that he might lose his right leg below the knee. But nature weighed in on his side; when he saw maggots crawling out through cracks in his plaster he was alarmed at first. But alarm turned to relief when the medical staff explained that his wounds had attracted blow flies, their eggs had become maggots and the maggots were busy consuming dead tissue, allowing both flesh and bone to heal healthily underneath. Many of the injured during the evacuation of Dunkirk three years earlier had enjoyed similar luck. In due course Peter was taken, complete with plaster casts, from Tripoli to Alexandria. The healing process, like the journey, was slow, tedious and kept him out of action for a long time.

At the height of Peter's fame in the Blitz, the press had kept a look out for him. Now he dropped off the radar; British newspapers which had covered his glory days in London and Gibraltar variously reported him dead or a prisoner of war. But by chance the surgeon who dealt with him in Tripoli was an acquaintance of Peter's brother Dick, by then a quali-

---

BACKGROUNDER:
MAGGOTS IN HEALING

Maggots active on a heel wound
Courtesy Live Science

The first modern use of blow fly maggots for therapy is credited to a confederate medical officer called Zacharias during the American civil war: 'Maggots in a single day would clean a wound much better than any agents we had at our command'. WW1 also provided plenty of scope for blow flies. An American orthopaedic surgeon, William Baer, recorded how the compound fractures and abdominal wounds of two soldiers left lying on the battlefield for a week swarmed with maggots. Their wounds were granulating and free from infection. This evidence also supported the idea that wounds heal quicker when moist. Baer used maggot management later as a civilian and in the 1930s it became quite popular. Larvae of the green-bottle fly, a species of blowfly, were available commercially. How do maggots achieve the cleaning of wounds? One theory is that maggots eat the micro-organisms that cause infection and destroy them during digestion. A neglected alternative is that maggots exude an antibacterial chemical—allantoin, today a common component of cosmetics, mouth washes and shampoos. When antibiotics became widely available after WW2, the use of maggots in medicine waned, but as bacteria have become increasingly resistant to antibiotics, interest in maggots has returned.

fied doctor, and he alerted the Danckwerts family. In keeping with his laconic and self-effacing nature, Peter rarely talked about the events on that Sicilian beach, even to his family. Indeed, he rarely mentioned his bomb disposal work to anyone. With a wry smile and characteristic raising of an eyebrow, Peter would claim later that he was fortunate the mine was Italian:

German explosive devices are dangerous but they are designed logically and therefore pose little threat to anyone disabling them with care. The Italians are whimsical in their designs, but their devices are not lethal.

By November Peter was sufficiently recovered to face repatriation. Still in plaster, he boarded a hospital ship heading from Alexandria through the Mediterranean to Liverpool.

The journey [was not] a pleasure cruise. We proceeded through submarine-infested seas, lit up like a Christmas tree. I did not have much faith in the inviolability of hospital ships; the argument about 'you sank one of ours so we'll sink one of yours' had already developed during the summer.

The paintwork of hospital ships—white with a broad green stripe along both sides broken by large red crosses—and their bright lights were supposed to give them protection under the Geneva convention.

HS Talamba with identifying paintwork off Alexandria in 1943
Courtesy Imperial War Museum, E 24034

But in spite of such clear identification, an Axis aircraft had sunk the *Tairea's* sister ship—*HS Talamba*—on D-day in broad daylight near Syracuse. Since then Axis planes, again during daylight hours, had deliberately bombed *HS Newfoundland* off Salerno just south of Naples in September, setting her on fire and smashing all the lifeboats bar one.

Peter's ship was packed, and he chatted to some of his fellow-passengers. One officer had been in charge of a gun battery on the Suez Canal. During a German air-raid the guns had opened up and the racket sent the local feral dogs into hysterics. One of them had bitten Peter's new acquaintance. The antidote to possible rabies was a course of twelve injections to the pit of the stomach which left him paralysed from the waist down. Peter commented sympathetically that he was one of the 'more bizarre casualties of the war'. Meanwhile, on board ship, boredom reigned:

There was nothing for the patients to drink and thereby alleviate the tedium of our eight-knot voyage. However, the commandant of the ship discovered that he had one naval officer aboard among a cargo of pongos or brown jobs [contemporary slang terms for army personnel]. The commandant had had some connection with the Navy, or possibly the Merchant Marine, himself and regarded us two as the elite of the passengers. I would be invited to drink pink gin with him each forenoon watch, while he regaled me with reminiscences of his 'rounds' in some former incarnation. He had gone round wearing white gloves and would draw his fingers along shelves, picture frames and so on. If a glove were soiled he would take it off without saying a word and throw it on the table, leaving the room in icy silence.

They finally docked just before Christmas 1943 on a typical Liverpool day—a sad contrast to the sunshine they had left behind in Egypt. Through the drizzle came the customs men...

...young and able chaps who might have been better employed at that moment in history, or so I felt in my bogus role as military hero. They delved in the sand bins on the upper deck where one of the crew, purely by way of a joke, had planted a potato stuck all over with razor blades. After that it was total war, and it took us 12 hours to clear customs. I feared that they might cut the plaster casts off my legs.

As a holding operation Peter was put into a local hospital. His youngest brother Micky was, by chance, stationed nearby with the Fleet Air Arm, and came to visit. He naturally asked Peter if he needed anything and Peter suggested a bottle of whisky. Micky duly delivered one and was later reprimanded by his commanding officer for 'exceeding the mess bill deemed appropriate for a junior officer'. The episode reminded Peter vividly of an incident in Evelyn Waugh's novel *Men at arms*. The chief protagonist, Guy Crouchback, is thrown together in the Royal Corps of Halberdiers at the start of WW2 with one Apthorpe, an old Africa hand. Apthorpe is plagued by an unspecified African disease (Waugh implies it is alcoholic liver disease) and when he is hospitalised during a recurrence Crouchback—innocently inspired by charity—brings him a bottle of whisky. It kills him. Peter, on the other hand, noted with appreciation that, unlike Apthorpe, he didn't die but 'benefited enormously as a result of [my brother's] misdemeanour'. Once out of hospital he travelled down to London:

> I was welcomed as a hero by my boss at the Admiralty—for no good reason, as my wounds had been practically self-inflicted. I was given plenty of leave and discovered the advantages of my 'wounded hero' appearance: crutches, medal ribbons, both legs in plaster. London taxi-drivers used to stop for me, even in preference to Yanks. I joined a queue outside a cinema and the manager came out and escorted me to a free seat in the best part of the house. My wounds seemed to appeal to some atavistic impulse in girls...

Peter's superiors didn't rush him back into service, and he continued extensive sick leave in the Portsmouth area, overseen by the Royal Naval hospital *Haslar* in Gosport and by his mother at the nearby family home in Emsworth. Eventually he was declared fit. Now 27 years old, he was ready to rejoin the war effort. Then, out of the blue, he got a telephone call taking him off bomb disposal duties and requiring him to report to the Admiralty.

Combined Operations 7

INALLY RECOVERED from his absent-minded oversight in Sicily and taken off bomb safety duties, Peter needed a useful job. As requested, he reported to the Admiralty. It was 6 March 1944; the tide of the war had turned but the conflict continued fiercely:

I went to the office of the director of personnel services at the Admiralty, where I was affably received. 'Light duties, eh, and a degree in chemistry. Now I think I have just the thing for you. Lord Ampthill is moving from Combined Operations Headquarters (COHQ) where he had been looking after Smoke [as a means of hiding manoeuvres from enemy eyes]. How would that suit you?' I thought that if it was good enough for Lord Ampthill it was good enough for me, and very soon I was sitting at his desk at COHQ.

Peter's new office address was 1A Richmond Terrace, London, off Whitehall and right opposite Downing Street. COHQ had been established early in WW2 to harass the Germans in Europe and deliver good news to prime minister Churchill at a time when such a commodity was in short supply. A well-known example is the Bruneval raid on Normandy when Combined Operations paratroopers backed by the Navy successfully stole state-of-the-art German radar equipment. By the time Peter got to COHQ, the general tone of the news was a little better and preparations for the Normandy landings were in full swing.

COHQ had been set up by Lord Louis Mountbatten in his usual princely style. It had assembled the prettiest girls from all three services and had a car-pool, a private cinema, the only all-ranks bar I encountered during the war and a canteen, run by society ladies, which served the best food in Whitehall. So passed an agreeable year, devising and attending trials often involving large quantities of explosives, which caused either apathy among the assembled brass when nothing went off at all or indignation when the intensity of the blast and missiles amongst the spectators was too great.

Uppermost in COHQ minds were the beaches chosen for D-Day landings. Peter was employed on experiments and trials connected with amphibious operations designed to overcome underwater obstacles which the Germans had laid to hinder beach landings. The secrecy surrounding COHQ activities—protagonists were sworn to reveal nothing even to their nearest and dearest—means that records are few. In contrast, rumours about what the participants were up to are plentiful. One tale suggests Peter was involved in a demonstration of tellurium sulphide gas as a possible deterrent weapon during the landings. Almost inevitably the wind shifted halfway through and gave the VIPs a dose; the story goes that they decided tellurium sulphide was not a proper weapon.

Another better-substantiated example of a COHQ revolutionary weapon upsetting its audience during trials (albeit this was a Danckwerts-free zone) was the Great Panjandrum. Allegedly designed by Barnes Wallis of Dambusters' bouncing bomb fame and certainly named by the novelist Neville Shute who worked on the project, it was intended for clearing beach defences. Two ten-foot wooden wheels with one-foot wide steel treads were joined by a central drum, the whole thing being propelled by small cordite rockets attached to the wheel rims. The idea was to carry the Panjandrum close inshore in a landing craft, drop the front ramp and ignite the rockets. The great wheel would trundle down the ramp, plunge through the shallows and roll up the beach at 60 miles an hour, crushing barbed wire as it went, until it encountered a more substantial obstacle. Then 4,000 pounds (almost 2,000kg) of explosive packed in the central drum would explode. In trials the rockets fired unevenly; at least one demonstration saw onlookers fleeing for their lives after the Panjandrum careered about, toppled over on its side and its rockets released themselves. It was judged as much a danger to its own side as to the enemy.

In July 1944 Peter's superior officers recommended him for promotion to acting lieutenant commander. The document, generous in its praise, was signed by his immediate boss Captain Thomas Hussey and Commodore Hugh Faulkner, director of Combined Operations (Naval). But in spite of their backing, no promotion resulted. Peter was not perturbed; he enjoyed working alongside some remarkable people. Much later, in 1983, he reflected on a few of them in *New Scientist*, responding to a piece by the magazine's weekly parliamentary columnist, Tam Dalyell MP, about people he had known in science.

Many of the famous and the notorious passed through COHQ. I remember particularly our genius-in-residence, Geoffrey Pyke, who invented Habbakuk, the giant iceberg ship, and the concept of power-driven rivers. The last time I spoke to him over tea, Pyke discussed the best techniques for committing suicide—to which he eventually resorted.

Pyke's Habbakuk was a project to create floating ice platforms, made from a resilient ice-and-sawdust mix christened 'Pykrete'. The idea was that fighter and bomber planes operating over the North Atlantic might be able to land on them and refuel. COHQ had a lot of development work carried out in Canada but the war was won before it bore fruit. Power-driven rivers were another product of Pyke's fertile brain. This imaginative concept would deliver invasion equipment (and even personnel!) quickly to shore from invading ships by placing them in cylindrical containers within pipelines. This one didn't get past the drawing-board.

Then there was the brilliant Marxist physicist Desmond Bernal. Peter nominally shared an office with him, although Bernal had so many irons in the fire that he was seldom there.

He was rather like a 12-inch gun with no gunlayer; he always turned up to meetings late and disorientated, although he had a middle-aged secretary who used to follow him about with his diary, and try to keep him in line. I was amused that Dalyell described Bernal in his physical decline as being "surrounded by marvellous women". He always could pull them in, although it was hard to figure out how he did it. The girls I knew at COHQ thought he was particularly unattractive, but I expect his own girls applied different criteria.

Bernal was greatly involved in the study of Normandy beaches where consistency would decide whether or not invading tanks and equipment-carriers would get bogged down. Quite late in the planning for D-Day it was realised that the surfaces of some beaches thought to have been of hard sand actually consisted of exposed peat which could be very soft. To get information Bernal consulted French geologists and ancient documents on peat-cutting. He was also involved in the micro-mapping of the invasion beaches. Timed aerial photographs of the water line plus tide tables gave contour maps of the beaches, accurate to metres horizontally and centimetres vertically. In the event the stormy conditions on D-Day undermined some of this work. All the same...

...there was very nearly a breach of security because of Bernal's desire to be on the beaches by D+1. His lank locks had to be shorn and he had to be kitted out with battle-dress and ammunition boots. All this took some time, and he was not a particularly military figure at the end of it.

One thing that Bernal did not get close to, Peter observed, was news of the atomic bomb. Sir John Anderson, who was in charge of matters atomic at COHQ and a major player in the bomb project, described Bernal as 'redder than the flames of Hell' and insulated Bernal as far as he could from news about the bomb. On one occasion when a major report on the US Manhattan project arrived in the office, Bernal was away. Peter dealt with it, and circulated his appreciation of its implications within COHQ. One of the responses he got was a serious enquiry about the likelihood of atomic hand grenades as a development. When Bernal returned he was, on this occasion, pressed to assess the report himself but merely responded: 'Danckwerts has said it all. There is nothing for me to add.' This sort of thing made Peter...

... forgive him for the many occasions on which he had unthinkingly made me feel mentally subnormal.

Another personality amongst the collection of talent at COHQ was Tom Gaskell. Born in the same year as Peter, the two men hit it off and became good friends. Gaskell was a lively and enthusiastic specialist in geophysics. His professor at Cambridge, Teddy Bullard, was a consultant to HMS Vernon on degaussing, and worked with Gaskell on means to combat acoustic mines. So Peter and Tom Gaskell had a lot in common. Gaskell worked for Bernal and when 'the Professor', as they called him, returned from brief discussions at Louis Mountbatten's far eastern head-quarters in Ceylon about possible beach landings against the Japanese, Gaskell was sent off to take his place. He arrived in Kandy in February 1945 and set about trying to organize the sort of research into Japanese-beaches that was carried out in Europe before the Normandy landings.

The end of the war in Europe—VE day—came on 8 May 1945. Peter's younger sister Diana was nursing at St Thomas's Hospital on the south bank of the Thames at the time; permitted to take two people onto a balcony overlooking the celebrations, she invited her eldest and the youngest brothers Peter and Micky. They judged the VE show amazing and emotional, with a mass display of fireworks from barges moored in the river. Micky Danckwerts was beginning a writing career as magazine editor for the Fleet Air Arm, the navy's search and rescue operation; later he was to move on to write publications for Shell and ICI.

In August 1945 Japan capitulated and Tom Gaskell returned to London. Looking for accommodation, he and Peter got together with Micky and another COHQ man in a flat at 1 Courtfield Gardens, Fulham. The two friends took short holidays together. They went climbing in the Swiss Alps and in the spring of 1946 they joined in a party to walk for a fortnight in Ireland. Peter enjoyed walking and climbing modest mountains and Gaskell was prepared to try anything once. With the war ended, they talked a good deal about what to do next: the prospect of switching from the occasionally intense excitement of wartime activity to a humdrum, nine-till-five existence was a challenge. They shared a range of relaxations: Peter was sufficiently recovered from his Sicilian injuries

BACKGROUNDER:
TOM GASKELL

From a school in Worksop, Thomas Frohock Gaskell won a scholarship to Trinity College, Cambridge. He got a first class degree in natural sciences in 1938. As part of his course he attended Lord Rutherford's last series of lectures, and the great man got him a research position with Teddy Bullard, an outstanding geophysicist who later provided compelling evidence for continental drift. During a summer vacation Gaskell had worked in Desmond Bernal's crystallography laboratory but preferred the outdoor work which Bullard offered. He got early experience of explosives and hydrophones investigating rock strata in the English Channel. The war caught up with him, and at first he was engaged in tracking underwater missiles and explosions. Rejoining Bullard in Portland and later in Edinburgh, the two worked on measures to combat acoustic mines. Gaskell got called to America for six weeks to advise on a new homing torpedo; on his return to Portsmouth, he acted as a consultant on counter-measures for mines in canals and collaborated on the COHQ study of Normandy beaches for D-Day. After the war Gaskell served as chief

petroleum physicist for the Anglo-Iranian Oil Company, multi-tasking as sports secretary, rugby player, horticultural adviser, duck hunter and mountain climber. In 1950 he became chief scientist aboard HMS *Challenger*, spending two and a half years circling the globe, making seismic surveys of the seabed and, covertly, monitoring contamination of the atmosphere caused by atom bomb testing. On his return, he continued a whirlwind life from BP's Sunbury research station, finding time to attend and address innumerable meetings of grown-ups and children, contribute to *New Scientist*, appear on radio and TV, edit journals and sit on committees. He wrote books on popular applications of geophysics like the recovery of North Sea oil and world climate well into retirement.

Tom Gaskell in the Swiss Alps

1946: Peter plots a route through Swiss glaciers... and then follows it
Courtsey Anthony Gaskell

to tackle Gaskell at squash on courts in Dolphin Square, and in May 1946 Peter took his friend to the family home in Emsworth on Chichester harbour for a weekend's sailing.

During the year following Germany's surrender, COHQ people took part in various mopping-up investigations in Europe. Peter carried out his own post-mortem on an operation COHQ itself had planned: an attack by Allied troops on the Dutch island of Walcheren, at the mouth of the Scheldt estuary. In late 1944 this island effectively denied the use of the port of Antwerp to Allied armies. The third largest port in the world at the time, Antwerp had been taken from the Germans almost undamaged. Walcheren had become a strategic obstacle out of all proportion to its size or the strength of its German garrison. Most of that garrison belonged to the 70th infantry, the so-called *Weissbrot* division, whose members were either so elderly or so delicate that they had to be fed white bread instead of the frightful German army-issue black bread. The submission of Walcheren had become a major Allied objective.

First the RAF bombed its dyke and let the sea in. Like most islands in the Low Countries it was below sea level and only the dyke and constant pumping kept the land dry and fertile. The effect of the bombing was not all that had been hoped; the garrison withdrew up onto the dykes and into bunkers which remained above water. They continued to dominate the estuary. So on 1 November 1944 COHQ launched a major amphibious assault—Operation Infatuate—to take the island, with attacks going in on either side of the breach in the dyke. It was, Peter recalled, a bloody battle, involving Canadians, Dutch and British attackers.

Photographic reconnaissance had shown that two German pill-boxes commanded the breach, one on each side. According to standard German practice, Peter wrote in a memoir, they were orientated not to fire out to sea but to enfilade the beach. Each pill-box was protected from fire from the seaward side by an extremely thick, reinforced concrete wall. At that time it fell to an inter-service Anti-Concrete Committee to deal with this kind of wall. One of the weapons its members had sponsored for the job was a 17-pounder gun mounted on the foredeck of a landing-craft known as the LCG(M) or landing craft, gun (medium). The intention was to beach the LCG(M) near a pill-box and destroy it with armour-piercing shells. The assault on Walcheren was eventually successful, but it was clear that something had gone wrong with the LCG(M) concept. Peter decided to take a closer look.

As I had a personal interest in beach obstacles and was deeply sceptical about the Anti-Concrete Committee, I wrote my own travel warrants and set out for Walcheren. With the co-operation of the Royal Netherlands Navy I got the use of a DUKW (the famous 'duck'), an amphibious vehicle equally at home on the road or afloat. My driver and I proceeded across the island, sometimes with wheels on the ground, sometimes free-floating and propeller-driven.

Peter found that the tide flowed strong and clear through the houses of Walcheren, in at the back doors and out at the front. First of all he and his driver got to the town of Middelburg at the centre of the island but slightly elevated ('like the isle of Ely in the fens', Peter noted). There they took on food and drink before pushing on to the pill-box which Peter had marked down from a photographic reconnaissance unit survey.

At last we found it. The seaward face had been hit by 20 to 30 rounds from the LCG(M) before it was sunk, but inside the gun emplacement there was no sign of damage. I picked up a piece of steel reinforcing-rod which had been left on the beach by the Organisation Todt [the Nazi engineering group engaged in large-scale construction] and probed the holes. All the shot had gone about two-thirds of the way through the concrete. The Anti-Concrete Committee had either underestimated the thickness of the concrete or overestimated the penetrating power of the [17]-pounder gun. I made them feel bad about it when I got back with my photographs and measurements.

Peter also noticed another curious thing. When the RAF breached the dyke and water poured into the island, the local cows had also retreated onto the dyke. Many of them had then trodden on anti-tank mines and were just heavy enough to set them off. Their bloated carcases lay on the dunes with three legs in the air, the fourth having been blown off by a mine.

The human population was more canny. I happened to be there on a Sunday and at about 9am figures emerged from the erstwhile German bunkers looking like a folk-lore exhibition. In spite of all privations the women were dressed overall in the Dutch/Flemish style—dazzling-white starched skirts and bonnets, with golden corkscrews in their hair. Before I left I saw the first preparations for repairing the breach in the dyke. The inhabitants of Walcheren did not wait for Marshall Aid or any other kind of subvention. They were using the technology of the seventeenth century—a raft of willow twigs weighted down with stones.

Immediately after the capitulation of Nazi Germany, the British had managed to persuade supreme Allied commander in Europe General Eisenhower that they should carry out some test firings of V2 rockets. For their trials the British had to put together rockets from recovered parts;

BACKGROUNDER:
V2-WATCHING

The V2 rocket was developed at the German rocket test centre at Peenemünde by a team led by Wernher von Braun and was first successfully launched in 1942. Over 3,000 of the missiles were fired at targets in Britain and the Low Countries in 1944-1945. The V2 was propelled by a rocket engine which used alcohol and liquid oxygen as fuel. When Nazi Germany capitulated, there was an existing agreement that the USA and Britain would share V2 hardware equally. By a series of lucky coincidences in May 1945 the Americans discovered, and without making a fuss were able to pack and transport by train to Antwerp and thence by Liberty (cargo) ship to the USA, some 100 unused V2 rockets and supporting equipment. The British eventually caught on but their protests came too

V2 rocket ready to launch at Cuxhaven, 1946

late. Some 150 German V2 technical personnel, von Braun and the V2 technical archive later followed the rockets to the States and formed the core of the Moon landings project in the 1960s. The British eventually assembled eight complete V2 rockets to evaluate and, in an event known as Operation Backfire, tested three of them in the first half of October 1945 on a range at Cuxhaven, in the British zone of occupation. Two performed perfectly. The tests required a lot of help from captured German firing troops and rocket scientists who—having just been aiming their rockets at London—were not at all enthusiastic about being handed over to the British for the trials. But in practice their assistance was vital, and the tests went off without aggravation. The test firings resulted in a five-volume report on V2s being deposited in Whitehall.

V2 spare parts arriving at Cuxhaven, 1946

the Americans had already come across V2 stocks and made off with them. Test launches of three rockets took place in early October 1945 from Cuxhaven, on the North Sea coast of Germany, from which the Nazis had carried out some of their earlier V2 launches on London. Peter was part of the British team observing the launches and later he was able to say a bit about it on the radio. The BBC Home Service, forerunner of Radio 4, ran a weekly half-hour show in 1946 called 'The world goes by'. Recorded on a Thursday, the programme was broadcast on a Sunday at tea-time. The interview with Peter about rockets went out on 12 May. Presenter David Lloyd Jones introduced him, explained his connection with V2s then invited him to talk about what would be needed to get a rocket to the Moon. Lloyd Jones launched things by saying he didn't think any of the Cuxhaven V2s hit the Moon. Peter agreed:

They fell in the North Sea, which was what was intended. All the real progress that's been made in the development of long-range rockets has been as a result of military demands ... Since then, the development of the V2 rocket has brought the project of such travel [to the Moon] much closer to fulfilment. The V2 reached a speed of something like a mile a second, and it only travelled about 200 miles before falling back to the Earth. A rocket which got up to five miles a second would have enough speed to circle the Earth endlessly, without falling back again and without using any more fuel. After [that] it wouldn't be a revolutionary step to construct a rocket that would reach seven miles a second and escape from the gravitational attraction of the Earth altogether.

Peter went on to describe some of the technical challenges of what, before the end of his life, became reality: a landing on the Moon and a sampling of the minerals to be found there. Lloyd Jones, ever keen to catch the attention of the listening public, raised the danger of sunburn on the Moon's surface and then the way that a lack of gravity would hinder the pouring of a pint of beer. Peter, clearly a less-than-enthusiastic admirer of 'The world goes by', responded drily:

I can't see you drinking it in an oxygen respirator. And in spite of the nasty cracks about materialism that have been bandied about in your Sunday postscripts recently, I don't see why scientists and technicians shouldn't have some non-utilitarian aspirations. They'll get to the Moon one day, whether it raises the standard of living or not.

Lloyd Jones was determined not to be outdone: 'Perhaps if you go yourself, you can make a report on the programme—or in a postscript', he suggested. 'And in any case,' he claimed, 'some people can drink beer standing on their heads, in complete defiance of the laws of gravity.'

In December 1945 Peter was granted a month's leave from the Navy in order to make a special trip. He was sent on a BIOS (British Intelligence Objectives Sub-committee) mission to visit three small German fullers'

BACKGROUNDER:
HOME SERVICE SPACE VISIONARY
In his May 1946 BBC Home Service interview with David Lloyd Jones on how man might get to the Moon, Peter described many of the challenges which were overcome in the following 25 years. After explaining how fast a rocket would need to go to escape from the Earth's gravitational pull, he continued:

If you made a still larger rocket, that carried rather more fuel, it'd be able to swerve round the Moon and return to earth, where it could be made to land by means of a parachute or folding wings. Clearly the first rockets wouldn't carry any crew; it'd be wiser if they were controlled by radio or automatic pilots. They should carry cine cameras or television transmitters, so we might be able to see what the Earth looks like from 240,000 miles away. Actually landing on the Moon, as distinct from crashing, would be a difficult business to arrange. As you get nearer the Moon, its gravitational pull becomes stronger than that of the Earth and the rocket would begin to fall towards the Moon. If it wasn't checked, it'd hit the surface at a speed of a mile and a half a

second. The Moon has no atmosphere, so wings or a parachute would be useless; the rocket would have to carry extra fuel for braking, which it might do by turning through 180 degrees so that the jet pointed towards the Moon not away from it. And if the rocket got damaged, it might mean that you wouldn't be able to get back to Earth again.

Getting back is going to be quite difficult, because you wouldn't have any launching-gear or ground crew. Fortunately, the force of gravity at the surface of the Moon is about one sixth that of the Earth, so the rocket would be fairly easy to handle. And, of course, it would need far less fuel for the return trip. On the way back it would start falling towards the Earth. I reckon it would have to be steered then, so as to graze the Earth's atmosphere and get slowed down a bit before it swung out into space again. This 'braking orbit' would have to be repeated several times, until the rocket was traveling slowly enough to open its wings or its parachute and land. Without it the rocket would be burned away by friction in the Earth's atmosphere. On this basis a trip to the Moon would take about 100 hours, and on the return trip you'd have to allow an extra 24 hours for braking orbits.

earth producers which had operated under the wartime umbrella of the Süd-Chemie company based in Munich. Peter knew a good deal about this activated clay used in oil refining; his employer in 1939, Fullers' Earth Union, produced it and had been the pre-war British agent for these German companies. Peter's task was to report on the current conditions of the sites, their production capacity and any technical advances compared with British practice. A second objective was to look at Süd-Chemie's production of artificial bentonite (an activated clay used for oil well drilling mud and foundry sand) and of Neuburg Chalk (a filler used, without refining, for things like paint).

The three fullers' earth mines—Rehbach, Daberg and Pörndorf—were north of Moosburg an der Isar, some 50km to the north-east of Munich. Of the three factories they supplied, only the Moosberg site turned out to be operational. The three Neuberg Chalk sites lay close together on the river Danube, familiar territory for Peter from pre-war vacations, about 75km to the north of Munich.

Stripped of my uniform, clothed in a shapeless battledress with no insignia whatever, I was sent off with a colleague [JN Wilson] to snow-bound Bavaria, in the American zone. We travelled in a Humber staff car driven by the stupidest service driver I have ever encountered—and that is saying a good deal. He lost no opportunity of driving into snowdrifts, sliding into ditches or just losing the way, although we had maps and tried to tell him which way to go. Our US friends and allies were just as much of a liability. Trying to get 'gas' from a US Army fuel dump with an illiterate corporal in charge was one thing. Being given the brush-off by officers of the US Army of Occupation, once our quarry sensed our presence and suggested to them that it was in their best interests to get rid of us, was something else...

Peter's report made clear that the two Süd-Chemie directors he interviewed often declined to produce technical details without authorisation from the military government. So he and Wilson had to work by observation and questioning. The fullers' earth mine sites were difficult to access in winter along tracks of poor quality. Still, Peter remembered with pleasure a visit they made to the Pörndorf mine, a small and isolated installation deep in a forest. The foreman and his wife were so relieved to discover that they were British and not American that the foreman went into their garden and dug up a few bottles of good Moselle.

We drank a number of toasts, including 'To hell with the Yanks—*ein Volk ohne Herz*'. I am sure that any American investigator would have met with a similarly warm reception in the British zone.

Peter discovered that their driver was as literal-minded as he was inefficient. He had orders that he was only to drive his passengers on journeys related to their official business, and was wary of side-trips. Nevertheless:

Just before we left Munich we fooled him into taking us to the concentration camp at Dachau. We were shown round by ex-prisoners still in their striped pyjamas. It was a deeply depressing experience. When we got back to the car, our driver said: 'Been joy-riding, have we? I shall have to report this.'

Peter also began to write short articles about his wartime experiences. He submitted them to the popular gentlemen's monthly publication *Blackwood's Magazine*, or the '*Maga*' to readers. The editor of the day—James Blackwood—reacted enthusiastically. The *Maga* had a terrific literary track record dating from the beginning of the nineteenth century, and had serialised books such as Conrad's *Heart of darkness* and Buchan's *39 steps*. Its heyday coincided with the glory days of Empire with colonial stories of military derring-do, hunting, shooting and fishing. As the Empire faded, so did the *Maga*. But war had a magical effect on circulation: readership surged during WW2 and by 1946 it had 30,000 readers worldwide. The Chief of Naval Information had to cast one eye over Peter's

Peter aged 29, just prior to discharge
from the RNVR
Photo: Basil Shackleton

drafts, keeping the other on the Official Secrets Act, but he made few if any deletions. James Blackwood published three articles (or 'papers' as he liked to call them) in December 1945, May and November 1946 and the publisher was generous: Peter received a royalty of two guineas (£2.10) per page of 700 words. He let the editor know that his friend Tom Gaskell was also keen to see his wartime adventures in print and was 'churning out a description of a walking tour in Ceylon' which he intended to submit. Gaskell took the hint and in September 1946 sent in 'Ceylon holiday'. Blackwood responded sympathetically but found the story 'somewhat lacking in drama'. He returned the manuscript, rather disappointed that 'you do not appear to have had any misadventures.'

On 20 May 1946 Peter was 'released to shore', still classed as a temporary lieutenant. He had collected two major medals and, between the long periods of boredom typical of war service, had been through his fair share of tense moments. The half-forgotten excitement of civilian life now awaited.

BACKGROUNDER:
PETER'S COHQ REPORTS
From 1944 to 1946, Peter was a COHQ experimental and trials staff officer. His boss, Captain Thomas Hussey, wrote in 1944: 'A most outstanding officer in all respects. He has been employed on experiments and trials connected with the solution of the problem of underwater obstacles in amphibious operations and has proved to have ability, initiative, drive and tact. He has great knowledge of explosives.' Already interested in explosives from childhood, Peter had now had five years' wartime exposure. In 1944 and 1945 Hussey gave Peter straight eights and nines out of ten across the spectrum of characteristics but a lowly seven for administration. In his 1946 report Peter was getting the hang of communicating with all ranks, and his trip to Germany had made his superior officers aware of his linguistic abilities: 'He is loyal and has a pleasant manner with his seniors and juniors. Speaks German fluently.' But even at this young age Peter's interest in administration was clearly flagging: a five. Nevertheless Hussey concluded: 'A most outstanding officer who is strongly recommended for immediate promotion.'

Side-stepping Regulation 18B 8

THE LAST TIME PETER HAD BEEN IN A CIVILIAN JOB was in July 1940. After graduating from Oxford a year earlier, he had taken a position with Fullers' Earth Union in Redhill, Surrey, as an industrial chemist. When war came his position had been classified as a 'reserved occupation' helping the war effort and so he was not required to join up. But he became weary of the phoney war, bored with his job and tired of his evenings with the Home Guard. He wanted to go to war. But then, just before he volunteered, an official letter arrived at his flat with the potential to divert him in an entirely different direction:

As a possessor of a funny name, I was invited to visit the Chief Constable of Surrey and explain myself.

Shortly afterwards Peter had found himself closeted with a member of the constabulary at the local police station. The constable wanted to know why his surname sounded Dutch and might even be German. The British government—egged on by a press campaign against a supposed fifth column, imagined as living in Britain and standing ready to assist the German invasion force when it arrived—had invoked Regulation 18B. This allowed for the internment without trial of anyone suspected of being a Nazi sympathiser. Originally introduced to deal with members of the Irish Republican Army committing senseless minor outrages in London, it suspended the right to *habeas corpus*.

Regulation 18B was a serious threat. Being judged undesirable meant not just an internment camp but the possibility of being exposed to considerable danger *en route* to wherever that camp might be. The *Arandora Star* tragedy was fresh in Peter's memory. Earlier that month, on 2 July 1940, some 1,200 suspicious-sounding individuals, mainly Italian long-term British residents who had been detained when Mussolini declared war on 10 June, were herded into the *SS Arandora Star* in Liverpool for transportation to Canada. A further 375 on board were guards or crew. Off the north-western coast of Ireland the *Arandora Star* met the ace

German commander Günther Prien in his submarine U-47. A single torpedo was enough. About 500 drowned, and bodies were being washed up on the Irish coastline throughout August.

The surname Danckwerts may have sounded Dutch or German but there were a number of mitigating circumstances. A summary of the adventures of a generation or two of Peter's forebears would have explained a good deal. Naturally the constable was a busy man and didn't have the time to hear very much; had he been able to spare half an hour, this is what he might have learned:

The Danckwerts did indeed come originally from Germany. There is even a coat of arms, first granted in Hanover in 1776 when various branches of the family were being ennobled as 'von Danckwerth': a sand eagle sits atop a knight's helmet with the usual surrounding drapery; the helmet is perched on a shield showing three green hills supporting a rose and on a green field below lie two crossed blades of forged silver. The family tree traces the Danckwerts name back to Lüneberg in the sixteenth century.

The British branch owes its identity to Peter's great-grandfather, Viktor Adolf Danckwerts. Born in Göttingen in 1821, the third of four boys, he studied theology and medicine. Then he took a post as an army doctor. His father—a bookseller and editor—died when his son was 20. Viktor married Ida Sophia Louisa Wilmans who was four years his junior and a doctor's daughter from Stade in Lower Saxony. She had already presented him with a son and baby daughter when, in the summer of 1855, relations within the family apparently reached a crisis; he upped sticks and took his wife and two children to Britain.

Why Britain? The British government was advertising for recruits to fight in the Crimean War. Its army was traditionally small: disease, casualties and apathy at home had led to a severe shortage of volunteers. So the government looked abroad for recruits: effectively to man a British Foreign Legion. In the event, the response from young men in continental Europe was so great that the British were able to form separate units based on nationality. Viktor joined the Anglo-German Legion as a doctor.

But he was too late. The war was ending, and this presented a problem as much for the British Army as for the new recruits. Assembled in Shorncliffe Camp on the outskirts of Folkestone, most could not return home, even if they wanted to, because they had sworn allegiance to a foreign power. Fortunately the Army was not the only British organization short of manpower. The Empire needed young settlers and many of these 'stranded' foreigners chose India or the Americas. But Viktor Adolf was one of four doctors amongst 2,362 frustrated would-be Crimean cannon-fodder who set off by boat, with wives and children, to sample the life on offer in the southern hemisphere on the borders of Cape Colony. In early 1857 they arrived in the East London region of South Africa.

Few of these ex-soldiers were ideally suited to the life of a frontiersman and farmer in Cape Colony. Many soon drifted on to India, to defend the Empire against the 'Indian Mutiny'. Viktor Adolf, however, stayed put. As a doctor he must have been a valued member of society. Along with all the other immigrants, he received a grant of land in the border area. His was in the largest of the new settlements, some 150km north west of East London on the Kubusie river, where he built a substantial house. His neighbour and good friend was the leader of the Anglo-German Legion's pioneering group: Baron Richard Charles von Stutterheim, who gave his name to the new settlement. The baron didn't stay for long, not even long enough to complete the ambitious turreted house he had planned. He returned to Germany at the end of 1857, gambled heavily and in a relatively short time lost a lot of money. His bank in Germany refused him further funds and he settled the debt by shooting himself.

Von Stutterheim's friend, now Adolf (or Adolphus) Victor Danckwerts, was made of sterner stuff. He had a responsible medical career and—what is more—took Cape Colony's population requirements seriously. He fathered a further four children with Ida, making six in all. The Danckwerts family prospered and moved house to Bedford, to the west of Stutterheim. There in 1866 Ida Sophia died; she was 41, and had given birth to her sixth child only a year earlier. Adolf married a second time— to Emily Painter. Although the Painters were also recent arrivals in the area, Emily's father Richard was already a member of the legislative assembly of the Cape of Good Hope. Emily's sister Amelia married a successful young captain in the Indian Army, William Henry Lowther of the Earl of Lonsdale's family, and that union was later to make a significant impact on the life of Adolf's eldest son. The Danckwerts moved on to Somerset East, west of Bedford, and about 100km north of Port Elizabeth. Emily bore Adolf four more children, further increasing the Cape's stock of Danckwerts. Adolf died in Somerset East in 1879.

Meanwhile Adolf Viktor's eldest son Wilhelm Otto Adolf Julius—born in Germany in 1853 before the emigration—had been getting an excellent local education. He attended Bedford School—under the eye of an out-standing teacher, the recently-recruited minister Robert Templeton, a Scot—and then Gill College in Somerset East, named after an immigrant doctor from Market Harborough who had left £23,000 in 1863 to establish the college. Leaving school at 17, young Wilhelm Otto Danckwerts took time off. He went to explore the diamond fields of Kimberley and worked for a while as a farm overseer. According to contemporary reports he took to the life very well, 'able to shoot a green parrot on the wing, and with a stock whip flick the ear of an ox in the leading span of a team'. But he wasn't destined to be a farmer.

To further his education he had won a place as a pensioner—he had to pay his way—at the Cambridge University college Peterhouse. He travelled to Britain and took it up in 1873. In spite of flak from his contemporaries for what they saw as his rural colonial background, he graduated in mathematics with honours four years later. He came only 82nd in a class of 96, so it was perhaps as well that he wasn't counting on mathematics for a livelihood. He had other ambitions: his Peterhouse obituary reveals that, during his time at college, 'William Danckwerts had kept terms at the Inner Temple'. He was called to the Bar in 1878.

William Otto trained as a lawyer in Lord Bowen's chambers in Brick Court. One of his contemporaries was Herbert Asquith, to become British prime minister from 1908 until 1916. Asquith wrote of his experiences at Brick Court that 'I learnt, I hope, many things there, and amongst the other lessons which I learnt was one which every man who aspires to practice with success at the Bar in these days has to learn sooner or later, and that was the dangers, the multiform and manifold dangers, of an encounter with Danckwerts.'

Early in his career William Otto married well. William Henry Lowther—the Lonsdale married to Amelia Painter whose sister Emily had become William Otto's stepmother—had by now retired from a heroic career with the Indian Army, mainly in Bengal, with the honorary rank of major-general. He and his wife had set up home in England on the Lonsdale family estates south of Penrith, Cumbria. With lineage from Saxon times, they stood well up the social scale and were exceptionally well-connected. Even though those connections stretched as far as the South African Danckwerts, the Lowthers were not particularly enthusiastic when—in the early 1880s—William Otto approached them for the hand of their daughter, Mary Caroline.

The Lowthers turned the young man away, suggesting that he might return when he could demonstrate an annual income in five figures. It didn't take him long; in 1885 he swept Caroline off to London. Together they produced three sons and a daughter. The first son, Harold, followed in his father's footsteps and even outshone him. The second son was Peter's father Victor Hilary Danckwerts. The other children died young—Elsa Gretchen, aged 11, of leukemia and Richard William at 22 in WW1.

From the start of his legal career William Otto was known for a quickness of temper and an irritable disposition. His opponents described him

A window of Great Salkeld church—a Cumbrian tribute to Peter's great grandfather William Henry Lowther: *Magistratus indicat virum*—the office defines the man

disparagingly as 'a Boer'; he had the reputation for being 'the rudest man in London'. But he was admired too. He showed a fine grasp of a brief, argued it lucidly and was considerate of those around him in court, which made him a favourite with the Bar. He joined the south-eastern circuit around London and his practice increased rapidly. He specialized in revenue and rating, showing an amazing memory for past cases and Acts of Parliament, particularly the National Insurance Acts.

William Otto's cases covered a wide range. In 1884 he took a small part in the case of the *Mignonette*, a *cause célèbre* involving cannibalism. He appeared for the police in 1898 to prosecute the secretary of the Legitimation League for selling a book it had published called *Sexual Inversion* by the well-known sex spcialist Havelock Ellis. The police claimed it was indecent. According to press reports of the time, William Otto noted in court that the object of the League appeared to be to do away with the current marriage system and to promote free love. He named certain 'practices' which he said it advocated, activities immediately refuted by the defence. Then William Otto indicated that he would like to read passages from the book. The judge addressed himself with concern to the women present: 'Things are now going to be read which no woman ought to hear; no decent woman will remain for one moment.' *Reynolds's Newspaper* reported that the only movement women in the court made was to lean further forward in their seats to hear more distinctly what was about to be read.

In 1900 William Otto 'took silk' as a Queen's Counsel. As the Spy caricature of him in *Vanity Fair* shows, he was 'one of the biggest men at

---

BACKGROUNDER:

THE CASE OF THE MIGNONETTE

William Otto Danckwerts was never one to duck a tricky case. In 1884, with a mere six years' experience as a barrister, he played a part in the case of the yacht *Mignonette* which set standards in the judgment of cannibalism. Putting the case in a nutshell, four men were contracted by the yacht's Australian owner to sail the *Mignonette* from Southampton to Sydney. In the Atlantic, 1,600 miles north-west of the Cape of Good Hope a large wave hit the yacht and sank it within five minutes. The crew took refuge in an open lifeboat with little food and no water. After three weeks adrift they were in dire straits and the young cabin boy in a coma; two of the three others decided their only hope was to kill the boy and eat him. A few days after the killing, the remaining three were picked up by a passing vessel and returned to Falmouth in Devon. Emotions ran high throughout the country at the prospect of determining whether necessity was a justification for murder. A junior Treasury counsel at the time, William Otto was briefed for the prosecution in the initial hearing before magistrates in Falmouth. He entered a hostile environment, for local opinion was right behind the crew in what they had done. William's son Victor later recorded, perhaps over-dramatically, that a conviction achieved could have put his father's life in danger. But William Otto showed himself to be fair-minded, making it clear that the prosecution sympathised with the three men. Getting evidence against them was a problem, as they exercised the right to silence. To overcome this, William Otto dropped the case against the third man who had opposed the murder and called him, and various others to whom the crew had told their story, as witnesses. Local press magnanimously described his handling of the situation as 'skilled'. The magistrates duly committed the two crew members for trial before judge and jury in Exeter, where a Queen's Counsel took over the prosecution. After many twists and turns, not to mention political interventions, the men were found guilty but were sentenced to a mere six months.

the Bar'. He became a freeman of the City of London, yet his temper and irritableness—or perhaps his earlier reputation for such things—meant that he was never made a judge. Nevertheless, William Otto and Caroline were comfortably off and lived initially at Chessington Lodge on the way to Leatherhead, where they kept five domestic servants and a gardener. Later they moved to South Kensington at 2 Brechin Place, off the Brompton Road. William Otto retained a certain colonial enthusiasm for life: at formal parties that he organized from time to time he is said to have enlivened the evening by firing a soda water siphon under the table at the ladies' legs.

William Otto Adolf Julius Danckwerts,
'Danky, a rude and learned personage',
Vanity Fair cartoon by Spy
(Sir Leslie Ward),
Man of the day no 716, 23 June 1898
Courtesy Peter Michael Danckwerts

To escape from the pressures of London, William Otto held a long-term lease from the Duke of Sutherland on a hunting lodge in Scotland. Syre Lodge lies deep in the Scottish highlands, north-west of Kinbrace. According to accounts in *The Scotsman*, which in those days ran listings of such things, he shot a lot of deer and, with his wife, fished for salmon in the nearby river Naver. William Otto's sporting interests in Scotland extended beyond slaying deer and catching salmon; the Danckwerts' oral history suggests he was responsible for at least one illegitimate Scottish son. The 1912 Sutherland register of motor vehicles shows the details of William's modest shooting brake, just below a Rolls and a Cadillac belonging to the Scottish-American steel magnate and philanthropist Andrew Carnegie. Today there are still reminders that the Danckwerts were active in the area. When the couple's third child, Elsa Gretchen, died from leukemia in 1902, William Otto had her buried in a small graveyard lined by yew and holly trees on a hill just behind Syre Lodge. The memorial—two large slabs of stone—is visible from the house. And the road crossing of the burn between the Syre and Badenloch estates, owned by Lord Leverhulme, is still known by locals as 'Danckwerts Bridge'.

William Otto Adolph Julius Danckwerts KC died in the spring of 1914 at the early age of 60. *The Manchester Guardian* suggested that 'those who knew him well had long suspected serious internal trouble,' and praised him as 'conspicuous for his learning and remarkable gifts of memory and exposition,' even if, in earlier years, he was 'self-opinionated and over-proud of his learning.' *The Manchester Guardian* noted that, 'he not in-frequently cited cases and *obiter dicta* [statements made in passing] which had escaped their lordships' recollection, and then it would be "Of course it is so, if you say so, Mr Danckwerts"'. *The Times* obituary commented: 'He made such a reputation as a lawyer that whenever a subtle question was likely to arise in an action his name was among the first, if not the first, to be mentioned as counsel to be retained.' Those who listened as he unfolded an argument 'knew well how he justified this reputation, and knew also with what gratitude the bench always received the reasoning which he contributed to the settlement of important points in many a modern statute.' A footnote recorded that 'Mr Danckwerts was a Fellow of the Zoological Society and had presented to the Gardens many valuable animals.'

His wife Caroline anticipated a comfortable widowhood. She was well aware of the success of her late husband's practice, although she may not have realized that he was earning around £20,000 a year, equivalent to more than £1.5million in 2010 money. So she was surprised to find, when probate was granted, that his bank account was almost empty. Caroline made further enquiries with the practice clerk. She had for many years been puzzled that she seemed unable to retain female domestic staff. One after another they had left service at the Danckwerts' London home in Brechin Place. Now she found out why. Her late husband had been 'picking them off', setting them up in rooms around the capital and visiting as the fancy took him. His practice clerk admitted to managing the whole undertaking. Danckwerts of the next generation were slightly more circumspect in their behaviour but still made their mark.

The British Danckwerts 9

WILLIAM OTTO ADOLPH JULIUS DANCKWERTS, Peter's grandfather, who established the British branch of the Danckwerts family, had been born in Germany and raised in Cape Colony. He studied in Cambridge and remained in Britain as a successful lawyer. His second son Victor Hilary, Peter's father, was born in 1890, educated briefly at Winchester College for two years until 1904 and then at *HMS Britannia*, the on-shore officer-training establishment overlooking Dartmouth. He specialized in gunnery and his early service record was exemplary. His commanding officers saw him as a man destined for high command: 'Able, zealous, promising,' were typical comments in his naval career log.

Scarcely twenty, Victor had met and was courting a girl two years his senior: the eighth daughter of a thoroughly naval family, the Middletons. One of Joyce Middleton's ancestors—Charles Middleton—had served briefly as First Lord of the Admiralty in 1805, the year of the battle of Trafalgar. Joyce and Victor were engaged by the time WW1 broke out. She later recounted to her own children how she had visited 2 Brechin Place in London to present herself as a potential daughter-in-law: Victor's father had chased her round the table offering a guinea a kiss.

Courting at a ball: Joyce Middleton and Victor Danckwerts pre-WW1

Victor was by now a lieutenant and assigned to the Monmouth class armoured cruiser *HMS Kent*. In 1914 the *Kent* was sent to chase German ships out of the south Atlantic. The trip culminated in the 'Battle of the Falklands' when Victor acquitted himself well. But he was less lucky with the local wildlife. He must have drunk water or eaten food contaminated with sheep's flukes. Parasites threatened his liver, but he recovered. On the way back to Britain, *HMS Kent* rounded Cape Horn and put in to Esquimalt naval base on Vancouver Island for repairs. Victor got shore leave; he cabled home to his fiancée, inviting her to travel to the Canadian west coast and marry him. She may have seemed a delicate and inexperienced young lady, with a maid to handle practical tasks, but she was also determined and resourceful. On the boat to Canada other travellers took her under their wing, and she continued by train across the continent. Later she told her children that she took off her engagement ring during the voyage to have more fun. Meanwhile in Vancouver, Victor and the whole crew of the *Kent* were being entertained as conquering heroes, with ample press coverage of 'the Battle of the Falklands'. The local Pantages Theatre showed film of everyday life on the *Kent* amongst its vaudeville acts and

---

BACKGROUNDER:
CHARLES MIDDLETON AND
COPPER BOTTOMS

The son of an eighteenth-century customs official in the Borders of Scotland, Charles Middleton served for 22 years in the Royal Navy. He was made comptroller in 1778 by John Montagu, fourth Earl of Sandwich. then on his third stint as first sea lord. It was the height of the American war of independence. The *Oxford Dictionary of National Biography's* entry on Montagu describes Middleton: 'An obscure Scot with an undistinguished service career, Middleton was a surprising choice, but none ever better vindicated Sandwich's judgment, for Middleton became the outstanding naval administrator of the century— though he was always a difficult colleague, cantankerous, egocentric and disloyal, and only Sandwich was able to manage him successfully.' Following the loss of America, Britain's naval supremacy was severely threatened: the French and Spanish fleets were expanding; Britain's was elderly and in bad repair. Middleton oversaw with urgency a controversial 'copper bottoming' of the fleet, extending the life of ships and giving them greater speed through the water. He is also credited with introducing the carronade, a small cannon with a large ball, which added greatly to the fleet's firepower. Able and energetic, Middleton served briefly as first sea lord at the time of Trafalgar. In this position, naval historian Sir Julian Corbett remarked a century later, he was 'the man who, for ripe experience in the direction of naval war in all its breadth and detail, had not a rival in the service or in Europe.' In the course of his career Middleton was knighted and later ennobled. As Lord Barham he played a creditable but passive part in helping opponents of the slave trade—for instance, by making premises available for meetings.

BACKGROUNDER:
THE FIRST BATTLE OF THE FALKLANDS

On the morning of 8 December 1914, a fleet of five German cruisers appeared off Port Stanley in the Falkland Islands. A British fleet of seven had arrived the previous day and was 'coaling', intending then to sail round Cape Horn and hunt for German ships in the Pacific Ocean. They were delighted when their opponents saved them the trouble. HMS Kent had steam up and was on guard duty in the outer harbour at the time; Lieutenant Victor Danckwerts was in his bath when the alert sounded. The Germans made off to the east, and the Kent shadowed while the other British ships got up steam and followed. There was a happy sense of anticipation amongst the crew, 'with all the hands who were not stoking on the forecastle watching the chase', Victor wrote. Rear-admiral Sturdee aboard HMS Inflexible invited HMS Kent's Captain Allen to deal with the cruiser Nürnberg. On paper, the Kent was half a knot slower; she had not taken on coal, and so was not ready for a long chase, but being lightly loaded could get up a good speed. The stokers worked wonders, and the Kent reached 26.5 knots, four beyond anything she had achieved before. As her furnaces were burning coal at an enormous rate the crew threw in anything combustible—wooden boats, spars, companionways and ladders, even deck planks. The action conjures up scenes from the film of Jules Verne's Around the world in 80 days. By late afternoon the Kent was within range. Her 6-inch guns opened up at 11,000 yards (nearly six and a half miles) and by a combination of fine marksmanship and good luck one of her shells struck the stern of the Nürnberg, disabled guns and seriously affected her speed and her ability to manoeuvre. The Germans' 4-inch guns fired more rapidly, and shells fell on and around the British cruiser. One entered a gun port and burst inside, causing a flash of flame to shoot down the ammunition hoist and into the passages below. A marine sergeant threw burning material aside and, seizing a hose, extinguished the fire, thus saving the ship and the 700 onboard. Damaged and on fire, the Nürnberg soon gave up and sank with the loss of some 350 men. The action had lasted just under two hours and the 'real fight' only 50 minutes. The Kent lost only six men, in spite of being hit 36 times, and suffered no vital damage. Lieutenant Danckwerts, who had spent several hours on the foretop—a platform at the top of the ship's foremast—directing gunnery in an icy wind, told his fiancée that 'I am somewhat the man of the moment, but the engineering commander is the blue-eyed boy, as our steaming was more than remarkable… and we all owe our scrap to him'. Four of the German cruisers—the Scharnhorst, Gneisenau, Leipzig and Nürnberg—were sunk; only the Dresden survived. By the time the Kent arrived in Port Stanley she had little more than sweepings of coal in her bunkers.

the Maritime Museum of Victoria retrieved and displayed mementoes of the
Battle. When the train carrying Joyce
finally arrived, Victor was waiting on the
platform to meet her. 'I never thought
you would come,' was his greeting.

They were married on 26 June 1915:
someone gave them 12 silver napkin
rings, 'one for each child'. Fellow officers
decked out the *Kent* appropriately and
presented the couple with a silver salver
inscribed with all their signatures. The
newly-weds honeymooned locally, and
discovered how competitive they were.
Bathing in a Victorian lake, Victor struck
out for a rock some way from the shore.
Joyce followed, even though she wasn't

Victor and Joyce Danckwerts on
their wedding day

quite such a good swimmer. The rock would only accommodate one and
she nearly drowned. Things did not improve: Joyce fell from a horse and
broke her collarbone, and Victor was hospitalised with pneumonia and
pleurisy. They spent the rest of the honeymoon in separate hospitals.

When the couple got back to Britain, Victor Danckwerts was seconded
from April to August 1916 to the Navy's headquarters for dealing with
mines—HMS *Vernon* in Portsmouth. A sharp mathematician, he worked
on the development of paravanes for mine-sweepers (see p100). After a
variety of commands and promotions, he became the Admiralty's assistant

HMS Kent off Vancouver Island flying the 'wedding garland' for Victor and Joyce Danckwerts

director of plans from 1932 to 1935. He was beginning to move in influential circles: on 12 November 1936 Captain Danckwerts received an invitation to dine on the royal yacht with Edward VIII, a month before his abdication. He represented the Navy at international conferences and commanded a destroyer flotilla in the Mediterranean during the Spanish Civil War.

By March 1938 Victor had become director of plans at the Admiralty. The *Daily Star* reported that, 'Big Dan… one of our brainiest naval men… is breezy, frank and pleasant'. 'His father,' the *Daily Star* continued, 'was a man who dealt with judges as frankly as his son does with sailors'. When WW2 began Big Dan found himself serving Winston Churchill at the Admiralty. Early in 1940 the two fell out over one of the First Lord's 'more bizarre plans': Churchill wanted to harass the Nazis from the Baltic but Victor thought it strategically inadvisable and dragged his feet. He also expressed reservations about Churchill's nocturnal habits which meant that, while Churchill might sleep by day and work by night, his staff had to work round the clock. Churchill sacked him. It coincided with further illness which the navy diagnosed as tuberculosis requiring 12 months in a sanitorium.

During 1941 the government sent Victor, fit once more, to Washington for talks with the Americans. He showed a tendency to make decisions off his own bat, and Churchill sidelined him for a second time. Then, in

Victor in Washington, 1941, making his own decisions

the spring of 1942, he became deputy commander of the Eastern Fleet based in Ceylon (Sri Lanka) with the rank of acting vice-admiral. At the end of 1944 he retired definitively and was made—oral history again—the promise of a knighthood when he arrived home. A long-held ambition to fish for trout in New Zealand persuaded him to go home the 'wrong way' round, but he never made it: the kidney problems he had suffered from throughout his life had left him with high blood pressure and, off New Zealand at the modest age of 54, he suffered a stroke. They buried him at sea. With Victor, Joyce had produced five children between 1916 and

1924: Peter, Dick, Hilary, Diana and Micky. She nursed Peter in early 1944, while he was recuperating from his Sicilian experience, and eventually outlived him by four years, calling it a day, deaf as a post, at 100.

Victor's elder brother by two years, Harold Otto Danckwerts, followed his father into the law. Just as gifted, he drew on his mother's genes for a milder disposition. He told Leslie Blackwell, his contemporary in the South African legal world, how a London newspaper described him as the gentle son of a violent father. As an illustration, he described walking with his father in Hyde Park at the age of 10 dressed in his Eton jacket and top hat. They came across Sir Edward Carson, then the English attorney general. Young Harold forgot to tip his hat to the great man, and his father laid into him there and then with a pair of leather braces. But Harold, with his relatively mild manner, succeeded where his father had not: he became both a judge and a knight.

Sir Harold Danckwerts in 1965
Photo: Portman Press Bureau

Sir Harold is best remembered beyond the legal profession as the Lord Justice of Appeal who was asked in 1952 to arbitrate on whether the recommendations of the 1946 Spens committee on general practitioners' pay in the new National Health Service had been fully implemented. Reporting, Pendennis wrote in the *The Observer*: 'Danckwerts is not one of those elaborately witty or irascible judges whose public fame is often won at the expense of their colleagues' respect. He likes quiet things—photography, whippets, making model ships. His family is Hanoverian in origin—he rather thinks he is a Count of the Holy Roman Empire—and he is a friendly and humorous man, with a comfortable Hanoverian appearance. Whatever he decides it is likely to be sensible.' In the event, his decisions allowed doctors to catch up on salaries after wartime deprivation, and rather took the government's breath away.

There were other idiosyncratic contributors to Peter's gene pool. His maternal grandfather Richard Middleton, born 1846, had an adventurous naval career. The grandson of an admiral, Middleton's first career choice was to join the Royal Navy. But he failed a medical, reputedly because of a bang on the head in the school playground. Sent on a long sea voyage to recuperate, he was successful the second time round and enjoyed a stirring naval career in the Mediterranean and the Far East. His final commission was in charge of a sloop-of-war named *Flying Fish*, based in Zanzibar. Patrolling the channel between Mozambique and Madagascar, his job was to intercept slavers and free their captives, an activity of which he was justifiably proud later in life. Returning to the UK in 1877 Richard Middleton married, left the Navy and settled down near Southsea to keep goats and tend vegetables—a life which he imagined would bring contentment. But it didn't; he was bored. So he moved to the busier environment of Blackheath, on the south-east outskirts of London. Joining the local Conservative club, he became its secretary and ran a constituency office for two local MPs.

Such a gifted organizer and fundraiser did Middleton prove to be that, by 1885, the national Conservative party asked him to be its chief agent, then a new concept. His efforts in the following years were said to have 'contributed in great measure to the party's electoral success in 1895'. The party's senior members were immensely grateful to 'Skipper' Middleton and organised a 'whip-round' for him amongst more than 4,000 well-heeled party members. With the proceeds they presented Middleton with a silver box, an album of good wishes signed by all the donors and a cheque for £10,000, or about £1 million in 2010 money. That large sum must have caught the eye of his eldest son Dick. When

Richard Middleton, *Vanity Fair* cartoon by Spy (Sir Leslie Ward), Man of the day no 806, 18 April 1901
Courtesy Jane Clarke

Richard Middleton died in 1905, Dick—known thereafter in the family as 'the Rotter'—tricked his gullible mother Elizabeth into signing away her late husband's estate to him. He took off for a life of wine, women and song, before syphilis brought him up short in Paris. She retired to a tiny thatched cottage in West Ashling near Chichester, Sussex, which her grandchildren recall as 'quaint' and having no electricity.

The Middleton family in the 1890s: (standing from r) Richard (second), Dick the Rotter (fourth), Elizabeth (fifth); Joyce is at the front, on the back of the cart

Of course, neither the policeman who dealt with Peter's case under Regulation 18B nor the Chief Constable of Surrey got to hear all this. But they heard enough. Peter wrote later:

It was a cheap triumph, but I was able to say to the copper who interviewed me, 'My father's an admiral and my uncle is a judge'.

They let him go, and off he went to join the Wavy Navy and meet his destiny.

Early life and prep school 10

W HEN HER HUSBAND is in the Navy, his wife doesn't get to see him too often. Family extensions generally coincide with leave. In January 1916 Lieutenant Victor Danckwerts was on leave at home: 25 Junction Road, Southsea, now gentrified as Wimbledon Park Road. Victor reported for duty again in March and set to work on the design of paravanes for minesweeping, a new concept in naval warfare. Joyce was pregnant throughout the spring and summer of 1916 and on 14

25 Junction Road
(now 25 Wimbledon Park Road)

October their first son was born. They had chosen his names some time before: Victor was a traditional one and went back five generations, but Peter was a first for the family. They chose the names carefully; Peter's mother explained later that they had deliberately given their son the initials of the ParaVanes Department in which Victor had been working that summer. During the rest of his life, which began in an era when to use first names was considered inappropriate or even cheeky, Peter's contemporaries referred to him as 'PVD'.

Before long there was more leave and another boy—Richard Evelyn. And then two daughters—Hilary Joyce and Diana Jean. The family urgently needed more space. In 1921, and by now a lieutenant commander, Victor took his family to nearby Emsworth on

Joyce Danckwerts with Peter and
baby Dick in 1918

Chichester harbour. They bought Merton Lodge, 75 Havant Road, on the main road about a mile west of the centre of Emsworth. It was a large two-storey building, dating from about 1900 with four bedrooms, two attics and a variable staff of a nurse, two maids and a part-time gardener. The garden was big enough to grow vegetables and to keep chickens; in those days shrubbery shielded the house from what is now the widened and rather busy A259 road. Joyce Danckwerts' last child before she called a halt in 1924 was Michael John.

BACKGROUNDER:

PARAVANES

'An underwater glider with teeth': that was the WW1 invention of two British naval men—Cecil Usborne and Dennis Burney—which became known as the paravane. It looked like an aircraft and was towed by a wire from the bows of a ship. As it was towed it moved away from the ship and downwards until, at a certain depth depending on design, the paravane reached an equilibrium position. In those days mines were contact weapons positioned just below sea level on cables anchored to the seabed. When a paravane wire snagged a mine cable, the cable was deflected down the wire towards the paravane, where it met and was severed by a powerful cutter blade. The mine floated to the surface and could then be destroyed by gunfire. Earlier WW1 methods of combating mines, like stretching a serrated wire between two vessels to snag and cut mine cables,

were primitive, inefficient and tied up two vessels at a time in a very inflexible formation. German mine laying in the North Sea posed a big threat, and Admiral of the Fleet Earl Jellicoe wanted a protection system which could be fitted to single vessels, allowing them to retain speed and flexibility of movement. The paravane developed by Victor Danckwerts and the team at *HMS Vernon* in Portsmouth provided a solution. A vessel towing two paravanes, one on either side of the bows, was deemed safe from moored mines. Paravanes were still in use during WW2.

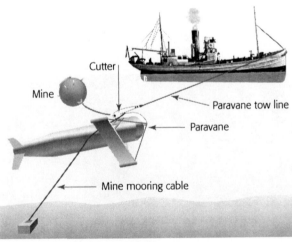

The children grew up in and left home from Merton Lodge. Peter's later assessment of the situation at Merton Lodge expressed, one suspects, for effect was 'a state of genteel poverty'. His mother, with some reason, took great exception to the phrase. A lieutenant commander's pay meant that there was enough money around but not a lot more. Victor was pragmatic: when he came home on leave he bought a car; when leave ended he sold it again. Emsworth lies on one of many inlets of Chichester harbour and Victor was a member of the Emsworth sailing club. He was a contemporary of Louis Mountbatten who served as the club's commodore in 1933. The sailing club was positioned to the west of the village waterfront, on the same side as Merton Lodge, and had a tidal swimming pool. The whole of Chichester harbour lay at the club's disposal— a large and relatively safe expanse of water on which to learn to sail. The Danckwerts children did just that. In due course Peter had use of a 16-foot (5 metre) dinghy lent to the family and spent many hours sailing.

There were other hobbies which claimed his time. Photography was one. He set up a dark room in one of the attics at Merton House and it allowed him to experiment with developing his own black and white film. Chemistry was another hobby:

I believe I was about seven years old when I first decided to become a chemist.

(l to r): Peter aged 5 with his siblings Richard, Hilary and Diana

That naval tradition: (l to r) Richard, Victor, Hilary and Peter

Merton Lodge, Emsworth

I had been reading a school story about an illicit 'stinks' experiment. It was necessary to turn out the reaction mass into a large vessel at some crucial stage. The only vessel available was the Junior Inter-House Hockey Challenge Cup. It turned black. I was fascinated and asked my mother what chemistry was about. Although she had no idea, she rose splendidly to the occasion: 'It's about things like making diamonds, darling.' What red-blooded boy could have resisted such a challenge?

Peter developed an interest in explosives and his brothers joined in as they grew older. As a young boy Peter often ended up with the debris from amateur bombs embedded in his face and around his eyes; his mother was called upon to remove it. The ingredients for bomb-making were, in those days, largely available from the local chemist's shop. While he found the making of gunpowder—from charcoal, sulphur and potassium nitrate or sodium chlorate—unrewarding, and nitroglycerine a step too far, Peter certainly enjoyed the pre-war schoolboys' favourite domestic option: nitrogen triiodide. The ingredients were iodine crystals and strong (880) ammonia solution. In the evening he would put the crystals in an egg-cup and cover them with ammonia solution. Next morning he would pour off the liquid and spread the residue—damp nitrogen triiodide—on the floor or on seats. There it dried and crepitated, or crackled, disconcertingly when the unsuspecting walked over it or sat on it. Peter also discovered to his embarrassment that it went off spontaneously during the night.

Peter's most outlandish experiment with explosives took place when he was a bit older: a bomb in a bucket. He probably would have used a mixture of sugar with sodium chlorate weedkiller as a cheap and readily-available fuel. Anticipating a biggish bang, he took the precaution of submerging the bucket under water—in the village pond. Triggering the fuse from a distance allowed him and spectators to keep a respectful distance. Up went the bomb and up jumped the bucket, right out of the water, creating a large wave which animated onlookers.

Another bit of technology which held a fascination for Peter throughout his life was the Wimshurst machine. The original design dates from 1883 when James Wimshurst assembled his 'influence machine'. He mounted two discs, each isolated from the other, on a pulley system operated by a handle which allowed him to make the two discs rotate in opposite directions; positive charge built up on one, negative charge on the other. Wimshurst collected the two static charges separately, stored them until they were big enough and then let them discharge between two metal rods as artificial lightning. Peter built a machine in his attic laboratory; it is a nice challenge for a bright schoolchild. Much later he described in *New Scientist* the 'formidable difficulties' he encountered waiting for shellac varnish to dry on two glass discs; it got easier when non-conducting plastic gramophone records became available. The Wimshurst machine wasn't just a plaything; in the nineteenth century it had been a workhorse of physics research.

Peter attended various dame schools (pre-school or kindergartens) and then, aged 8, was sent off to Stubbington House preparatory school for boys. Stubbington is a leafy village just west of Portsmouth and not so very far from Merton Lodge. Some indication of everyday life at Stubbington House can be gleaned from another of Peter's contributions to *New Scientist*—a letter in 1982 sparked by a debate, then raging, on the safety of urea-formaldehyde foam for cavity-wall insulation. There was concern, but not much evidence, that formaldehyde vapour might leak out of the foam and threaten the health of the insulated. He wrote:

At my prep-school it was the practice [at the start of each term] to herd all the 120 boys into a sealed room and fumigate them with formaldehyde generated by heating Meta[formaldehyde] tablets over a stove. The more elderly members of the

---

BACKGROUNDER:

FOSTER'S NAVAL ACADEMY

Stubbington House school is not a household name, but a few Old Stubbingtonians have made it into public consciousness. The best remembered is Captain Robert Falcon Scott, he of the doomed trip to the South Pole in 1911. Naval veterans and historians may also recall various sea lords: for a brief period in 1916 two Old Stubbingtonians ran the Navy as first and second sea lords: Sir Henry Jackson and Sir Somerset Gough-Calthorpe. The reverend William Foster founded Stubbington House in 1841 as 'Foster's naval academy'. With the Solent just down the road, it became 'the recognised place for coaching towards a naval cadetship'. By 1926 the school had been significantly expanded and the proprietor was William Foster's grandson, Montagu (Jr), who was knighted for his services to education in 1928 while Peter and Dick Danckwerts were pupils there. After WW2, the school's clientele was changing and in 1961, chasing the market, the school moved to a building in Ascot. Eventually, in 1997 and finding new competition too hot, it closed its doors. Back in Hampshire the local council, with great wisdom, acquired the original site and re-developed it as a community centre. Stubbington House itself, an attractive but impractical Queen Anne mansion, was demolished but the assembly hall, a classroom block and the lovely surrounds survive.

Stubbington House assembly hall, 2010

staff seemed unmoved by this, but at the end of the prescribed 20 minutes the boys would be scrabbling at the doors, waiting to be let out of this mini gas chamber. My point is that if the medical histories of the several hundred boys who survived this treatment could be followed up (name and address of school to *bona fide* researchers) we might have some hard facts in place of the usual uninformed guesses.

Peter's brother Dick joined him at the school for a year, before he was diagnosed with rheumatic fever and asked to leave. According to Dick conditions at Stubbington House were Spartan, and the discipline fierce. 'There was an institution called the "Pub": if a boy did something terrible, like cheating in exams, he was put across a vaulting horse and beaten severely before the whole school. No right of reply or anything like that.'

But some foundations of Peter's character were laid in Stubbington. In an autobiographical note written fifty years later Peter recalled:

Between the ages of 10 and 14 I wasted many dreary hours on Latin and even Greek, not to mention the dates of the kings of England. However, amongst the masters there were a few to whom I owe a debt for a thorough grounding in the writing of English and a basic appreciation of the principles of mathematics. The academic standards were not high, and it was regarded as little short of a miracle when I got a place at Winchester College.

Then, as now, Winchester College was an expensive place to educate a son. Grandfather William Otto's annual income had been quite enough to send all three of his sons there but he had passed nothing down. Victor, with five children and a captain's salary, would have found the fees impossible. But help was at hand in the nick of time from an unexpected quarter. Peter's godfather Alec Neilson KC—a friend of, and long-time collaborator with, Peter's grandfather William Otto—died in April 1929. A good friend of the family, he had kept an eye on William Otto's widow Caroline in the fifteen years since her husband's death. In a 1929 codicil to his 1928 will, Alec Neilson left her a picture, 'The pass at Bramber', which she may have admired in his drawing room at 11 Upper Berkeley Street in London, and added £1,000 'in memory of a long and happy friendship from which I have learnt all that there is to know of the goodness of a good woman'. To Peter he left £5,000, the equivalent of more than £200,000 in 2010 terms and enough to cover his education. When he was told, Peter burst into tears.

Winchester manners 11

Peter arrived at Winchester College in September 1930, aged almost 14. Founded by William of Wykeham in 1382, Winchester had the reputation of providing the best all-round education in the country. Staff encouraged pupils to develop their sporting abilities as well as their intellectual powers; old Wykehamists, as they are known, have a reputation for what sometime poet laureate John Betjeman once called a 'broadness of mind'.

Winchester College is walled and fortress-like. The central, flint-walled Chamber Court provides accommodation for those fortunate enough to have won scholarships; the ordinary fee-payers live in outlying 'houses'. Houses carry, both formally and informally, the names of their early housemasters; Peter was assigned to Du Boulay's, also known as Cook's or 'C' house. It stands on higher ground above the school, amongst the houses of ordinary citizens, and pupils streamed, and still stream, down the hill in the morning to classes.

Each Winchester house was in the charge of a housemaster. They were very much hands-on, and had a good deal of autonomy in setting the rules. At Du Boulay's Peter found Murray Hicks in charge. Recognised as 'a great outdoorsman', he made his mark with many of his pupils because he found time for them and trusted them. Trained as a lawyer, Hicks was an extremely tough character with masses of energy and a fine physique. In his youth he had represented Oxford at swimming. A serious leg wound suffered during the WW1 Gallipoli campaign caused him pain for much of his later life but it didn't hold him back. In an environment where such things were previously left to individual choice, Hicks had introduced daily cold baths or showers for all. He led from the front, never asking anyone to do anything he was not prepared to do himself. If games were cancelled because of the weather, Hicks would lead his boys on a gruelling cross-country run. He introduced his charges to his own favourite activities like skiing, swimming, sailing and walking. And he encouraged them to observe nature; for instance, to collect and get to know wild flowers. Calling on his own extensive record collection, he invited them in to listen to music. The sailing trips with his boys on Southampton Water were legendary.

As it had been at Stubbington, life at Du Boulay's was spartan. The main communal parts were centrally heated, but the 10-bed dormitories were not: at night the windows remained open and in the morning boys set off for their cold shower on stone floors. In the winter the water often froze in the wash-stands. Much of everyday life was built on tradition: attendance at chapel was obligatory but offered nothing more threatening than a relaxed form of Church of England service. Peter, like his father before him, was sceptical about religion. One family story has his parents kneeling beside their bed to pray before retiring for the night. When Peter's mother turned to look at her husband she noticed his lips weren't moving. She asked him why. 'I'm just keeping you company,' he replied.

On Sundays boys were required to dress in top hats and tails to attend chapel and often spent the afternoons walking through the Winchester water meadows. Nicknames and slang proliferated, to such an extent that Wykehamists old and new can talk together in a dialect of English, known as Winchester 'notions', which is unintelligible to the outside world. Peter took his meals in Du Boulay's but scholars ate at lunchtime in College. Peter told his family how, instead of eating off plates, they used traditional Winchester College wooden platters, or trenchers, ten inches square and an inch thick—conceivably the origin of the term 'a square meal'. The platters were battered, prone to split and they must have been fearsomely unhygenic; Peter described how scholars found it necessary to construct mashed potato dykes to stop gravy running into their laps.

Du Boulay's House, Winchester, today

These were the days when corporal punishment was accepted as character-building, perhaps as much for the deliverer as for the victim. Older boys were occasionally required to dish it out to their juniors. Laurie Pumphrey, in the year ahead of Peter but only three months older than him, found himself cast in the role of senior. (Pumphrey went on to a memorable WW2 as an involuntary guest at Colditz Castle and later became one of Her Majesty's more distinguished diplomatic servants.) He recalled having to administer a beating to the young Danckwerts for some misdemeanour. 'Perhaps he was caught eating in the wrong place or at the wrong time,' he explained. 'The weapon was a short cane about three feet long. I was not well practised in the art and missed him with the first shot—out of four, I think. Peter was entirely unimpressed: I can conjure up his saturnine expression even now. He had, I suspect, a general contempt for the conventions of public school life.'

It was, however, a fair cop. Every boy knew what was, or at least what was not, expected of him. The school rules at Winchester are displayed on the walls of the main hall; another panel spells out the rewards appropriate when pupils transgress. All in Latin, of course, to improve the boys' familiarity with the language. They are 'a fascinating exposition of the imperative mood' according to one Wykehamist. A lot of the discipline in Peter's time was couched in language appealing to the boys' supposed better natures. As an example: '[The Headmaster] begs the School to remember that grass (especially in spring) is too valuable to be carelessly walked over, and that hedges and hatches cost their owners money, and are not things to be injured in thoughtlessness. He feels that it only needs to be pointed out that such thoughtlessness would make free extension of bounds [areas where pupils were permitted to go, the opposite of 'out of bounds'] to a large School intolerable to the neighbourhood, and he trusts that the boys' good feeling, and their care for the character of the School, will prevent any occasion for complaints of mischief done by the School.'

Looking back in 1982, Peter appreciated his experience at Winchester:
This was a public school which was unique in valuing intellectual at least as highly as athletic attainment. In spite of its notable classical tradition, it provided an excellent scientific education. Despite the long hours devoted to the ineffectual study of Latin and Greek, I got a remarkably good grounding in physics and chemistry, but for some reason made no progress in mathematics. However, my training in English was reinforced.

Wykehamists are notorious for being laconic, taciturn, even silent. Rather than add nothing substantial to the discussion, or perhaps cast pearls before swine where the intellectual adequacy of the company is in doubt, a Wykehamist chooses to hold his peace. Peter was by nature shy and a man of few words; Winchester reinforced it. All the same, he regularly took part in school debates. *The Wykehamist* records him as speaking early in his final year against the motion 'That this house has no objection in principle to tyrannical government' and contributing, in a discussion on German and French points of view then current, a debating point that Nazism 'has a mission in civilisation'. In a discussion on whether arms manufacture and sales should remain in private hands,

---

BACKGROUNDER:

THREE STRIKES AND OUT

The final section of school rules or 'laws of the masters' loosely translates into the vernacular as:

In omni loco et tempore

(At all places and times)

• He who is an inferior must obey the prefects.

• He who is a prefect must govern according to rule (he must be without fault and an example to the rest).

• Both must refrain from all bad words and deeds.

• These or other offences, whenever they are reported, we punish.

• When the holidays are over, no one stays at home with impunity.

• Boys who go outside College without leave, we expel on the third offence.

The table of the laws of the masters: a chance to practise one's Latin
Courtesy Winchester College

Winchester College rewards: 'Either learn, or leave or be beaten',
Courtesy Winchester College

Dental hygiene at a German-English youth camp near Heidelberg, 1933:
Peter (far left, aged 16) alongside Michael Mackenzie Smith

he argued strongly against private manufacture and sales in what the editor described as 'an interesting and amusing speech'. And in November 1934 he briefly addressed a matter with some resonance for someone who would later live in a house reputed to be haunted. The motion was 'That this house believes in ghosts'; Peter spoke against.

At Winchester Peter enjoyed a very happy reinforcement to his childhood interest in chemistry in the hands of a brilliant and long-serving schoolmaster, Fred Goddard.

In a one-hour period he demonstrated the nature of chemical change by inducing explosions and under-water fires, burning through a steel plate by the Thermite reaction [using aluminium to reduce iron oxide, or rust, to iron in a reaction giving off a lot of heat and taking the temperature to around 3,000°C] and effecting a dazzling variety of colour changes. I knew then that my early inclination had been correct.

Heidelberg: the camp's entrance gate
Courtesy Julia Mackenzie Smith

The Burma Road
Courtesy Julia Mackenzie Smith

During the holidays there were group trips abroad. In the summer of 1933 he and a few contemporaries including Michael Mackenzie Smith took part in a *Deutsch-Englisches Jugendlager* (German-English youth camp) in Heidelberg. That year the new German chancellor Adolf Hitler had merged all youth organisations into just one—*Hitler Jugend*—and the German contingent at the camp wore the required uniform, reminiscent of boy scouts: white stockings, short corduroy trousers, a brown shirt, a scarf and a leather woggle. They indulged in communal tasks, like construction work and the preparation of pathways—photographs of them in action summon up thoughts of the Burma Road. Peter wrote home that the English contingent escaped from the camp every evening with the objective of getting drunk. He also claimed they deliberately set out to demoralise the Germans by not getting up until 10am.

Peter did not show much enthusiasm for sports involving physical exertion. It is true that he took part in that most ritualistic of Winchester sports, the peculiar form of football which is unique to the College. A cross between soccer and rugby (with rules that differ from those of the better-known Eton wall game), it was played in the 1930s between three teams—one drawn from the 70 scholars who lived within the College and two from boarding houses outside the College walls—at 6 or 15 a-side. The

15 a-side encounters were major events. But in two quite detailed reports from 1934 of the cut and thrust when Peter took part there is no mention of any noteworthy contribution from him.

In contrast, Peter excelled at rifle shooting. In his final year, 1934, he was at his best and represented Winchester in the annual public schools Officers' Training Corps (junior division) competitions at the National Shooting Centre in Bisley, Surrey. The main competition—the Ashburton Challenge Shield—took place on 12 July. Using a service rifle, each member of a team of eight had one sighter shot and then seven shots to count, over both 200 and 500 yards, each shot carrying a maximum of five points. Winchester's team did tolerably well to take fourth place out of 81 entrants; Peter contributed his team's best score of 66 out of 70. But the day before he had done even better. The individual competition, with a field of more than 200, involved seven shots over the same two distances. Peter recorded scores of 34 and 34, totalling 68 out of a

Peter sailing on Chichester harbour

possible 70. He won by a single point and took home the silver Gale and Polden Cup (value: £15 15s 0d) plus an unspecified item worth £5. There was great celebration at Merton Lodge.

The holidays from school were also an opportunity to continue experiments with photography and explosives at home and to sail on Chichester harbour.

Petty rivalry has always been a tradition between leading English public schools, that is to say, the ones which charge fees; it's usually about which one claims to provide the best education. Envy is a major aspect, clear in the anonymous saying: 'You can always tell a Wykehamist, but you can't tell him much.' When, in 1982, Winchester celebrated the 600th anniversary of its foundation there was a re-enactment of this rivalry for all to see. It took place in the columns of the newspaper which, in those days, was the country's leading intellectual organ (and the one which most Wykehamists might be expected to read)—*The Times*. Under the title 'The shameless elite', the editorial writer sermonised briefly on the characteristics of a Wykehamist. He acknowledged that for 600 years Winchester had 'given the most intellectually excellent education available'. And he emphasised that Winchester showed elitism 'in the good sense of that word'—that is, preferring the best to the second-rate in the world of ideas. Frustrated by lack of space, the writer left an interpretation of William of Wykeham's famous motto for his school 'Manners makyth man' to another time. Instead he told a short story, which he claimed to be a 'defining characteristic anecdote':

A group of three enviously-schooled men—an Etonian, a Wykehamist and a Harrovian—were standing together in a room, so the story went, when a lady entered. The leader writer then claimed the following sequence of events: the Wykehamist called for a chair, the Etonian fetched one and the Harrovian proceeded to sit down on it. There was outrage amongst Wykehamists. One wrote incredulously to correct the heresy: 'It was the imperious and patrician Etonian who commanded that a chair be brought; it was the unobtrusive, efficient and—dare I say it?—well-mannered Wykehamist who provided it.'

And there was more to come. 'The Wykehamist did not 'call out for somebody to fetch a chair", wrote another sympathetic reader. 'He asked the lady, very politely, whether she would like one. Surely that makes all the difference.' And a Harrovian, claiming without evidence to be neutral, wondered: 'Is your leader writer perhaps an old Etonian? In the correct version the Etonian says 'This lady needs a chair', the Wykehamist fetches one and the Harrovian sits down with the girl on his knee.' Finally Peter Danckwerts entered the fray at a characteristic tangent:

I suggest that the comparison... be extended beyond the question of manners to that of pragmatism. During the first Atoms for Peace conference at Geneva in 1955, the Mr Big of the international energy scene was confined, raging, to his room by a cold. He instructed me and a fellow delegate to go to a pharmacy and get him some blackcurrant syrup. My colleague (an Etonian) demanded *courants noirs*. I (a Wykehamist) waited for the pharmacist to ask us (in English) what we required.

Winchester had a tradition of supplying undergraduates to New College, Oxford. But there were other options. Peter wanted to study science, and Winchester offered a recently-established scholarship, 'perhaps more closed than any other' Peter recalled, for a Winchester scientist to study chemistry at Balliol College, Oxford. The scholarship was founded in memory of a young Wykehamist, John Ewan 'Jack' Frazer, who had died in tragic circumstances. Peter recalled with his customary modesty:

There were only two candidates in my year [1934] and I was the rank outsider. Nevertheless I won it, to universal dismay, and became a Scholar of Balliol, a notable title. I must say that I have been somewhat cynical about honours and awards ever since.

So Peter left Winchester in the summer of 1934. He was now equipped by nature and education to broaden his horizons, but didn't go straight to Oxford. The attractions of a 'gap year' proved irresistible.

BACKGROUNDER:
THE JACK FRAZER SCHOLARSHIP
Jack Frazer was a Wykehamist and Balliol all-rounder. After a star-studded career at school, he got a first class degree in science and played both football and cricket for Oxford. He stayed on at the university as a junior staff member of the science faculty. Skiing was another passion, and New Year 1927 found him in Klosters taking part in the 'Ski championship of Great Britain'. On 2 January, on a free day between the two legs of the cross-country race, he took off on his own to explore a nearby mountain slope and ran into a half-hidden rock. Frazer died later that day in hospital in Davos of internal injuries and was brought back to Klosters for burial. Jack's father Ewan, a surgeon, established several memorials at Winchester to his son. One was a fund to raise money for a new sports pavilion, administered by Douglas Jardine, later captain of the England cricket team on the infamous bodyline tour of Australia in 1932-33. Another was the memorial scholarship to Balliol College. Established in 1929, it was worth around £100. Immediately there was a row at Winchester, because of the well-established New College, Oxford, tradition that the best scientist at Winchester took a scholarship there. But the headmaster, Alwyn Williams, calmed things down and the new award continued as planned.

To Balliol
via Salzburg 12

GERMAN WAS THE LANGUAGE OF THE 1930S, especially for a
budding chemist, and the Danckwerts family clearly felt that they
knew of an acceptable country in which to learn it. Peter took
his 'gap' year before university in Austria, arriving there in the early
autumn of 1934. His siblings pursued their own interests in Austria and
Germany later in the decade. Both his sisters found romantic Austrian
attachments until the 1938 Anschluss put a stop to them.

Peter himself recorded that:

Happy to escape from school, I spent a year in Austria, that charmed country,
living in Vienna and Salzburg, learning a fair amount of German and conceiving a
passion for the Baroque and for the mountains.

Much of what Peter got up to during that period is lost in the mists of
family memory, but snippets emerge, largely from postcards and annotated
photographs. He became acquainted with skis for the first time and was
hooked. And a great facility to 'get to know people' emerged in the
course of the year. Friends and colleagues in later life would marvel that
Peter seemed to know someone everywhere he went. He was very
presentable and, in spite of an apparent shyness, people were impressed
by him, warmed to him and liked him. In Salzburg he lodged in *Hellbrunner Allee*, a very long street stretching out into the suburbs south of
the city, on which most of the dwellings are mansions or even castles.
Although the addresses on this high-class street may have suggested considerable income, some of the inhabitants were not in such good economic
shape as the appearances which they tried to keep up. They would have
happily provided unofficial accommodation for a respectable young
Englishman.

Salzburg has its full measure of tourist attractions, and in the 1930s they
pulled in visitors of all kinds, Peter included. In the 400-year-old water
gardens of *Schloss Hellbrunn*, water drives a lot of gadgets and visitors are
likely to be unexpectedly soaked to the skin. The salt mines, from which

the city gets its name, are drier but equally fascinating. Opportunistic photographers, on hand to take pictures as mementos, offered a fast turn-round. They caught Peter with some British sightseers on the traditional train which took visitors on a tour of the salt caverns. A couple of seats behind him sat a minor Hollywood film star: the Barratt Browning relative Edward Moulton-Barrett, a bit part player under the screen name Charles D Waldron.

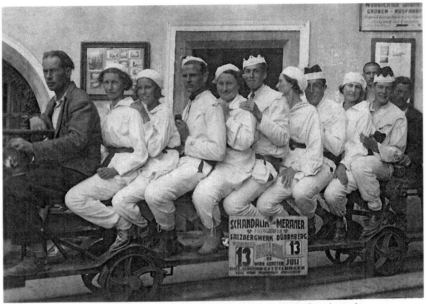

All aboard for the salt mines (from l): Peter aged 18 (fourth) and Edward Moulton-Barrett (sixth), Salzburg 1935

Peter also met local minor aristocracy, including a branch of the Vetsera family. The Vetseras had been caught up in the Mayerling tragedy in 1889 when Crown Prince Rudolph, the heir to the Hapsburg throne, took 17-year-old Mary Vetsera to a hunting lodge outside Vienna. It turned out to be a suicide pact. In the wake of the scandal, the Vetseras kept their heads down. Mary Vetsera's younger brother Feri, a *Freiherr* or baron, married Margit von Bissingen who was of Hungarian stock and born a countess—a bit of a step up on her husband in the ossified Austrian class structure. He fought in WW1 and drowned in Volhynia (now north-west Ukraine, on the borders of Poland and Belarus) while on the Russian front in 1915. His widow with her three daughters Nancy, Alitschi and Nora lived on in Vienna until 1927 when they all moved to Auerspergstrasse 27 in Salzburg to be near friends. Peter must have been impressed to meet the three youngish baronesses. They photographed him in the Austrian Tyrol in the summer of 1935 beside the *Seebensee*—almost in the shadow of Germany's highest mountain, the famous *Zugspitze*. Peter appears with two baronesses Vetsera; the third may have been behind the camera.

After a year in Austria, able to communicate in

Peter with two baronesses Vetsera contemplating the *Zugspitze* and the *Seebensee*, Austrian Tyrol, 1935

BACKGROUNDER:

THOSE INFAMOUS VETSERAS

Just before dawn on 31 January 1889, a carriage left the royal hunting lodge at Mayerling, just outside Vienna. Inside a 17-year-old woman sat bolt upright between two uncles. Their destination? The *Heiligenkreuz* graveyard four miles away where Mary Vetsera, already dead and held upright by a walking stick down her spine, would be secretly buried. 24 hours earlier she had been shot dead by Crown Prince Rudolph, the heir to the Hapsburg throne, in a suicide pact. The Hapsburgs had long been the dominant European dynasty and in 1889 Emperor Franz Joseph commanded the Austro-Hungarian empire from Vienna through a court which was distinctly decadent and corrupt. His son had been schooled to rule but saw no prospect of it happening soon. Hungarian friends enthused by the prospect of independence pressed him to further their separatist cause, so

that the court tracked his every move. Denied divorce, Rudolph and his wife Stephanie were trapped in a loveless marriage. He sought refuge in sexual adventures and morphine. His fantasy was a 'romantic suicide pact' with a lover. An unlikely Romeo, racked with venereal disease, he eventually found a willing Juliet in Mary Vetsera—the naïve and impulsive daughter of Eleni Baltazzi-Vetsera, a dedicated social climber. Egged on by her mother she took the trip to Mayerling with Rudolph. Afterwards, the court imposed a total cover-up, beginning with the secret burial for Mary and a lying-in-state following 'heart failure' for Rudolph. The Vetsera family, while not persecuted, was quietly forgotten about. Franz Joseph had Mayerling partly destroyed and a monastery built in its place. The court was greatly aided in its cover-up by the rapid growth of a romantic myth surrounding the episode, expressed later in films and a ballet.

German and bearing his notable title of scholar, Peter arrived in Oxford for the 1935 autumn term. Since its foundation in 1206 Balliol College has changed in many ways; standards of presentation have certainly risen. To-day the availability of sumptuous grass has led to a replanning of quads, and gardeners with scissors trim the edges of the sward. But a Balliol man from the nineteenth century on would still recognize the general archi-tectural shape of the place. Peter lived rather splendidly on the first floor of the (now demolished) seventeenth century Old Coffee House. Being a scholar he was allowed to spend his four years lodging within the main

college, rather than finding rooms outside in the town. He had a male 'scout' who provided his basic requirement in the morning—a bowl of hot water in which to wash and shave—and would have dealt with the laundering of bed linen and tow-els. He could even fetch lunch, although a sandwich from the buttery was better.

Peter's digs at Balliol: the Old Coffee House

Balliol had a reputation for preparing men to rule, both as politicians and as executives. Some of Peter's contemporaries later became household names. Future prime minister Edward Heath was organ scholar and started a choral society: Peter at some stage sang in his choir, known as Mr Ted Heath's Glee Club. He also noticed Denis Healey—later described by some as the best prime minister Britain never had—but they never spoke. The chemistry course took four years, and Peter found the atmosphere 'singularly bloodless':

My mentors were engaged in the investigation of the properties of solutions of salts so dilute that they were referred to as 'slightly polluted water'. However, it was nec-essary to work one's way through the various teaching laboratories. I was a slovenly experimentalist. In [Balliol] we had a ceremony at the end of term known as 'hand-shaking'. One passed the length of the Hall to High Table where one's tutor sat with the Master [Sandie Lindsay, moral philosopher and founder of Keele University]. 'Well, Master,' said my tutor, 'I have a report from the inorganic chemistry laboratory which says that Mr Danckwerts fails to achieve in two hours what an intelligent lab boy could do in ten minutes.' The Master sighed: 'This life of cultured leisure, Mr Danckwerts...'

14 schoolmistresses on the Danube: Peter's charges as courier

There were skiing trips during the vacations: the first, in the winter of 1935-36 was frustrating: 'practically no snow' Peter recorded. During the long summer breaks from college Peter, like many of his colleagues, looked for ways to earn some pocket money in an agreeable way. The summer of 1936 saw him back in Austria—as a travel courier shepherding a party of schoolmistresses on a canoeing trip down the Danube to Vienna. He invited his brother Dick and sister Hilary, both keen canoeists, to come along as well. This was a smart move: he managed to delegate most of the work to Dick. The party crossed the Channel and caught the train in Ostend towards Bavaria. Somehow or other they found they were two or three tickets short. So every time the inspector came along it was Dick's job to shepherd a few ladies into the toilet where they locked themselves in until the inspector had passed. Once they got to their canoes they found that at every stop the menu was the same: *Wiener schnitzel* followed by *Apfelstrudel*. On arrival in Vienna, Peter abandoned the party on the spur of the moment, leaving them to find their own way home, and went off with Dick to the Salzkammergut for more canoeing. Surprisingly, the

Peter, brother Dick (r) and sister Hilary (l) as canoe couriers in 1936

same company engaged Peter the following year. The party crossed to Ostend as before, but some of them went astray at the station. By the time Peter found them the train had departed; they had to wait 24 hours for the next one.

Back at Balliol, the all-important final examinations were held in the summer of the third year. This was the Danckwerts approach:

I cannot claim that I was particularly diligent during my first two years… so that in my third year it was necessary to do some serious work. I acquired some excellent textbooks, which I read from cover to cover, I studied the predilections of the examiners and tested myself on previous examination-papers. As a result I did rather well, a result [a first class degree] which I believe gave my tutor the idea that something was wrong with the system.

Peter at 21 in Köln, Germany

His tutor was Ronnie Bell, a greatly respected but rather crusty don of the old school. He was not long returned from Copenhagen, where he had studied with one of the great men of experimental chemistry—Johannes Nicolaus Brønsted. Bell was a charming man. Small in stature and wearing a characteristic pair of wire spectacles, he nevertheless had a reputation—said to be based on his Scandinavian experience—for proving irresistible to the women. One story has him conducting an oral examination with a young lady. Looking to end the interview on a pleasant note, he asked her what her plans for the future were. Blushing, she whispered that she had been planning to go home for supper…

Bell's major interest in chemistry was the study of acids and their ability to catalyse reactions. As the rather esoteric Balliol couplet about Bell had it:

My job is to refute the fallacies
In Brønsted's acid–base catalysis.

In their fourth year, after the final examinations, students had to carry out research work, and Bell set Peter to work on a project in one of his areas of interest. Peter called it a 'light-hearted piece of research', an interpretation which probably went some way to underwhelming his tutor. He performed adequately, producing what was required: a brief paper on the matter in the *Journal of the Chemical Society*. But Bell was not over-impressed; he told others later that Danckwerts was the last of his students that he would have expected to be a success in research.

In the 1930s there was no single site for laboratory work in chemistry. Peter's research for Bell was classified as physical chemistry and so took place in the Balliol-Trinity laboratory. It occupied subterranean cellars on either side of the wall separating Balliol from Trinity College next door. Ventilation must have been pretty poor; whenever Peter referred to his research in the Balliol-Trinity laboratories in print he comments on the prevailing fog of cigarette smoke.

Peter's experience at Balliol had been valuable but not defining. Chemistry was essentially a literary subject at Oxford at the time, and he was taught no physics and no mathematics. He wrote later that:

> It was said that if an Oxford chemist encountered a differential coefficient in a book, he turned the page; if he saw an integral sign he shut the book. Perhaps the most valuable part of my intellectual heritage from Balliol was a sense of scientific rigour combined with scepticism—the latter summed up in the imputations of the word 'bogus'.

BACKGROUNDER:
THE BALLIOL-TRINITY LABORATORY

Balliol had been the first college to provide some sort of chemistry facility for its students, starting in 1848 in a small cellar. As the subject developed, Balliol negotiated an agreement with Trinity College to combine their efforts in both laboratory facilities and teaching. A larger and updated laboratory was inaugurated in 1879, entirely underground and on Balliol's side of the wall which separates the two colleges. Balliol was responsible for upkeep, and Trinity made a contribution. In 1882 the Duke of Bedford, whose two sons had recently been at Balliol, funded a lecturer to teach their students. By 1897 space was very tight and, in spite of the fierce rivalry between the two colleges, they agreed to extend the shared laboratory under the wall on to the Trinity side.

One Harold Hartley, later instrumental in getting Bell his placement with Brønsted in Copenhagen and a significant force in British science and technology, was elected a Balliol fellow in 1901 and served as Bedford lecturer. By 1904, to help space demands further, the University designated separate laboratories for four branches of chemistry—analytical, inorganic, physical and organic: Balliol-Trinity got physical chemistry. Things remained like this until Peter's time. By 1941 the University had finally completed a unified building for chemistry and the Balliol-Trinity laboratory was closed. Some of the laboratory rooms are still used: the Balliol teaching laboratory is now a music and exhibition room and the corresponding Trinity facility a workshop. Balliol's original 1848 laboratory is now part of the undergraduates' beer cellar.

Later he was to encounter the pragmatic use of mathematics to solve everyday problems in chemical industry and, by applying them himself to unsolved problems, he would go on to make his reputation as a scientist. But that was some way off. As the fourth year was spent on 'light-hearted research'...

...in one's final summer one was free to punt on the river, go to balls and otherwise lead the *dolce vita*. The year was 1939.

BACKGROUNDER:
BALLIOL-TRINITY RIVALRY
In Oxford, the colleges of Balliol and Trinity sit alongside each other on Broad Street and, like many who share a boundary, they are bitter rivals. The two, as in Peter's time, may have happily co-operated in a shared chemistry laboratory, but former students reminiscing about their youth find it hard to resist recalling that florid manifestation of Balliol-Trinity rivalry— *The Gordouli*. Balliol students wrote *The Gordouli* in the 1890s. The antics of the young, newly released from the control of parents or school and experimenting with alcohol in the early hours, are often harmless, but the lyrics of *The Gordouli* sound like a vulgar form of inverted racism. At the time Balliol was then one of the few colleges to let non-whites in; Trinity prided itself on keeping them out. The song was inspired by Galetti di Cadilhac, an Anglo-Italian Trinity man whose skin, weathered by the gentle breezes of the Mediterranean, suggested to Balliol undergraduates that he might be coloured. Its name was derived from *Gordoulis*, a brand of Egyptian cigarettes popular at the time. The spirit of the song could be described as playful, but essentially says: 'Anglo-Saxon all-white Trinity, what on Earth is Gordouli doing amongst you?' Here are some of the gentler bits:

Bloody Trinity, Bloody Trinity,
  Bloody Trinity!
If I were bloody Trinity,
I'd go into a public rear,
I'd pull the plug and disappear.
Bloody Trinity, Bloody Trinity,
  Bloody Trinity!
I don't give a f*** or damn,
Who the hell or what I am,
As long as I'm not a Trinity man!
Bloody Trinity, Bloody Trinity,
  Bloody Trinity!

Sometimes an opportunity for revenge arose. When the 1935 movie *Sanders of the River* arrived in Oxford, there was animation in the audience during one sequence featuring a canoe manned by 'natives' and steered by Paul Robeson. Then someone shouted, 'Well rowed, Balliol!'

That reserved occupation 13

THE SHADOW OF WAR hung over Peter's graduation in June 1939. But it wasn't until September that Nazi Germany invaded Poland and Britain felt obliged to take up arms. By then Peter had a job: he was employed 'as a raw chemistry graduate' with the company Fullers' Earth Union in Redhill, Surrey. The centre of his activities was the Cockley Works on Nutfield Road, just east of the town. Written histories describe it as being fitted out with up-to-date technology during the 1930s; Peter was soon to discover just what that meant.

A naturally-occurring clay, fullers' earth has a long history. The Roman historian Pliny the Elder mentioned it in AD29, and records go back well before that. It comes close to the surface at Redhill and had been dug up there since time immemorial; a great adsorbent of dirt from oils and cloth, its bleaching properties were widely used in the Middle Ages. In the twentieth century fullers' earth had been made more potent—treatment with acid made it more active. By 1939 it was in great demand; most significant for the war effort was its value as a purifying agent in the production of fuel for motor vehicles, tanks and aircraft.

The Fullers' Earth Union logo created from various grades of the product
Reproduced by permission of Surrey History Centre

In the inter-war years Fullers' Earth Union had agents all around the world and one joint venture—with the Tannenberg Company in Saxony-Anhalt, Germany—which started production in about 1931.

I went to the Fullers' Earth Union to do 'research'. Working on my own, with no experience of industrial research, it seems in retrospect that I was unlikely to have achieved much; in any case the matter was taken out of my hands when the war broke out and I was sent into the factory to 'boost production'. This probably did me a lot of good, but not the factory. The job really called for a chemical engineer rather than a conductivity-water chemist [a reference to Peter's mentors at Balliol College and their interest in very dilute solutions]. The factory was a witches' kitchen which had been designed by the managing director, an ex-colonel with no training in chemistry or engineering.

Nevertheless, a job boosting production of fullers' earth, and hence fuel production for the war effort, qualified as a 'reserved occupation' and protected the holder from involuntary call-up to the armed services.

The managing director of Fullers' Earth Union at the time was a decorated WW1 lieutenant-colonel named Wilfrid Taunton Raikes and known to all as 'the Colonel'. In designing the company's acid-activation process he had some help from his wife Gentle, who had prior experience

---

BACKGROUNDER:

A SHORT HISTORY OF CATS' LITTER

Fullers' earth—the apostrophe is a moveable feast and sometimes even dispensed with—has been mined around the globe for some seven thousand years in those places where the geology brings it to, or close to, the surface. The name reflects its use in history: in fulling, or cleaning textile materials—for instance raw wool—by kneading the two together in water. Adding fullers' earth helped because it is a good absorber of dirt and lanolin. Smectite is another old-fashioned name, montmorillonite is what a chemist might be tempted to call it, and bentonite is a shorter and easier version for industrial technologists to get their tongues round. They are all roughly the same. In Peter's day the Cockley Works of Fullers' Earth Union in Redhill activated fullers' earth by treating it with sulphuric acid. The surface of the clay is covered with negatively charged sites, each one holding by attraction an ion of one of the metals sodium, calcium, magnesium or iron. Acid treatment replaced the positively-charged metal ions with positively-charged hydrogen ions, and this made the clay even better at absorbing impurities from oils. Fullers' earth has had a myriad of uses: as a catalyst, as a household cleaner and in cosmetics, toothpaste and skin treatment—even, in an emergency, to absorb poison in the stomach. Cream containing it is an obscure but effective way of protecting a baby's bottom against soreness. To cap it all, fullers' earth also absorbs odours, and the largest current use is as cats' litter; its natural earth colour is said to make the animals feel at home while defecating. The company Fullers' Earth Union consolidated operations of several small companies in or near Redhill and Bath in 1890; the Laporte Group bought it in 1954. Demand gradually fell away as better alternatives were found, and Laporte finally abandoned the Cockley Works in 1996. The adjacent quarry, an unusual landscape used in the 1970s as a location for filming episodes of the popular TV serial Dr Who, is now a landfill site.

of the chemical industry. Peter was initially involved in laboratory activities at Cockley Works. There he joined Barbara Emödi, another new arrival. Emödi had written to the company from her native Hungary in 1939 claiming to be able to improve the quality of its products. The management, acutely aware that the Colonel's process had never been optimised, took her on. But hostilities intervened, and the company had to make special arrangements to get her to Redhill. In the laboratory Emödi had the help of a small team of teenagers who were learning the business. Peter set the scene:

Peter boosting production at Fullers' Earth Union

Raw fullers' earth quarried in a neighbouring field was stewed up with sulphuric acid in gigantic lead-lined autoclaves to activate it for use as an adsorbent in oil refining.

Emödi was looking at ways in which the activating of fullers' earth might be improved; she and Peter would have talked about the challenge. Emödi tried out a process in the laboratory using superheated steam. There were all sorts of advantages: it avoided two expensive stages—there was no need to dry the raw material before activation or use a high pressure autoclave. Repeated repairs to the pressure vessels were a major part of activation costs. Even better, the level of activation achieved by the 'Emödi method' made the product a much better bleaching agent. The best results came at a slightly higher cost: using three times the amount of acid normally employed. After treatment with superheated steam for a short period the slurry was held for a day or so at a high temperature. The Colonel was quite excited by it all. He wrote to his engineering consultant and directors at the end of April 1940: 'Even if the total costs of Dr Emodi's process come out heavier than our present process, it must be remembered that the earth is very much more efficient.' In due course Emödi became a leading authority on the product.

After this low-pressure activation the process would follow existing practice for which equipment was already in place:

The mixture was decanted, the residue washed, and the acid effluent neutralised with lime. This produced a gelatinous precipitate of hydroxides of aluminium and iron,

Fullers' Earth Union's Cockley Works, Redhill, in the late 1930s
Photo: Aero Pictorial

plus some calcium sulphate. The effluent was fed to a tank where a precipitate was supposed to flocculate, settle out, be thickened and disposed of.

The flocculation tank was a very fine one and quite new, according to Peter. The writer and journalist Jack Trevor Story, who at that time was earning a living as an engineer, made a passing reference to it in *New Scientist* 44 years later. Story was reminiscing about working with Fullers' Earth Union at Redhill in 1942, a little after Peter's time there, but Peter recorded that what he had to say 'sent a shaft straight through my vitals'. In the winter of 1939-40 it seems that the Fullers' Earth Union flocculation tank was only capable of dealing with about one quarter of the factory's throughput. Why? As far as Peter was able to discover, the tank had been sized and commissioned in summer months, when the temperature of the water inside was fairly high and its viscosity low. In winter the water was colder, its viscosity was higher and this slowed up the rate at which the flocculate settled. So in cold weather the tank was simply unable to cope with the projected throughput. This was one of several pragmatic lessons which Peter learned at Redhill and which came in useful later. Peter described another odd practice then current in the plant:

Most of the vital connections in the plant were made with rubber hosepipes fastened with the baling wire used to tie up the bags of finished product. Every so often one of these impromptu joints would give way and send a shower of dilute sulphuric acid into the lower regions. It would be hard to think of a greater contrast to the pure chemistry I had learned at Oxford (or the highly sophisticated chemical engineering I was taught to admire later).

In January 1940 the Colonel added a graduate chemical technologist called Henry Hathaway to the payroll. When he arrived Peter showed him around the Works and must have been interested to find that the new boy was better prepared than he to tackle large-scale processes. By the summer of 1940 the sheer strain of life was beginning to tell. Already a member of the Local Defence Volunteers, which became the Home Guard, Peter was engaged in patriotic action in the evenings. He put his general disenchantment down to the extra strain of this, a lack of job satisfaction, long working hours and boredom—plus a degree of bellicosity which most citizens had acquired by then (encouraged by popular slogans of the times like 'You can always take one with you').

We were all keyed-up for a German invasion, and it seemed that we might be better-equipped and better-organised to take a bash at the Hun, who (we assumed) was about to invade us, if we were in uniform and had reasonably effective weapons. My few months in the Home Guard had been discouraging. Working all day and hanging around half the night with a Short Lee-Enfield 0.303 rifle with five rounds in the magazine and no replacements, watching the Surrey hillside for an enemy who did not appear, became depressing. We had one burst of excitement when a member of the platoon let off his gun indoors, and the bullet went right through a 9-inch brick wall with plenty of momentum to spare.

So Peter set about getting himself de-reserved. Before leaving his job, he had to make peace with his employer. The Colonel, then approaching 50, was a minor member of the aristocracy and had a background not dissimilar to Peter's—public school, Oxford, brother an admiral, cousin a major-general. But for Peter, at half his age, Raikes was 'an old fire-eater' and, Peter guessed, probably glad to see him go. In any case there was a ready-made replacement already on the staff—Henry Hathaway.

Henry Hathaway photographed by Peter exploring the Redhill plant in 1940

By a cruel irony I survived the war with only peripheral injuries, while my replacement at Redhill, a bright young chemical technologist from Imperial College, was killed while fiddling with an unexploded anti-aircraft shell he found one day.

Such were the uncertainties of the times. At the end of July 1940 Peter was on his way to war, and he didn't return to the civilian world until May 1946.

BACKGROUNDER:
HATHAWAY'S END

A Londoner and a year older than Peter, Henry Hathaway had studied chemistry and geology at Imperial College London. He too graduated in the summer of 1939. Before arriving in Redhill he had spent a few months with Murex Welding Company which specialized in the technology of the metal tungsten. At Fullers' Earth Union he became works chemist. Like Peter he joined the drive to streamline production and showed an aptitude for the development of processes. The treatment of effluent was an outstanding problem at Redhill, and in due course Hathaway helped design, construct and operate an improved plant. It cleaned up the effluent but brought another challenge: what to do with the gel-like residue. In July 1943 he moved on to Witton, Birmingham, as chief chemist with Bound Brook Bearings. The UK arm of a US parent company, BBB made bearings using graphite, a product vital to the production of all manner of mobile war machinery, from guns through four-wheeled vehicles to tanks. So it was another reserved occupation. Hard-working, versatile and something of a perfectionist Hathaway married in April 1944 and set up home in the Birmingham suburb of Harborne, not far from work. At the end of June he took time off to revisit Redhill. On 4 July, a warm Tuesday scarcely a month after D-Day, Hathaway went to get some fresh air on the North Downs. He headed out of Merstham and up Shepherds Hill, just north of where the M23 now crosses the M25. At some stage he came across an unexploded 20mm anti-aircraft cannon shell lying in the grass. Perhaps he tripped over it; maybe he unwisely tampered with it. Whatever the case, the shell exploded and a fragment pierced his heart. At the inquest two days later the coroner concluded that Hathaway's death had been accidental. He was 28.

Practice in Massachusetts   14

DEMOBILISED FROM WAR SERVICE on 20 May 1946 Peter was, by his own admission, 'on the beach'. Almost six years had passed since he left Fullers' Earth Union, and he would soon be 30. He was disappointed in prospects of a career in films, in spite of his physical attractions and the earlier encouragement aroused by women confronted by an injured war hero in uniform. He was wondering what to do about a job when a number of chance events ensued. First he bumped into a man in Oxford who had spent the war in Washington; while there, it seems that this academic had heard of a profession called 'chemical engineering'. Its exponents appeared to be equipped, as a result of their training, to carry out with conviction the sort of thing that Peter had been trying to do by common sense and intuition at Fullers' Earth Union in 1939. And by all accounts the Americans were rather good at such things.

Not long after Peter was describing that chance encounter to a friend. The friend knew about something called the Commonwealth Fund. He suggested that Peter might apply for a fellowship. The Fund had been set up in 1918 by Anna Harkness, the widow of one of the original investors in the US Standard Oil company, with a mandate to 'do something for the welfare of mankind'. By 1925 it had begun an international programme, including support for British graduates to study in the USA. Administered from New York but with an office and staff in London, the Fund was offering fellowships which looked extremely attractive, and especially so in the context of run-down, post-war Europe: two years' study in the USA, including six months' travel around the country getting to know Americans whom a British student might otherwise never meet. Ownership of a car was obligatory. To the British it looked like a Rhodes scholarship in reverse.

Peter applied for a fellowship, attended an interview at Harkness House in London and, as often happened when interviewers were exposed to the force of his charm and intellect, got what he wanted. The fellowship was to study chemical engineering on a 'practice course' at the Massachusetts Institute of Technology (MIT) in Cambridge, Massachusetts. The fellowship money covered tuition fees, travelling expenses including a return ticket by liner across the Atlantic and living expenses. During two years he received and spent just over $7,000.

Peter arrived in Cambridge, Massachusetts, in September 1946. The Institute had begun life in Boston some eighty years earlier; a technological university, its aim was to achieve the same intellectual level and respect as traditional universities. In 1916 it had moved to an imposing site on the banks of the Charles river where Peter found it. He was allotted a room in the six-story 'Graduate house' on Memorial Drive. Facing the river, this unusual structure, with an H-shaped footprint and strange castellations on top, stands immediately across Massachusetts Avenue from the great dome of the main MIT building and is still designated for student accommodation. Peter's first impressions on becoming a student again were not encouraging; soon after he arrived he wrote home that:

Life is a bit of a grind at this place, and it lacks many of the more picturesque and carefree features of a typical American university.

BACKGROUNDER:
THE COMMONWEALTH FUND
Anna Harkness, widow of Standard Oil investor Stephen Harkness, founded the Commonwealth Fund in 1918 with an initial endowment of $16.3 million (about $250 million in 2010 terms). Others made contributions and in 1950 the next Harkness generation added another $36 million (about £230 million now). In looking for ways to boost the welfare of mankind in the early 1920s, the Fund initially selected four areas: international, health care, children and youth. Financial fortunes have waxed and waned since then and the Fund, either from desire or necessity, has adjusted its programmes. The international programme has survived and the Commonwealth Fund fellowships, known since 1959 as Harkness fellowships, are the Fund's oldest continuously operating programme. The Harkness family originated in England, and British graduate students were a favoured group. The mission here was 'to advance international understanding through education and travel for young men and women of character and ability', and to promote a special relationship between the United States and the United Kingdom. Fellowships lasted two years in any field of study, and always required some time to be spent travelling around the United States. In 1930 the Fund leased Harkness House at 35 Portman Square as its London headquarters, and operated there until 1960. It currently rents space from the Nuffield Foundation at 28 Bedford Square.

In this and other ways MIT in 1946 was a considerable cultural shock and quite unlike his earlier experiences:

At Oxford I had been a scholar and a gentleman, in the Navy an officer and a gentleman. At MIT I was one of a multitude who were being processed by a vast and somewhat impersonal machine working double shifts. I was bewildered by every-thing–'material balance' [or 'what goes in must come out'] meant nothing to me. I didn't know what a differential equation was and (at the age of 30) I had never used a slide rule.

MIT Graduate house under conversion, 2010

Adjusting to his new surround-ings he could be naïve. One day the Americans on the course got to-gether and suggested to Peter that he come with them on a snipe hunt at the weekend. Peter got kitted out and borrowed a gun. At the last minute they revealed that the snipe was a protected species; he took the joke well. Peter was to spend two terms from September 1946 to May 1947 attending classes at the Institute itself before joining the chemical engineering practice school on location at industrial sites for six months. His reward at the end would be, if all went well, the title of Master of Science in chemical engineering practice and the letters SM after his name. Reflecting later on the classwork involved:

Since I had not been called upon for serious mental exertion since Oxford in 1939, I expected to find the course a testing experience, but there was an element of severity in it which exceeded my expectations. I think anyone with a conventional scientific training in an English university might feel the same way.

The teaching programme consisted of a number of courses covering chemical engineering, some mathematics and electrical engineering, and a small amount of research. In most courses the students had to work through a succession of numerical examples designed to bring out the principles of the subject while accustoming them to accuracy and diligence. The descriptive and literary approach, so familiar to Peter at Oxford, was almost completely absent.

Many of the class periods were devoted to discussion of these problems, the students—again, how unlike Oxford!—being expected to recite and justify their own solutions on demand. Exchanges between instructors and staff were often lively. Some of the notable characters on the staff could, with nicely calculated insults, raise the most torpid class to babbling indignation, or hold them attentive with anecdotes of factory and board-room. In each course there was a fortnightly 'quiz', consisting of a one-hour examination covering the past two weeks' work.

Peter had made, on his own admission, a fumbling start, overwhelmed by homework and struggling to adjust his ideas. But he found his feet quite quickly and soon managed, as he put it, to keep up with the crowd. He discovered on closer examination that it wasn't going to be that hard:

I scored zero on my first test and 100% on my last one. The principles were not so difficult; the main things I had to learn were how to apply them and put numbers into them, and how to adopt a commonsense and pragmatic attitude towards problems rather than a purely intellectual one.

He came to savour the pragmatic approach of an engineer when compared with that of the scientist:

My chief practical weaknesses were in mathematics and in the art of arithmetic. My formal deficiencies were in mechanical engineering (a minor part) and in humanities, since at Oxford one had been supposed to acquire culture by osmosis rather than formal instruction. The course did not provide insuperable intellectual difficulties for anyone with a good grounding in the basic sciences. The difference in attitude can be summed up in the statement: 'Scientists solve the problems they can, engineers solve the problems they have to.' An engineer uses scientific methods as a valuable tool, but knows that few engineering problems are capable of a rigorous formal solution. He also knows that it is often better to have an answer within a factor of two tomorrow than to have it within ten percent in a year's time.

The special virtues of MIT, he found, lay in the enthusiasm and intellectual liveliness of the staff, their ability to correlate the scientific and technical aspects of the subject and their personal experience of industrial developments. In addition they clearly spared neither themselves nor their students when it came to sheer hard work. MIT's unofficial slogan was 'Tech is hell'.

One day in early 1947 an aesthetic-looking visitor entered the classroom and sat down next to Peter. It was Terence Fox, the recently-appointed first Shell professor of chemical engineering at Cambridge University, England. Fox had studied mechanical engineering, in which he excelled, and wanted to get up to speed in a discipline in which he had had no formal training.

So, with his new department due to open in October 1948, he set out on a tour of North American departments. During the MIT class Peter made an impression on Fox when, in Peter's own words, he 'answered smartly a question flung at him in the disconcerting American style', so different from his pre-war experience at Oxford. He felt that he owed to this incident his subsequent good fortune when he returned to England.

In May 1947 the theory course ended and there was a break before the practice course began. Peter invested $550 of his Commonwealth Fund grant in a 1937 Dodge and set off to fulfil some of the travel commitments of the fellowship. His itinerary took in Richmond, Knoxville, St Louis, Chicago, Detroit and Niagara Falls before he returned to Boston and took a side trip to New Hampshire for some salmon fishing. While Peter was driving round the States, back in the UK his younger sister Diana and her fiancé John Garson had scheduled their marriage for 20 September 1947. Peter was, of course, unable to be there. But he supplemented his habit of posting food parcels home by sending the happy couple a 'fantastic' celebratory cake. The parcel arrived to spontaneous excitement in ration-restricted Britain and expressions of great credit to Peter for having thought of it. Then they opened the box. Inside were the consequences of something Peter hadn't considered. The cake had become mouldy *en route* and was inedible.

To prepare for the arrival of practice school students at the industrial sites, MIT had sent out junior staff members of the department in advance.

BACKGROUNDER:
THE DEVELOPMENT OF MYTHS
Versions of Peter's meeting with Cambridge's new professor Terence Fox in an MIT lecture room abound. The original story, told in the main text from Peter's own words, has been subtly changed by Chinese whispers, or exaggerated as a form of respect. These enhanced versions followed him around later. In one, Fox missed something the class leader said and asked Peter what it was. The private exposition he gave was so good that even Fox, the ultimate intellectual engineer, was impressed.

A later version is that not only did Peter answer the question flung at him but phrased his answer in such a way as to suggest that the questioner's intelligence might somehow be in doubt. This sounds very much like the addition of someone bruised by the playful style in which Peter could put someone down. A more ambitious—and frankly ridiculous—version of the same story has it that following the episode no member of the teaching staff dared to fling a question at Peter again. This one is not at all in the MIT spirit, and Peter would have been the first to disown it.

At each site there was an MIT assistant professor, known as the director, and an MIT instructor who assisted him. Their mission was to get the hang of the processes. By the time students arrived the staff would have been there for six months to a year and the director, with the help of a company go-between, would have chosen the projects the students were to tackle. The student groups (and there were two in Peter's intake) rotated between the three sites. In June 1947 Peter found himself in a group of six along with five American citizens: Dick Asmus, Ray Baddour, Walter Benzing,

Peter on the practice course...

John Kellett and John O'Neill. As the only one with a car, he was the obvious choice to drive the group from place to place. They all piled into the Dodge. Ray Baddour recalls Peter driving quite well on the unfamiliar right-hand side of the road, but riding kerbs whenever he had to turn left. Their first stop was Parlin, New Jersey, where Standard Oil was based at the time. They were set to work on the site of American Cyanamid, part

Peter's MIT practice course group:
(back row, l to r) Peter Danckwerts, Walter Benzing, Dick Asmus;
(front, l to r): Ray Baddour, John Kellett and John O'Neill

..and relaxing (second left) during it

of the Standard Oil group. At this stage Baddour had little to do with Peter who, he recalled, 'while not antisocial was not outgoing'.

After eight weeks and assessment, they moved on to Bethlehem Steel in Buffalo, New York. On the way Baddour recalls that they passed through New Brunswick and crossed the heavily-polluted Raritan river. During wartime few questions were asked about the waste that industry dumped into rivers, and there must have been a lot of oil floating on the surface of the Raritan. It had caught fire and Baddour noticed Peter got a huge kick from it.

Big though the American Cyanamid site was, Bethlehem Steel's Lackawanna steel works in Buffalo was enormous. The team was restricted to peripheral activities like coke-making, heat transfer and fuel efficiency. The main process—the making of steel—was off limits. On arrival in Buffalo the six students clubbed together and bought some crates of the local beer, which was quite inexpensive. They put the bottles in the fridge and, to log consumption, stuck a piece of paper to the fridge door. Each of them could then place a tick against his name every time he took a bottle. Baddour soon noticed that, while the number of checkmarks against American names was modest, Peter's row already stretched right across the page. Whereas Peter had got a top grade at Parlin, Baddour got the assistant professor's top mark in Buffalo. From then on, Peter sought out his company. Baddour could see that Peter was quite different from the others. It wasn't just his educational background; at 30, he was older than the rest of the group and had 'a bearing about him'. In the group, Peter was respected as a 'no-nonsense person'

The six moved on finally to Bangor, Maine. Alice Robertson, a journalist representing a local newspaper group, was lying in wait. The editorial office had got a tip-off about a 'British admiral's son in Bangor for training' and her report appeared on page one under that heading. An experienced reporter adept at extracting information from visiting personalities, she persuaded Peter to tell her, amongst other things, a bit of what had happened to him when he tripped over the anti-personnel device in Sicily. Her scoop appeared next day in the *Bangor Daily Commercial*. The six MIT students lived in the Bangor House Annex, 17 May Street, in downtown Bangor. Home to all groups of passing MIT students, those leaving would pin to the Annex wall, for the benefit of those yet to come, a list of all the girls they had met in the town, complete with contact telephone numbers. Then new arrivals knew whom to call in for parties. There was some competition: on one occasion Peter introduced Baddour to a girl friend of his who was, in Baddour's memory 62 years on, 'really beautiful'. Baddour began dating her and Peter was not amused. 'It didn't happen again,' Baddour admitted.

Two local paper mills were available for the course: Penobscot Chemical Fiber Company in State Street, Bangor, on one side of the Penobscot river, and Eastern Pulp and Paper Corporation in South Main Street, Brewer, on the other. A paper mill was quite unlike any industrial plant the group had seen so far. The long pipelines and big distances of an oil refinery had created a rather impersonal atmosphere, while the lay-out of a paper mill and the proximity of one part of the process to others gave a feeling of intimacy. Here the six students could get into everything. Peter and Ray Baddour worked at Eastern Pulp and Paper, a site that had been built up over a long period. Paper-making uses a lot of water; there were no plans showing how the water lines had been laid out, so the two of them were set to trace the lot. It took them into all the nooks and crannies of the plant.

A paper-mill takes wood chips as raw material and extracts lignin, the glue which holds wood together, leaving clean cellulose fibres for paper-making. Eastern Pulp and Paper operated what was called the 'sulphite process'. This yielded a nasty waste liquor containing all the unwanted materials, including lignin. In those days such waste was dumped straight into the Penobscot river: the liquor was full of wood sugars, looked like molasses and made heavy demands on oxygen dissolved in the water. Deprived of oxygen, fish suffocated. Even in the 1940s the operating company was looking for ways to treat its effluent to make it less objectionable. Finding uses for waste made a lasting impression on the two of them. When Baddour was employed at MIT four years later, he had the idea of growing *Torula utilis* yeast on this waste liquor, harvesting the yeast and feeding it to cattle. The very same yeast made an appearance, under its older name *Torulopsis utilis*, in a couple of articles Peter wrote at about the same time in the UK magazine *Food*; they showed how the yeast could be used in an industrial process to produce fat.

The visit to Maine coincided with Thanksgiving—the fourth Thursday of November. Here Baddour discovered another of Peter's major characteristics: that he seemed to know people everywhere. Somehow Peter got them both invited for a deer hunt on the first day of the season. 'We were outfitted with bright orange jackets and two powerful rifles, and dropped off at the edge of a forest. When we started moving amongst the trees, we kept spying overfed city dwellers noisily crashing through the woods and carelessly dangling dangerous firearms. This so frightened us that we hid behind a tree whenever one of them came into view. Finally we gave up in frustration. That night we ate with our hosts: venison from their freezer.'

The practice course ended early in February 1948. The group returned to Cambridge, Massachusetts, and were put up in a dormitory to wait for their results. When they were published (see p142) Peter's grades showed a respectable 'honor' (or credit) in most subjects, including the practice trips, although only a pass in 'differential equations'. How did he see the whole experience?

It was a unique opportunity to become acclimatised to a country and its people, and I have felt at home in the USA ever since. The course was brilliantly conceived and executed. In general, but with exceptions, chemical engineering in 1948 in the USA (which dominated the international scene) was a non-intellectual subject. Most of its leading exponents and teachers were concerned primarily with practical problems and methods of solving them using the minimum of analysis. At each process plant, apart from carrying out measurements and experiments, we learned about the history and the state of the art of the process, and about scientific topics related to it. In the Lackawanna Steel works, for example, I learned more about the theory and practical aspects of [heat] radiation than I could have done in any other way. Even Boltzmann [a famous name in the theory of heat radiation], I like to think, could not have had a better intuitive appreciation of 'black-body [perfect emitter] radiation' than I acquired by peering into furnaces.

Peter's MIT teachers were an impressive group. On that part of the course conducted in the classroom, Peter had found himself sitting at the feet of Warren 'Doc' Lewis, the grand-daddy of US chemical engineering. Lewis had first arrived at MIT as a student in 1901, soon after the chemical engineering course began. Before WW1 he studied briefly in Breslau (now Polish Wrocław) and returned to a life-long career on the MIT staff spiced with a lucrative side-line as a consultant to industry. His lectures were legendary for their combination of beautifully organised material and Socratic exchanges with his students. His 1923 textbook, written with William Walker and William McAdams, underpinned the

---

BACKGROUNDER:
DOC LEWIS IN ACTION
Doc traditionally taught 'Course X' at MIT and when he formally retired at 70 Tom Sherwood organised a special project to celebrate. He invited all Course X alumni ('who started by cursing Doc and then came to love him') to send in stories. They made up a small book entitled A dollar to a doughnut—Doc's catchphrase for a bet with the students. A doughnut cost a lot less than a dollar in those days. Doc was always trying to make students think for themselves and not simply to follow textbooks or quote other people's conclusions. And he was a great firer of questions. Once, when a student gave him an unsatisfactory answer, Doc asked a corollary question and then several more, getting an answer each time, until he arrived back at the original question. The answer now was obviously the opposite of the student's original reply. Leaning over the student's desk, he simulated nose-pinching and

yelled at him in a high-pitched voice, 'I can lead you around by the nose. Why the hell do you let me do it?' In his younger days Doc encouraged a student by bellowing, while shaking a finger in his face, 'Now, Bonehead Brown, how would you solve that?' His colleagues gradually persuaded him that 'intellectual processes were not lubricated by these tactics, plus the bristling eyebrows, snorts, puffs and groans'. He was also a great thumper of the table or his adversary's shoulder. But he wasn't all sound and fury. After one of those demanding quizzes, a depressed student came to Doc's office saying that he couldn't cope any more. 'Doc practically took him apart on the spot, told him he didn't have time for any "snivelling weaklings" and finished with a terrible dressing-down.' The boy left, white as a sheet. Doc dug out the kid's quiz sheet, added a point here and there, and boosted his grade. Lewis continued to be an influence on the department until he died in 1975, aged 92.

Doc Lewis caught in charactistic full flow
Courtesy MIT Museum

fledgling discipline. In Peter's time at MIT, Lewis was approaching formal retirement, although he was still bubbling with energy and went on teaching for much longer. As a believer in an adversarial system, he was renowned for firing some of the more outrageous questions at students.

His unscrupulous arguments gave us a good warning of the unjustness of the outside world, but he knew how to galvanise his class by saying 'Do you want to know how I made my money, fellers?' Even Judson, his yellow-soxed feet propped on the chair in front of him, showed signs of animation.

BACKGROUNDER:
ARNOLD JUDSON MISJUDGED?
Peter may have underestimated Arnold Judson, his classmate in yellow sox. Judson doesn't recall the sox, but didn't forget Lewis's advice. A 'straight As' student at school, he had found chemistry and maths so easy he had time for, amongst other things, playing a Beethoven piano concerto with the MIT Symphony, accompanying the glee club on piano and timpani, and studying musical composition (for which he later became well-known). He was on the 1947 chemical engineering course, he says, 'because he found it exciting; more marks were given for approach and method than for the right answer'. He switched to organisational behaviour after his degree, feeling that there were similarities between what an organisation does and what happens in a chemical process. He ended up as a management consultant, author and composer.

One of many ways Lewis made money as a consultant turned out to be through an exclusive contract he negotiated with Exxon.

Tom Sherwood was another professor giving classes at MIT when Peter arrived. Quite unlike Lewis, Sherwood was urbane and reasonable. He was, Peter realised, one of the first all-rounders to bring science and mathematics to bear on a variety of chemical engineering problems. Peter admired him and assessed him as wise and charming:

I particularly treasure a paper in which he discussed experience, experiments and explanations of the fact that the ice at the top of an ice-flow contained less salt than the sea water from which it was formed, so that it could be melted and used for drinking water. This was an early illustration of the diversity of situations to which so-called 'chemical engineering principles' could be applied.

Sherwood later shared an office suite with Ray Baddour at MIT. Baddour recalls how good Sherwood was at writing: he could write a manuscript straight out in ink, while Baddour 'wrote with a pencil, and kept an eraser in the other hand'. Peter shared Sherwood's facility with words and was soon to share another distinction—that of having a useful parameter in chemical engineering named after him. The dimensionless Sherwood number was useful in calculations involving mixing. A graduate of McGill University in Canada in 1923, Sherwood had then studied at MIT with McAdams and Lewis. He stayed on, and like all the best MIT staff, worked widely with industry.

Another giant of the 1940s professorial staff was Ed Gilliland. Gilliland had arrived at MIT in the mid-1930s, studied under Sherwood and after the war developed the mythical status of a brilliant no-nonsense teacher,

---

BACKGROUNDER:
DIMENSIONLESS NUMBERS

A lot of numbers in life have 'units' attached to them. For instance, Michael Schumacher wasn't driving that Ferrari on TV last night at 100, he drove it at 100 miles per hour. Dimensionless numbers don't have units attached. Suppose we take the ratio of the speed of Schumacher in some unspecified form of transport in miles an hour against his walking speed (say 3 miles per hour): the result is a dimensionless number because the units cancel out top and bottom. If he's walking, the Schumacher number is 1 (because 3mph/3mph=1). If we hear that the Schumacher number changes to 3, he's probably on a bike (at 9mph). If it becomes 33, he could be in that Ferrari (at 100mph). So the size of a dimensionless number gives information about transitions: the (fictional) Schumacher number would be about transitions between forms of transport. It's OK to measure in kilometres but be consistent throughout.

author and researcher: probably the brightest on the faculty. Baddour was very keen to get away into industry but Gilliland, seeing how sharp he was, kept persuading him to stay on and satisfy his enthusiasm for industry as a consultant. When Gilliland stepped down as head of department in 1969, he presented his recommendations for a successor: Baddour on a list of one.

Peter was to see quite a lot of these men in the future. Indeed, so impressed was he by his experiences at MIT that they informed his views on academia and student courses thereafter. But for now he took off in his Dodge on the second part of his Commonwealth Fund tour of the country. It was required that he take in the US west coast. He had told Alice Robertson—the reporter he had encountered in Bangor, Maine, a few months earlier—how he wanted to see the southern and south-western

---

BACKGROUNDER:
SHERWOOD MAKES GOOD COFFEE

The useful chemical engineering tool named after Tom Sherwood is the dimensionless Sherwood number. So what is it? Sherwood looked at the way two components mix with each other. Take preparing a cup of coffee. The idea is to get enough flavour molecules in the coffee beans to mingle with molecules of water until they make a tasty cup. The technical phrase is mass transfer. It happens in two ways: one is called diffusion—the molecules have energy of their own and this naturally leads them to move from the bean to the water; the other is convection—the heat of the water adds energy and creates currents to speed up the rate of molecular mingling. Rate of convection divided by rate of diffusion is the Sherwood number (Sh). Why is it useful? Knowing the size of the coffee particles, the Sherwood number can be calculated from data in reference books. Diffusion depends on material properties which are also in the books. Then, knowing the rate of diffusion and the Sherwood number, the rate of mass transfer due to convection can be calculated. What about practicalities? Pouring hot water onto coffee beans is no good; the area of the bean surfaces is too small for good diffusion. Grind the beans first—this vastly increases the surface area and gives much more opportunity for molecules to mix. And they don't have to travel too far. So diffusion increases to an extent dependant on how fine the beans are ground. Convection increases as the temperature goes up: you can use boiling water but super-heated steam in one of those Italian machines is better. Stirring the coffee could increase convection further. Theoretically this data can reveal the speed and length of stirring to make the best cup of coffee. In real life: suck it and see. But attention to a single variable—the Sherwood number—potentially provides the insight to make an ideal cup of coffee.

states 'because he hadn't seen them yet, and because the weather is reasonable.' He drove south to Charleston in South Carolina and on to the Everglades National Park in Florida. Then he headed west via New Orleans and across the southern states to the Grand Canyon. Finally he arrived in California, with 20,000 miles on the clock. But time was up, and he sold the Dodge. Peter shared a few other thoughts with Robertson, and she reported that 'he would like also to visit Hollywood, but he is afraid he might never get away again'. The thought of film stardom seemed to linger at the back of his mind. But away he did get, and on his return to Britain found that his special charms had an audience after all...

BACKGROUNDER:
PETER'S MIT GRADES
Pass grades at MIT in 1948 ranged, in descending order,
from 'Honor' through 'Credit' to mere 'Pass':
Industrial chemistry (180 hours): Credit
Colloidal chemistry (90 hours): Honor
Chemical engineering 1 (150 hours): Honor
Chemical engineering 2 (150 hours): Credit
Differential equations (135 hours): Pass
Thesis (375 hours): Credit
Parlin station (180 hours): Honor
Buffalo station (180 hours): Credit
Bangor station: no record

Cambridge bohemia 15

RRIVAL BACK IN BRITAIN from the United States in the late summer of 1948 was quite a shock. After two years in a relatively unrestricted and affluent country, Peter had to cope again with austerity: ration books, limited transport and a country on its economic and physical knees. With his new master's degree in chemical engineering, he cast around for an industrial job but found that potential employers weren't sure what a chemical engineer was. When Peter explained, they felt they could get by without one. So he got in touch with Terence Fox, the head of the new Cambridge department of chemical engineering, whom he had bumped into the previous year at an MIT lecture. There was an interview in Cambridge; the department's recently-leased site was nothing more than a rectangle of hardcore, awaiting the imminent arrival of temporary buildings. Fox invited Peter to join three other new members of staff: Kenneth Denbigh, Stan Sellers and John Kay. The salary seemed to Peter to be 'ridiculous' and further shocks lay in store.

Nowadays it's hard to get about in the centre of Cambridge for crowds of tourists; in 1948 it was almost exclusively the domain of university students and the colleges. For many a student the abiding memory of the city in the late 1940s was of a primitive college room which became ice-cold in winter. There was no heat other than a bowl of hot water delivered by the 'bedder' in the morning and a small gas fire, suited to little more than making toast. Standing before it, the front roasted while the backside froze. The smell of boiled cabbage wafted gently from college kitchens. Most teaching members of established university departments enjoyed access to slightly better conditions than these. They were invited to become 'Fellows' of one college or another, an invitation which carried attractive free dining rights. But Peter and his new colleagues did not qualify; the University had yet to recognise the new department of chemical engineering as a serious intellectual activity worthy of endorsement. That made things that bit more of a struggle. At the advanced age of 32, Peter might have expected something better. Peter's new colleague and fellow lecturer Kenneth Denbigh made an approach to Gonville & Caius College; the master—Sir James Chadwick, discoverer of the neutron and head of the British mission attached to the Manhattan project which applied nuclear physics to the development of the atomic bomb—took him aside and explained that he didn't think Cambridge was a suitable place for applied science. To reinforce this opinion he also confided in Denbigh that many people feared the department would emit corrosive fumes.

Only Fox, as his department's professor, had a college fellowship and that was a hangover from his previous position on the staff of the University's general engineering department. He was a Fellow of King's College. As a bachelor Fox had the chance to rent a desirable suite of college rooms at a modest rate. His suite, Y7, was on the second floor in the building close to King's College's bridge over the river Cam. His living conditions there were quite comfortable, and must have seemed palatial to Peter. The suite commands magnificent views: to the north the vast green sward stretching from the college's celebrated chapel down to the river, and to the east a quiet stretch of the river upstream of the bridge. Peter was envious...

...we felt like second-class citizens. At the end of my six years [1948-54] I had achieved dining rights in one college 'once a week in full term at my own expense'.

Attachment to a college would have at least provided the four chemical engineering staff members with a convenient 'lunch club'. But rumour had it that in fact high-table conversation, far from being high-minded, was disappointingly provincial. So they may not have missed much. As an alternative the four bought sandwiches at the Eagle pub in Bene't Street, where once horse-drawn carriages from London ended their journey. It was soon to become more famous as the place where James Watson and Francis Crick talked out their ideas on the structure of deoxyribonucleic acid (DNA). At the time organisation at the Eagle was poor and the experience, in the opinion of a contemporary, somewhere between Breughel and Kafka. So the chemical engineering four moved on to the Little Rose in Trumpington Street opposite the Fitzwilliam Museum. It was right next door to the department and had the reputation of being, in Peter's view...

...possibly the seediest pub with the surliest landlord in Cambridge.

In late 1950 the University made a gesture of a kind to unfortunates like the chemical engineering four: it opened a 'combination room' for unattached academics where they could take lunch, read newspapers and magazines, and meet people from other disciplines, most of them also 'outsiders'. The address was Regent House, a substantial building in the middle of Cambridge next door to the Senate House. The surroundings were splendid but space could be tight and, at lunchtime, contentious. An entry in the suggestions book of February 1953 over Peter's name complains to the management committee about guests behaving as if they were members: reserving blocks of seats and hogging the papers to which they had no entitlement. The committee replied in opaque and timeless officialese: 'Appropriate action is being taken'. A later suggestion from the same source proposed the installation of a coin-box telephone. It got short shrift.

Besides interaction with his colleagues, there were social highlights in Peter's life beyond the department. George Porter was a lecturer in the physical chemistry department at the time. Four years younger than Peter, his practical work on photochemistry—study of the behaviour of molecules during very fast chemical reactions—was to win him and his professor, Ronald Norrish, a Nobel prize in 1967. Porter hit it off with Peter and dropped in on him now and then at work; they had both spent the war in the Royal Navy Volunteer Reserve, Porter as a radar officer, and their contemporaries noticed that they both had aspirations for a lifestyle somewhat grander than the average lecturer. Porter was a natural performer and pioneered the presentation of science on television to a British public which had been becoming increasingly sceptical. He was later knighted and ennobled.

Peter also saw something of geophysicist Tom Gaskell, his good friend from Combined Operations days. At the end of November 1949 Gaskell had returned from two years in Persia with the Anglo-Iranian Oil Company; the two dined at one of Gaskell's favourite watering holes—the Pheasantry in King's Road, Chelsea, popular with film stars, artists and politicians—to celebrate his return. Gaskell had been brought up in Cambridge at 39 Hills Road and his parents still lived in the house. So, on professional visits to confer with his friends—the Cambridge geophysicist academics Teddy Bullard and Ben Browne—Gaskell had a convenient bed for the night. He and Peter would meet in Cambridge for a meal at the 'KP' restaurant on King's Parade, a visit to the theatre or a drink. New Year 1950 found them as guests of a large house in Norwich with excursions for a hunt ball and clay pigeon shooting. Then in April

1950 Gaskell took off on a two-and-a-half-year trip around the globe as chief scientist aboard *HMS Challenger*, making seismic measurements of the ocean floor. The expedition identified the deepest known point—in the Mariana trench near Guam—36,000 feet (11km) down. At the same time Gaskell had been instructed, with great secrecy, to monitor the atmosphere for fallout from atom bomb tests. When he got back in early 1953, he and Peter resumed occasional meetings. Both men seemed unsure of how best to progress their lives; in April 1953 Gaskell recorded in his diary that they 'talked of what we should aim at in the future, but we both drift and so are not much use to each other'.

On arrival in Cambridge Peter had found lodgings in a room on the ground floor of a house in Jesus Lane, next door to the grounds of Jesus College. A landlady may have looked after some of the basic necessities, like bed-making, but otherwise he had to fend for himself. The top floor room was rented by 24-year-old Bryan Robertson who had just taken up the job of curator of the art gallery established by the Heffer family, owners of the eponymous Cambridge bookshop. Robertson regularly met local artists, sculptors and their circle; it was probably he who introduced Peter to their world, although Terence Fox also had contacts with local artists and may have been responsible. Throughout his six-year stay in Cambridge Peter kept contact with artists, writers and composers, although he remained largely on the fringe. They regarded him as a bit inhibited, even 'square', but he enjoyed their company and was enough of a renaissance man to be interested in what they did. And during the early 1950s this contact led to emotional and physical consequences.

Robertson was very friendly with Elisabeth Vellacott, then a designer and printer of textiles with an interest in stage sets and costumes. For a while in the early 1950s, after Robertson fell out with the Heffer Gallery and lost his job, he became just as impoverished as she and the two met every day for lunch in her Ram Yard studio on Round Church Street (now replaced by a concrete car park). Sensing her talent, he encouraged her. In due course they both went on to greater things: Vellacott won a national reputation as an artist and Robertson, after a move to London and the Whitechapel Gallery, became the country's most respected art impresario and critic. Vellacott's circle of artist friends in Cambridge was extensive. It included Cecil Collins, a painter in the Blake tradition whose work interested the Tate Gallery, and his wife Elisabeth, who also painted. Nan

Youngman, a painter, and Betty Rea, a sculptress, were also part of Vellacott's world. An older member of the circle was Cecil Lady Taylor, whose first-floor flat at 15 King's Parade commanded an enviable view of King's College Chapel. She had been married to Major General Sir George Taylor, director of Army bomb disposal from 1940 to 1942. After the war she wanted to paint but he needed her as his social secretary and hostess. So she left him and set up on her own in Cambridge. Much younger, and at first still students, Jasper and Jean Rose had a studio just next door to Vellacott in Ram Yard. They lived, in the judgement of a contemporary, 'on the smell of an oil cloth' in one room with a bedroom up a ladder to the loft.

Not everyone in post-war Cambridge was living on a shoe-string. Peter Mitchell became a student at Jesus College in 1939, graduated and began research in biochemistry. Mitchell had private means: money from wealth created by his uncle who developed the engineering and construction company George Wimpey between the world wars. According to his biographers Prebble and Weber, Mitchell was a bohemian and had the strong conviction (with which Peter Danckwerts sympathised) that

---

BACKGROUNDER:
BRYAN ROBERTSON

Brought up in the deprived Battersea of the 1930s, Bryan Robertson had no formal education beyond school. He learned to appreciate art in bed, where he was often confined with asthma, devouring every art book he could get his hands on. He spent 1947 in Paris and met modern masters like Brancusi, Braque and Giacometti. His career took off in 1949 when the Heffer family made him curator of their new Cambridge art gallery. That first job was short-lived; by 1951 he was unemployed. He accepted Peter Mitchell's offer of a flat to tide him over, and moved into a flat at the Mitchells' accommodation in Grange Road. Robertson saw great potential in Vellacott's work and encouraged her to draw and paint. They forged a lifelong friendship. In 1952, probably through the influence of his friend the art historian Kenneth Clark, Robertson became curator of the Whitechapel art gallery in London and launched a series of exhibitions which opened British eyes to the international art world. It turned the gallery from a 'cultural soup kitchen' into Britain's most influential contemporary art gallery. He revived interest in Turner, Stubbs and Hepworth and introduced Australians like Drysdale and Nolan. Realising that New York was replacing Paris as centre of the art world, he showed Pollock and Rothko. Bryan Robertson became the most influential and revered art 'impresario' of his generation.

scientists needed similar freedom to artists and composers in order to do their best work. In his student days Mitchell had enjoyed such company at the Peacock, a small restaurant in All Saints' Passage. He knew, amongst others, Vellacott, Robertson and the Roses. Marian Sugden was another member of the group: a competent artist, she was married to Mitchell's college friend, the physical chemist Morris Sugden. Gabor Kossa, owner of the antique shop in Trumpington Street opposite the Fitzwilliam Museum, joined in from time to time. A Hungarian émigré, he had been involved in his home country with a ballet company and was reputed to have known Diaghilev. Beyond his business interests, Kossa put on plays and

---

BACKGROUNDER:
PETER MITCHELL

A charismatic figure, often to be seen about 1940s Cambridge with shirt open to the waist and a purple jacket offsetting his shoulder-length hair, Jesus College student Peter Mitchell was said to have looks resembling the young Beethoven. Extraordinarily bright, his all-consuming research interest became the mechanism by which energy to power the human body flows from food into the bloodstream and thence into the cells of the body and its muscles. Competing scientists ridiculed Mitchell's 'chemiosmotic' theory (a combination of chemical reactions and osmotic pressure in which hydrogen ions carry energy into the cell through the cell wall) when he proposed it in the 1960s. But he persevered. While other theories failed to convince and lost support, Mitchell's survived and won him the 1978 Nobel Prize for chemistry, even though some of the detail was unresolved. There was controversy; a rival, no doubt jealous, suggested Mitchell got the award for bio-imagination. Mitchell had begun

his career with staff positions in biochemistry in Cambridge from 1951 to 1955 and then in zoology in Edinburgh. But he was plagued by gastric ulcers and retired from Edinburgh to Bodmin in Cornwall. There, with the help of his uncle Godfrey's money, he renovated nearby Glynn House and turned it into a private research laboratory, inviting a team of faithful colleagues to work there with him. It was quite extraordinary that he was successful, even though 'the big battalions' were ploughing huge resources into the same field of research. Like his uncle, Peter Mitchell too had a generous heart. One example: when fellow biochemist and avowed communist Joseph Needham returned from China ready to tell Cambridge about the wonders of ancient Chinese technology, their head of department Sir Frederick Hopkins denied Needham use of a lecture theatre on the grounds that the subject matter was too political. Mitchell threw open 60 Grange Road for the event. Mitchell was tolerant, gentle and definitely odd-ball.

was something of an amateur musician. Once he hired London's Wigmore Hall and gave a silent piano recital; reviews were mixed.

As a student during WW2 Mitchell had several girlfriends before being attracted to Eileen Rollo. This relationship became serious, and they married in December 1944. She worked as a nurse at a rest home for wounded servicemen at 60 Grange Road, a large Cambridge house with a big garden. In 1945 twins—Julia and Jeremy—were born to Eileen and Peter Mitchell but, as recorded by Prebble and Weber on the basis of extensive interviews with Peter Mitchell, the parents were gradually growing apart. In February 1945 Mitchell acquired the lease of the Grange Road house where his wife had nursed during the war—it had 60 years to run—and the family moved in. It was a large building, but in general Mitchell wasn't ostentatious with money. He let out flats in the big house—probably at a low rent.

Mitchell had at least one extravagant weakness: a 1926 open-top Rolls-Royce tourer in which he and his wife used to drive theatrically around Cambridge. In the summer of 1950 it played a small part in creating a defining rift between the Mitchells. They took off for Monaco in the Rolls to visit

60 Grange Road in 2010

Elisabeth Vellacott who was holidaying in a villa she had rented there for the summer. With them went Bryan Robertson, John McNeil (a bright but rather nervous young medical student said to be moonstruck by Eileen Mitchell) and Pat and Helen Robertson, no relations of Bryan. Pat was a theatre set designer; Helen had a reputation as a beauty. On the way Peter Mitchell and Helen Robertson fell into each other's arms. Eileen Mitchell was very upset; Mitchell was her man and it was extremely distressing to have him taken away like that. Their relationship, already strained, headed towards divorce, in those days a lengthy business. Perhaps as a distraction, the Mitchells were very active socially and ran frequent parties for their artist friends and other enthusiasts. Former participants say that, while champagne bottles may have been on hand, the atmosphere wasn't excessively boozy and the parties certainly didn't go on all night.

On the other hand, Peter Mitchell was known for his enthusiasm for an element of racketyness. The less rackety commented that there was something strange at Grange Road parties in the unsettling fluidity of relationships between the sexes. Francis Crick was a regular partygoer who shared Mitchell's enthusiasm; the artists in the group had no idea how celebrated both Crick and Mitchell would shortly become, each in his own field, as Nobel prize-winning scientists. John Gayer Anderson, a literary dilettante and college contemporary of Mitchell's, also joined the 60 Grange Road fray. Peter and Sheila Stern—he a lecturer in German, she in English—were lodgers in the house and so found themselves on the spot. Another keen participant was sometime surrealist and socialist Hugh Sykes Davies. He showed amazing wit, grace, verve and social charm but while he was good at so many things, he didn't seem to specialize in any of them. The Cambridge circle knew at least that Sykes Davies was a fair

---

BACKGROUNDER:
HUGH SYKES DAVIES

If your father was a teetotal Yorkshire clergyman and your grandmother's middle name was Abstemia, you might be forgiven for rebelling. So most forgave Hugh Sykes Davies, a Cambridge don notorious in his time. Graduating in English in the early 1930s he became an early member of the Apostles, along with Anthony Blunt and Guy Burgess (although he was a socialist then, rather than a communist). He also spent time in Paris; a gift for friendship helped him get to know Salvador Dalí and he became an organizer of the 1936 London International Surrealist Exhibition at the Victoria & Albert Museum. It attracted huge public attention. Dalí addressed the meeting dressed in a diver's suit, leading a greyhound on a leash. When someone stepped on the air pipe, Dalí was in trouble but Davies came to the rescue, loosening screws at his neck for

ventilation. Davies was a natural member of the Mitchell set. His *Times* obituary refers to great intellectual power allied to the graces of wit and social charm. Everything he accomplished was excellently done, yet he was too versatile to become pre-eminent in any particular field. Davies was keen on new things, including wives. His first, the poet Kathleen Raine, ran away. In 1947 he made prim, proper and aristocratic Fay de Courcy his third wife. They divorced in 1962 but he married her again as his fifth wife in 1982. He pointed out that that this too was new, as he'd never remarried anyone before and, anyway, this wife got him into *Debrett's (Peerage & Baronetage)* each time. When Peter Danckwerts died a few months after Hugh Sykes Davies in 1984, Fay Davies wrote touchingly to his widow recalling that 'when Hugh was in hospital Peter wrote the best and most heartfelt letter, advocating gin.'

pianist and capable of rendering to them acceptable versions of Chopin mazurkas on the piano accordion. This was Cambridge 1950s' bohemia: a post-war generation which was escaping from ration books, austerity and all the restrictions of wartime, and felt at liberty to create a new society with its own rules for the future.

One of the most memorable Mitchell parties was held in the spring of 1953. It was Peter Mitchell's idea: 'Clergymen and Goddesses'. There was quite a crush, which reminded the more excitable guests of Valhalla, and couples roamed in the gardens of 60 Grange Road, according to one participant 'probably debating theological dilemmas'. Eileen Mitchell dressed as Psyche; Mitchell himself chose to appear as the Devil, with horns and a tail fashioned from laboratory rubber tubing. Most men simply turned their collars round, and amongst them Sykes Davies was reckoned to look by far the most pious. Cecil and Elisabeth Collins were there: on that evening she was wearing a peacock blue wrap intended to cover a bare back. Clergymen present—who regretted that so many of the other ladies opted to dress as ancient rather than modern goddesses and were taking cover under curtains and dyed sheets—enjoyed watching local sculptor John Smith running his fingers down Elisabeth Collins' spine as though playing the xylophone. Jean Rose went as Diana, goddess of the chase, in a pair of outsize jodhpurs and a bowler hat, with a bow and arrow prop made from a coat-hanger. Sinophile Joseph Needham, another good friend of Vellacott and Mitchell, effected a Chinese priest in a huge blue robe with shoe trees on his head.

Other members of the group ran parties for their artist friends but of a rather different kind. Those organised by Cecil Lady Taylor, for instance, were of the tea and Battenberg cake variety. She knew both Peter Danckwerts and his boss, Terence Fox, and of their sympathy for the idea of freedom to realise their full potential. Nan Youngman and Betty Rea also hosted parties at their home, Papermills, an old and beautiful house on Newmarket Road just beyond the leper chapel. From time to time the artistic communities of Cambridge and London gathered there. Rea, an acquaintance of Desmond Bernal, left a lasting legacy outside Cambridge's Parkside swimming pool: a group of three bronze swimmers in conversation at the poolside.

In 1954 Peter Mitchell took off to pursue his scientific work in Edinburgh. Parties at 60 Grange Road ceased. The Cricks, having returned from a year in America and living in a five-storey house in Portugal

Place, tried to fill the gap. There were bright spots: Jasper Rose, a junior fellow at King's College and made a proctor (a sort of university police-man) by King's president Noel Annan 'as a way to discredit the role', remembers turning up to an event at Portugal Place late one evening to overhear whispering on the ground floor. 'Up there,' people pointed towards the stairs. 'Real models from London. Starkers!' But attempts at themed parties, like the vaguely sadomasochistic-sounding 'Planters and Slaves' evening, weren't—surviving participants claim—really the same.

Throughout his first period in Cambridge, which lasted six years, Peter was on the fringes of this bohemian world. He was interested in what its members did but wasn't fully attuned to their world; after overhearing a conversation between Bryan Robertson and the artist Francis Bacon one evening, he admitted that he 'couldn't understand a word they were say-ing'. Seen from the perspective of his day job, these were perhaps minor social distractions. Above all Peter was busy making his name in his area of choice—chemical engineering. But both his involvement with Cambridge bohemia and his abilities as an innovator in his subject area led to tangible results.

The Shell endowment 16

AMBRIDGE UNIVERSITY'S new chemical engineering department, which opened to students in October 1948, was unusual in many ways. For instance, of all the staff members only Peter had actually been a student on a chemical engineering course. And its governance was separate from the normal University arrangements. The department was a gift to the University, or more correctly to the country, from the Shell Group of oil companies. It was the brainchild of two individuals—John Oriel and Ronald Norrish. Peter wrote later:

Norrish alleged that the whole scheme was cooked up between him and Oriel in London pubs during the [wartime] blackout. To anyone who knew them, the milieu seems entirely credible.

Norrish was a formidable professor of physical chemistry. He specialised in photochemistry—that is to say, very fast reactions stimulated by short and intense flashes of light which are relevant to what happens in flames. Later he shared a Nobel prize with his colleague George Porter. Oriel was a Shell man through and through. A chemistry graduate, he had an encounter with mustard gas during WW1 and eventually became blind, but it didn't stop him continuing as a busy Shell executive. He knew the company intimately. During WW2 both he and Norrish could sense that, if Europe were to keep up with worldwide progress in the petroleum industry, it would need a good supply of competent chemical engineering graduates. Until then British and continental European industry had required chemists (skilled with test tubes) and mechanical engineers (adept at designing machinery) to put together plant in which chemical reactions might be carried out on a large scale and with efficient results. To achieve this they would 'knock lumps out of each other'. The arrangement survived on accumulated experience and flair; the early German chemical industry had thrived on it, but the British were less well served.

Things happen in big chemical reactors which chemists and mechanical engineers don't get taught about; they may learn a bit about them from experience but, as Peter always maintained, once they were immersed full-time in industry they rarely had the time or inclination to take an interest in theory. For instance, the pattern of flow followed by a liquid through a large reactor and the variation of temperature across it both affect the yield of a reaction. In addition, designing for the efficient separation of products from each other on a large scale requires special knowledge and skills. Chemical engineers learn how to make a multi-ton process successful; such a profession was, with a few glorious exceptions, almost unknown in Europe in 1945, whereas the Americans—notably at Massachusetts Institute of Technology—had been forging ahead for years.

Norrish and Oriel not only persuaded Shell directors to fund a professor and a department of chemical engineering but to do it on a grand scale. They were aware that it might prove impossible to attract the right individuals as staff members, because in industry they would already be receiving larger salaries than the University would ever be able to offer them. And it was vital to equip them properly for practical experimentation. In January 1945 Shell wrote to the vice-chancellor of Cambridge University offering a huge sum: £435,400 over seven years, the equivalent of some £50 million in 2010, and no strings attached. It also provided £2,500 each year (£300,000 in 2010 terms) for scholarships until further notice, to encourage students. Within two months—in spite of rumoured efforts by Oxford University and Balliol College, in the form of science and technology guru (by now Sir) Harold Hartley, to hijack the arrangements—Cambridge had accepted the offer. For governance of the new department a chemical engineering syndicate was created, unique in that the department was thus independent of the University's normal control by faculty boards, whose requirements could create a sense of frustration.

In the summer of 1946 Terence Fox had been announced as the department's first professor. Fox had graduated in mechanical engineering at Cambridge in 1933 with stunning effect. Peter noted:

He swept the board, not only achieving the remarkable distinction of 'starred first class' honours but winning all the possible prizes as well.

Fox then went to work for ICI at Billingham on Teesside before returning to Cambridge as an engineering department lecturer. As such, he became a Fellow of King's College. By 1946 he seemed an obvious, if slightly idiosyncratic, choice as Shell professor of chemical engineering:

Using the rule of thumb normally applied at Cambridge, the person appointed should be a Cambridge graduate who has served an apprenticeship elsewhere, such as a professorship at a provincial university, and on no account anyone who has actually applied. Fox was somewhat unusual; he had never been a professor and never attempted a doctorate.

But he was known to the academic world, was incredibly bright and it was time for him to take a step up. He was well-equipped for the main objective of the post: to make chemical engineering intellectually respectable to the University. Fox arranged to spend the academic year 1946-47—his first in the job—on a tour of departments in the United States, where chemical engineering was taught competently. In that way he could acquaint himself with the discipline and how the Americans did things. As it happens, several departments he visited had recently been wrecked by disastrous fires, and the trip left Fox somewhat neurotic about safety in his own department's laboratories.

Back in Cambridge it was proving hard to find a site on which the new department could operate. Eventually Paul Vellacott, the master of Peterhouse, the oldest Cambridge college, decided that he should encourage this fledgling department and arranged the lease of a temporary site between Tennis Court Road and Trumpington Street. The rectangle of hardcore which Peter had seen at interview was soon covered with single-storey huts, and they remained the department's home until 1959.

Peter's new colleagues were a mixed bag. In view of their lack of formal

Terence Fox in 1952: rumpled linen jacket and frequent cigarette
Courtesy Don Katz

training in chemical engineering, Fox insisted they all attend each other's lectures. Fox himself was a brilliant teacher but his lectures were of a frightening intensity, even for the most competent students. Furthermore, Fox wanted to make his graduate and undergraduate students into rounded individuals, and he often invited them to his rooms in King's College to meet artists from different disciplines. From time to time he played classical gramophone records to them and tried to winkle out their opinions. This may well have enhanced his daunting image.

Stan Sellers came from a background in oil. After working for Anglo-Iranian in the Middle East and then at Manchester Oil Refinery, he was at home with processes in which oil ran through pipes. Big, burly and an experienced chemical engineer, it was easy to imagine him out on the refinery getting things done. He taught that part of the course which dealt with what chemical engineers actually do when they design and operate a plant. Always imperturbable, he made a wonderful front-man for the department.

In contrast John Kay appeared to some of his colleagues to be quiet and studious, even stuffy. But his sense of humour could be tickled, and he was certainly one of the better lecturers. He had a degree and a doctorate in mechanical sciences, and between the two had served an apprenticeship in the mechanical engineering tradition. Peter remembered that:

He used to point with pride at the permanent way near Cambridge station and proclaim that he had cut the threads on the bolts which secured the fish-plates which held the rails together.

Kay had worked with Rolls Royce and so knew about aerodynamics. This was useful because the way air flows around aircraft bodies relates to how fluids flow along pipes, past stationary walls and around fixed barriers—'fluid mechanics' in a chemical engineer's professional jargon.

The 1948 department (l to r): John Kay, Stan Sellers, Terence Fox,
Kenneth Denbigh, Peter Danckwerts
Photographer: Edward Leigh

The fourth member of the quartet was Kenneth Denbigh. A grammar school boy and short in stature, he often wore a trilby hat with the front turned up, allegedly to appear taller. But in intellectual terms Denbigh had nothing to feel inferior about. He was quite exceptional; indeed, the Fox-Denbigh-Danckwerts period in Cambridge chemical engineering was, and is still, considered by many as the ultimate flowering of the profession. Denbigh had been at ICI Billingham with Fox. An industrial chemist with theoretical leanings, he had worked on continuous processes during WW2—the manufacture of two chemicals vital to the war effort, ammonia and the explosive hexanite. In the lecture theatre Denbigh had the reputation of being able to achieve what generations of students (and most lecturers) have considered quite impossible: he made thermo-dynamics— that most difficult of courses—interesting. Thermodynamics explains how energy is converted into heat and work, and how that con-version depends on temperature, volume and pressure. Denbigh achieved his miracle partly by drawing on examples from real life and partly by deploying his lively sense of humour. Some in his audiences recall an experience familiar to many a student following inspiring lectures: when they got back to their rooms in college and the euphoria faded, they couldn't quite remember the detail. So it was back to the books to go through it again, but at least knowing that it could be understood.

Students found Peter a refreshing and less demanding option after the

BACKGROUNDER:
LECTURING DUTIES IN CHEMICAL ENGINEERING, 1950-51

Years one and two
 (taught elsewhere): Mechanical sciences or Natural sciences

Years three and four:

Fox:	Mathematical methods, Kinetic theory of gases, Thermodynamic properties;
DENBIGH:	Chemical thermodynamics, Reaction and phase equilibrium, Surface chemistry, Colloids, Continuous systems;
KAY:	Fluid mechanics, Applied thermodynamics, Heat and mass transfer, Control;
SELLERS:	Unit operations, Control, Chemical processes;
DANCKWERTS:	Mass and thermal balances, Reaction kinetics, Electrochemistry.

stresses brought on by Fox and Denbigh. Peter's lecture material was generally too new to appear in the textbooks and so his lectures at that time could be brilliant, eye-opening experiences. All the same, he gave the distinct impression that he wasn't cut out to stand up in front of an audience and teach. Bill Wilkinson, an undergraduate from 1951 to 1953 who then stayed on until 1956 to do some research, remembers Peter as a bit disorganised and dreamy at the best of times. But Wilkinson was sympathetic to and appreciative of Peter's style. To keep himself on track in those days, Peter kept his lecture notes in a hard-bound notebook rather than loose sheets favoured by many. Even so the thread could be lost. One day in mid-lecture it happened to a class of 20. Even Peter became aware. He glanced up, looked at his notes, turned back and then said:

I'm terribly sorry. I flicked over two pages together by mistake. Let's go back...

Thanks to his background, upbringing and nature, Peter was very sure of his position in society. His wartime exploits and bravery were well-known in Cambridge and inspired awe in colleagues and students alike. His height and good looks made an impact socially, although colleagues quickly found that he wasn't an open personality. Shyness and reserve made him hard to get to know, and that was easy to mistake for aloofness. But once he spotted a topic of interest, or a spark of intellectual imagination, colleagues and friends alike found they had brief access to sunny uplands of delightful conversation spiced with an agreeable irony. He was a disciplined thinker and rather an unusual academic because of his concern to make usefulness to industry a major feature of his work. Those six years in Cambridge showed both his profession and industry just how valuable that unusual nature was going to be.

Life on Tennis Court Road 17

THE 1948 UNDERGRADUATE COURSE in Cambridge's new chemical engineering department had only ten students, but a number of graduates had signed up to do doctorates and they swelled the numbers. One of those applying for the undergraduate course was Sydney Andrew, who already had qualifications as a mechanical engineer. He arrived from ICI Billingham where his boss knew Fox and Denbigh well. At Andrew's interview for admission, Fox sat on the panel with representatives of other departments. In such a mixed group of university disciplines there is an irresistible, some might say juvenile, tendency to score points almost without realising it. In this case, though, it wasn't the professors doing it. Senior chemistry professor Alexander Todd was sitting in; as the applicant remembers it Todd asked: 'You don't seem to have much chemistry, Andrew. Will this be a problem?' 'I don't think so,' responded Andrew with the nonchalance of youth. 'It's mainly memory work and I have a good memory.' Todd glowered and Fox suppressed a smile. As he left the interview Andrew heard footsteps hurrying after him down the corridor. It was Fox. 'Don't you know who that was on the panel?' he demanded. 'I might agree with what you said to him, but you shouldn't have said it.'

The main access to the department site was from Tennis Court Road, but there was also access from Trumpington Street, which runs parallel to it, by a passage opposite the Fitzwilliam Museum and beside the Little Rose public house (now converted into a restaurant). Students and staff had to pick their way past Captain Canham's riding stables. Peter recalled:

Sydney Andrew in 1948
Photographer: Edward Leigh

Horses put their heads out of their stalls and eyed us as we went by. On several occasions I saw the Shell professor chasing Captain Canham's hens off our site.

The department's temporary buildings provided space for both lectures and research laboratories. There was also room for a few peripatetic scientists. Tom Bacon, whose fuel cell research was then funded in part by the University, was one. He was allowed space just outside Peter's office. Fuel cells generate electricity directly from a chemical reaction; Bacon's operated at 200°C and about 40 times atmospheric pressure, not huge by industrial standards but enough to frighten academia. Peter was somewhat apprehensive of the equipment at first, but eventually felt reassured when he saw that Bacon spent most of his time close to it. Bacon was the man whom President Nixon—at the Apollo 17 launch in December 1972—clasped around the shoulders and declared: 'Without you, Tom, we wouldn't have gotten to the Moon'. Bacon's fuel cell provided a light and compact source of electricity and clean drinking water in the spacecraft. In 1948 Bacon was working on a small scale at Tennis Court Road and with larger equipment at Marshall's engineering establishment on the eastern outskirts of Cambridge. He and Peter became firm friends.

Those who 'got inside Peter's shell' were enthusiastic admirers. Peter Gray, who joined the chemical engineering staff in 1951, remembers Peter's conversation as 'humorous, on the edge of cynical'. Bill Wilkinson,

Terence Fox (centre) and John Kay (right)
among the chemical engineering department's temporary buildings
Courtesy Miles Kennedy

A technician attends to Tom Bacon's fuel cell equipment
Courtesy Cambridge University Department of Chemical Engineering and Biotechnology

BACKGROUNDER:

TOM BACON'S FUEL CELL

Francis Thomas Bacon trained as an engineer and became fascinated by the way fuel cells were able to convert chemical energy into electrical energy. Throughout the 1930s he worked alone on the design of a fuel cell which could use hydrogen and oxygen as fuels and inexpensive nickel as the catalyst for reaction, rather than the costly metals used by other designs. Bacon's cell operated at high temperature and pressure. By 1948 an inheritance —he was a descendant of the famous champion of science Francis Bacon— had allowed him to work full time on his project; he organised small interdisciplinary research teams and by 1948 was receiving funds from Cambridge University and the Central Electricity Generating Board. The University allowed him space in the new chemical engineering department's temporary buildings. By 1959 his fuel cell design had become the benchmark, and he successfully demonstrated a 6kW version which produced cheaper electricity than other designs. Pratt & Whitney licensed the design for the Apollo spacecraft: it appealed for several reasons: light and compact, it produced clean drinking water as a valuable by-product of the cell reaction and worked on the two raw materials —hydrogen and oxygen—which were already carried by the spacecraft to fuel its main rockets. The need for pure hydrogen and oxygen gases limited the use of Bacon's design in more mundane areas, but his success spurred research into other designs.

even though he was only an undergraduate at that time, also found Peter very sociable and not at all stuffy. He learned things that students perhaps didn't expect to learn from their lecturers: in Peter's case that he was not married but was 'looking for a blonde with long legs.' Wilkinson is sure that it was Peter who introduced him to pink gin. But above all Peter was a shining example of how to write. He was so precise, Wilkinson recalls. And Peter had strong views on how scholarship should be attained. As many Cambridge colleagues found then and later, Peter didn't like swots. It was poor form, he thought, for students to let it be seen that they were studying hard. He himself cultivated and exemplified the art of apparently effortless scholarship—although, like the swan, he may have been paddling away furiously under the surface.

Those early years at Tennis Court Road were heady times for young staff and graduate students alike. Fox was a great believer in getting team members to chat over tea in a relaxed atmosphere. He considered that ideas could best be fertilised by what might today be called 'lateral thinking', exchanging ideas in a relaxed atmosphere. Researchers were encouraged to join staff for both mid-morning and mid-afternoon tea-breaks. They met standing up; Fox didn't believe in sitting down. One morning when a group of industrialists arrived on a visit to inspect the department, Fox explained his tea-time philosophy and, glancing at his watch, suggested: 'Well, it's about that time now. Let's go along and see what they are discussing.' They did, and it turned out to be their impressions of the latest Marilyn Monroe film.

In their fourth year all undergraduates on the new Cambridge course had to undertake a project, to give them some sense of work in industry. Peter supervised several of them. One—Peter King, later director of ICI research—investigated the production of the algae *Chlorella* by a continuous process. At the time there was widespread concern about shortages in the supply of food as the total world population boomed; *Chlorella* was popular as a possible means for producing synthetic protein. King's equipment was a four or five-foot Perspex column about six inches in diameter with stirrers down the middle. *Chlorella* multiplied quickly in the right conditions. The idea was to feed a steady stream of *Chlorella* in at the base of the column—with water, carbon dioxide and nutrients—let the algae multiply as it rose and watch green goo emerging at the top. A vital necessity was very bright light supplied from lamps around the column. Things got very hot as a result and one day the lights

Laboratory activities on Tennis Court Road
Courtesy Cambridge University Department of Chemical Engineering and Biotechnology

BACKGROUNDER:

CHLORELLA: DASHED DREAMS

Chlorella (meaning small and green) is a single-celled member of the algae family. Spherical and between 2 to 10 micrometres in diameter, the cells contain the green photosynthetic pigment chlorophyll. Chlorella multiplies rapidly, requiring only carbon dioxide, water, sunlight and small amounts of minerals. Scientists studying Chlorella have won Nobel prizes: the German Otto Warburg (1931) worked out its cell function and the American Melvin Calvin (1961) showed how it helps plants take up carbon dioxide. In the 1950s Chlorella was thought to be a good bet as a potential source of food and energy. Its ability to fix carbon dioxide seemed to be as good as that of highly-efficient crops like sugar cane. It has a high content of good things: when dry, it is about 45% protein, 20% fat, 20% carbohydrate, 5% fibre, and 10% minerals and vitamins. But development for large-scale production revealed unforeseen problems. Its photosynthetic power did not live up to initial assessments, harvesting Chlorella on a large scale proved difficult, and the cells containing vital nutrients were found to be indigestible by animals and humans. Peter's student working on Chlorella, Peter King, took his thoughts on synthetic protein production with him on graduation from Cambridge and eventually his employer, ICI, invested in excess of £100 million in synthetic protein production. It finally came to market in 1980 as Pruteen, but production costs were high and alternative protein sources had by then become cheaper, so the process proved uneconomical. Recent work has overcome some of the problems: Chlorella is now cultivated in large, shallow, circular ponds and shows potential as a source of biofuel.

caught fire. Fortunately Fox was away at the time for, with his neurotic attitude to safety, he would have been very upset. But he was due back soon. All the students rallied round to rebuild the rig over the weekend so that, when Fox returned, it seemed as though nothing had happened.

Teaching undergraduate students meant formulating questions for examinations. Fox was a stickler for thoroughness and took the view that, as examination papers were published to the world, they indicated the department's intentions and standards. When the American chemical engineer Neal Amundson, something of an intellectual celebrity in his own country, spent time in Cambridge in the 1950s, Fox showed him the department's examination papers. Amundson said he thought they must be quite tricky for the students. 'Students?' Fox retorted. 'They're not for the students, they're for posterity.' When, after enormous effort and sweating of blood, the four staff members provided Fox with a draft of their 1949 examination paper for the new students, Fox retired to his room with it. Emerging shortly after, and holding the papers like dirty handkerchiefs, he suggested: 'I should burn these if I were you'. Peter recalled that:

> He demanded complete rigour from all of us, but was amazingly patient and kind in helping us to attain it.

Fox's faith in what he was doing was bolstered when the first set of students graduated. Sydney Andrew turned in a first class examination performance and it gave Fox the assurance that his department was on the right academic lines. He would encourage his colleagues by commenting: 'The young are at least as clever as we are because they are younger'. But Fox didn't get it all his own way. Peter was delighted that others had a chance to look at Fox's papers:

> Fortunately we had a robust external examiner, a practical engineer, who commented on 'these trick mathematical problems' to our silent cheers. However, I am sure that Fox was right. No-one who looked at our Tripos [examination] papers could have dismissed the subject as an easy option, or even a descriptive one.

Peter's MIT friend Ray Baddour visited in 1951. He had been lured back to Cambridge, Massachusetts, from a job at Oak Ridge National Laboratory in Tennessee by Ed Gilliland to tackle a doctor of science course. By the summer of 1951 he had completed it and accepted a job on the MIT staff. He decided, against unsolicited advice, to take the summer off and boarded the SS Mauritania to Europe. When it docked in Southampton he collected ration books, like all visiting foreigners at the

A roof-top view of chemical engineering's Tennis Court Road buildings
Courtesy Cambridge University Department of Chemical Engineering and Biotechnology

time, and set out for Cambridge. During his stay, Peter showed Baddour some of Fox's examination papers. Baddour thought them 'very highly theoretical, really applied mathematics. They seemed very difficult to me.' Peter had, however, great respect for and loyalty to Fox, and continued to feel the right approach was being taken:

The counter-argument was heard from industry that the atmosphere was too cerebral; but there are some intellectually-demanding subjects, like mathematics, fluid mechanics and thermodynamics, which one would never learn once caught up in the day-to-day urgency of industry. Worse, one might not realize one's ignorance and that in the long run leads to disaster.

During the vacations the staff rubbed students' noses in the practicalities of industrial life by sending them off to refineries and chemical plants.

[There] they learned something about the application of their theoretical knowledge and the extraordinary difficulties in the way of making practical studies on continuous flow plants without drilling holes in pipes or otherwise upsetting the plant manager.

When his visit came to an end, Baddour unloaded on Peter his surplus sugar coupons and his last memory of the trip, as he took a taxi for Cambridge station, was of 'Peter buying candy and tucking into it as he left the shop'.

Peter's departmental chores didn't stretch him unduly. Others arrived to play a part. Peter Gray had a physical chemistry background, and in 1952 Denys Armstrong and John Davidson joined the team—Armstrong with a penchant for teaching and administration, Davidson with experience in mechanical engineering. Peter wrote later:

During the years 1948-54, although there was plenty of teaching to be done, I sat on no committees and supervised [the research work of] only one graduate student. I look back on this as my period of 'academic indolence', during which I had time to shove my feet up on the desk and actually think. Such 'insights' as I have experienced originated mainly in this period, stimulated by summer visits to chemical plants. The scene was very different from that of today [1982] when (dare I say it?) too many academics are pursuing too many non-problems. We had an almost virgin field to plough and very few competitors in launching some important new ideas.

That academic indolence made this period of Peter's life the richest of his innovative lifetime.

The fruits of
academic indolence

18

WITH HIS FEET UP ON THE DESK, thinking, Peter had been putting his time to good use. Always happier this way than talking formally, he had few social duties and plenty of time to pursue his thoughts on the process challenges facing chemical industry. He began to write those thoughts down and send them off for publication in a range of scientific journals. In the major areas he addressed—the absorption of gases in columns, what happens in chemical reactors, the mixing of solids—his work could be described as the first concerted application of science to chemical engineering.

Peter had been turning over in his mind a large mismatch between laboratory chemistry and what happens in industry. It concerned chemical reactors: raw materials come together in the reactor and, during their time inside it, new compounds are formed. Then these new compounds, plus any unchanged raw materials, leave the reactor to be separated. Chemists generally do all this on a relatively small scale—in test tubes, beakers or other small glass vessels. They work in batches, making a few grams, or at most a kilogram, of product at a time. Industrial chemical engineers are required to produce large amounts of useful chemicals, perhaps thousands of tonnes a day. So they have to use much larger vessels than chemists and—to make them strong enough—get them constructed from metal rather than glass. That makes it difficult to see what is going on. But there is another, rather more fundamental, challenge: the plan of campaign. If large quantities of C are required from raw materials A and B, it would be very tedious and expensive to work in batches; for instance, mix A and B to make a kilogram of C, stir hard, separate the product and then do it over and over again. What's more, the product quality might well differ from batch to batch. Production schedules and product uniformity generally require a different approach: continuous flow.

Chemical engineers might arrange for A and B to flow continuously into a mixing tank. Out of the tank would flow product C together with unconverted A and B. Then they would separate C, also continuously, and recycle unchanged A and B to the start. One way to produce more of C might be to use not just a single tank but several, one after another in a cascade, or to feed A and B into one end of a tubular reactor. Peter was grappling with how to predict mathematically the most appropriate arrangements to get the best yield of C in a reasonable length of time, the best volume for the tank (or diameter for the tube) and the best speed at which A and B should flow. In the nineteenth century these things were done by trial and error, taking each manufacturing process and each product as a special case. By the 1940s chemical engineers had just begun to think about what was happening inside the tank or tube, and how to maximise the yield of C. Ideally they needed an analysis that would hold good across lots of different processes.

Peter came up with that analysis. To throw some light on it, imagine a tank which holds 1,000 litres. Suppose that A and B enter the tank together at a constant 10 litres a minute, and material leaves the bottom of the tank at the same rate. The average time one molecule of A or B spends in the tank is 1,000/10 or 100 minutes. That's called its 'residence-time'. But some molecules find their way to the exit pipe almost immediately, and have a short residence-time; others hang around for a long time in 'dead' parts of the tank where there is poor mixing and have a longer residence-time. In this case there is a broad 'distribution of residence-times' for molecules of A and B, and this can have a profound effect on how much C is made from them. If A and B flow quite quickly down a tube, all the molecules go at more or less the same speed. This is called 'plug flow', and the distribution of residence-times is very narrow. If the flow rate down the tube is slower, some molecules are slowed down by friction against the walls of the tube and the distribution of residence-times gets broader. Peter provided mathematical analysis of the amount of time raw materials spent in a reactor—the residence-time distribution—and related this directly to the amount of C produced.

His contribution here was in the tradition of the great seventeenth century polymath Robert Hooke. Intrigued by the behaviour of a spring balance which extended as weights were placed in its pan, Hooke explained what happened to the spring in terms of mathematics. In about 1660 he formulated an equation, now known as Hooke's Law, which related the extension of the spring to the load applied to it. This kind of treatment was a great departure; until then explanations had been descriptive (rather like Peter's Oxford chemistry course). It was also enormously useful, because the equation allowed prediction of the behaviour of many springs in all sorts of situations. Following, as it were, in Hooke's footsteps but using calculus rather than algebra Peter was tackling a much more complicated problem. There are other factors besides flow rate and reactor volume which have an influence on the amount of product: the temperature and pressure in the reactor, for instance, and the nature of the reaction which is going on between A and B. All this adds complications, but does not detract from the simplicity, elegance and usefulness of the analysis of residence-times.

Peter's analysis allowed industrial engineers to work out the best combination of equipment and parameters for the job. It was just the sort of problem which attracted him; its solution provided tools to help engineers in industry. He kept in mind that the analysis had to be relatively simple because otherwise, if the mathematics were too taxing, industrial engineers would simply switch off. He was always pragmatic. The way to approach the problem had come to him suddenly. He wrote later:

I had been revolving the subject in my mind for some time and one day the solution crystallised almost immediately. I remember saying to my colleague [Stan] Sellers during our morning tea-break, 'I see it all now, Stan.'

Peter himself described this moment as similar, but on a minor scale, to the German chemist August Kekulé's vision in 1865 of the structure of a benzene molecule as a ring of six carbon atoms. A ring was an unprecedented structure at the time. Kekulé was, according to the version of the story, either dozing in a chair before the fire or lost in thought on the top deck of a horse-drawn bus when he visualised a chain of dancing atoms, one end linking to the other. It was like a serpent biting its tail, a popular motif of the times. Looking back on his own 'Kekulé moment' later, Peter recorded that:

I attribute the inception and success of the idea first of all to the fact that chemical engineering was at that time a wide-open field to which a limited (although very fruitful) amount of analytical thought had been applied. It was relatively easy to open up a far-reaching new topic.

Although Peter was acutely aware that industrial engineers would not welcome complex maths, he himself—after his rather late and stimulating start at MIT—seemed to find it quite easy. Going any further into his mathematical solutions would involve a revision course in calculus, equally unwelcome to those who might need it as to those who don't. Here instead is an interesting bit of logic about residence-time distributions. One industrial solution to get the best yield of C might be to set up a cascade of, say, three or even five tanks in series. If the mathematical analysis of these perfectly-stirred tanks in series is then extended to deal with an infinite number of such tanks, the mathematical expression becomes equivalent to that of a tubular reactor with plug flow. A tubular reactor is equivalent to an infinite number of perfectly-stirred tanks one after another.

The research Peter directed at residence-time in a reactor didn't just stop with academic analysis. In this case he really did try things out on a large scale, making use of the Cambridge department's excellent contacts with Shell. In 1953 he took two Cambridge undergraduates—Geoff Place and John Jenkins—to Shell's Stanlow refinery in Cheshire. There he had managed to negotiate the carrying out of an experiment on part of a real, full-scale operating plant: the catalytic cracker. This piece of equipment takes low quality fuel from oil and enhances its octane number. The Cambridge team chose—or more likely was restricted to by the plant management—the section where the catalyst was regenerated; here a flow of air passing upwards through a bed of catalyst particles covered in a film of carbon creates a hot, heaving mass which behaves like a liquid—a fluidised bed. The regenerator burns off the carbon film to refresh the catalyst for reuse. It was big; about 30,000 cubic feet of gas swept through it in around 30 seconds. The Stanlow project lasted about six weeks. Peter was on site with the two students at the start, took a great interest and helped design the tests:

[It was an] experiment on the heroic scale to which I have always aspired but which it is so difficult to attain in practice—namely, tracer tests in full-scale operating plant. We injected five cubic feet of helium gas at the bottom of the regenerator and took a series of ten samples in 60 seconds from the stack at the top. The samples, which contained less than three parts of helium in 10,000, were analysed (with a mass spectrometer). The variation of concentration with time yielded a residence-time distribution.

One unexpected discovery was that, in some parts of the fluidised bed of used catalyst, air and helium actually moved backwards, in the opposite direction to the overall flow.

John Davidson, who had just joined the Cambridge staff team, could see that Peter really enthused the two students. 'This test of theory under

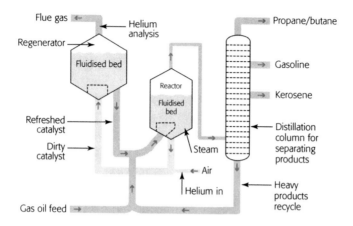

Catalytic cracking showing the catalyst cycle and the points where Peter's students injected helium and analysed for it in exit gases

BACKGROUNDER:
FLUIDISED BEDS

'Catalytic cracking' of gas oil, a hydrocarbon fraction obtained by distilling crude petroleum, was the first large-scale application of fluidised beds. Cracked gas oil provides the lighter and cleaner products of petroleum which are well-known in everyday life: propane and butane for stoves, gasoline as petrol and kerosene for jet and rocket fuel. In breaking large hydrocarbon molecules up into smaller ones, catalytic cracking also encourages the formation of compact molecules which give petrol a higher octane number. The process, a continuous one, involves two fluidised beds: the reactor and the regenerator; Peter explained what happens in the reactor:

The [catalyst] is in the form of a coarse powder and the [hydrocarbon vapour] is blown upwards through it at a fast enough rate for the particles to be suspended in the gas stream without being blown out of the reactor. In this condition the solid behaves very like a boiling liquid; it can be run off through pipes, so the technique is useful for handling solids in a continuous process.

The cracking of the heavy hydrocarbons to smaller ones produces heat, but everything in the fluidised bed is at more or less the same temperature. This makes it relatively easy to stop things getting too hot, something which might reduce the efficiency of the process. The products of cracking gas-oil exit the top of the reactor to a distillation column where they are separated at different points in the column. The heaviest material collects at the bottom of the column for recycle to the reactor. No process is 100% effective, and in this one some of the gas oil burns to carbon which progressively coats the catalyst particles. So the particles have to be circulated to a regenerator where, in another fluidised bed, air blows through them to burn the carbon off before they return to the reactor.

real practical conditions was Danckwerts at his best', he considers. Peter was modest about it:

It was successful only because of the energy and executive ability of [his student Geoff] Place.

Peter's ability to perform practical experiments like this in industrial plant in order to test his theories was particularly admired in the USA and raised him in following years to almost god-like status. This kind of experiment became quite fashionable, as industry tested the performance of its equipment against theory. Americans took to injecting helium enthusiastically into their catalytic crackers.

Geoff Place showed his energy and ability again when he collaborated with Peter in a second trial. This time it involved the variation of residence-time of material in a spray drier—the equipment which produces washing powder as dry, fragile spheres easily soluble in water. It is done by spraying a shower of a concentrated solution of the washing powder into a rising flow of hot air. The trial took place at the premises of Thomas Hedley (later Procter & Gamble) in Grays, Essex. Peter struck Place as a manager who could organise things, and very different from your average Cambridge academic. 'He didn't seem cut out to be a teacher, but with a small group he was very articulate; he knew what he wanted to say and had learned it for himself.' Through these projects Peter quickly became viewed as both brilliant and influential. His innovative thinking on

---

BACKGROUNDER:
GEOFF PLACE

With no ambitions to be an academic high flyer, Geoff Place was just as keen to apply his energies at university to learning bridge, or launching rockets on King's Parade, as he was to advance his supervisor's research. All the same, the Shell projects he had carried out with Peter stimulated his interest in mixing and the behaviour of flowing materials. After graduation he spent eight years as a chemical engineer with Procter & Gamble, where he got used to mixing cakes rather than catalyst particles. His get-up-and-go continued to serve him

well. Moving into management with P&G in Cincinnati, he rose as high as vice president, research and development. 'The faculty of the department in 1952-54,' he reminisces, 'I now recognise as memorable and truly outstanding. It had an enormous impact on my personal development, subsequent career and even now [2010] as a volunteer in paediatric health care.' (There was spin-off for Peter too: he acted as a consultant to P&G in the 1960s and felt able to tell his nephews and nieces that, at that time, he knew more about P&G's Daz detergent than anyone else in the world.)

residence-time distribution, although it turned out not to be infallible in predicting the performance of reactors, nevertheless opened a whole new field; a vast literature on residence-times developed over the following years, carried out by researchers inspired by Peter's ideas. Peter himself took very little part in it as his attentions were directed elsewhere.

Gas absorption is a technology widely used by industry in those days, and even more so today, usually by bringing the gas into contact with a liquid. It may be just a way of cleaning one gas from a mixture of others, but more often there is a chemical reaction going on at the same time making some new and useful product. Yet in 1948 there didn't seem to be what Peter called 'a rational method of design' for equipment to carry out the process. Tom Sherwood, Peter's teacher at MIT, may have turned his thoughts that way. Writing much later, Peter recalled that soon after arriving in Cambridge he had read the record of a 1943 discussion

BACKGROUNDER:
SOME GASES ABSORBED IN SIGNIFICANT INDUSTRIAL PROCESSES

Carbon dioxide:
In solutions of barium, calcium and magnesium to make carbonates;
To remove it from synthesis gas in solutions of potassium carbonate or amines.

Oxygen:
To oxidise aldehydes to acids and acid anhydrides;
To oxidise cyclohexane to adipic acid (Nylon manufacture);
To oxidise toluene to benzoic acid;
To oxidise cumene to its hydroperoxide (phenol manufacture).

Chlorine:
In trichloroethylene to give pentachloroethane (dry cleaning fluid manufacture);
In aromatic compounds to give chlorinated derivatives.

Nitrogen dioxide:
In water to produce nitric acid.

Sulphur dioxide:
In sodium nitrite with zinc dust to make hydroxylamine.

Sulphur trioxide:
In sulphuric acid to make oleum.

Hydrogen:
To saturate unsaturated organic compounds in the presence of catalysts.

Hydrogen sulphide:
To remove it from domestic natural gas by reaction with amines in water.

Unsaturated hydrocarbons:
Isobutylene in sulphuric acid solution to make tertiary butanol and for polymerisation;
Ethylene in benzene using a catalyst to make ethyl benzene;
Acetylene in cuprous chloride solution to make vinyl acetylene.

Hydrogen chloride and bromide:
In higher alcohols to make the alkyl halide;
In vinyl acetylene to make chloroprene (Neoprene manufacture).

during which Sherwood had commented: 'Personally I am getting pretty discouraged about work on absorption with chemical reaction.' Whatever the stimulus, it set Peter working on a theoretical basis for the design of gas absorption equipment. The work developed into a sustained research programme—the biggest of his career—pushing his ideas as far as they could be developed. It stimulated more than 50 'contributions to the literature' with Peter's name on them; these papers often involved his students, many of whom then took up aspects of the same research field when they left him, swelling the literature further.

Industry generally does gas absorption in a column. A typical column may be 10 to 15 metres high and a metre in diameter. In many cases it contains some random packing, like cylindrical ceramic rings or, once upon a time, crushed coke; typically a liquid trickles down from the top of the column and the gas bubbles up from below. The objective may be to absorb gas in the liquid, or perhaps to allow it to react with something dissolved in the liquid, or both. The development of a theory to explain what was happening—and, equally important from an industrial point of view, to predict the size of column needed to carry out a specific task and obtain a predictable yield—had been slow going. Conventional wisdom of the time explained gas absorption in a column by reference to a so-called 'two-film' theory, published by Walt Whitman (an MIT chemical engineer, not the father of American poetry) and his colleague Doc Lewis in 1924. This theory envisaged a thin and undisturbed 'stagnant' film of liquid on the surface of the packing and a similar undisturbed film of gas in contact with it. Gas would then be absorbed by diffusion into the film of liquid. This allowed the design engineer to make calculations using some well-established data. But at the same time assumptions had to be made—for example, that the absorption of gas took place at a uniform and steady rate. All this allowed the engineer to describe the absorption rate with a differential equation. This equation, in what was then standard practice, could be integrated over the whole column to find the overall rate of absorption in the column. This was the key to estimating the necessary dimensions of the column for a particular job. But... Tom Sherwood again, in 1952: 'it is curious that after 28 years the [two-film] theory itself has never been checked experimentally. Until further data... become available, the validity of the theory must be accepted'.

But Peter didn't accept it. After intense thought he decided that it was 'no more than a convenient fiction'. In a real packed column, he couldn't help noticing, a liquid film flows briefly down one piece of packing but then drops off and hits another piece of packing below, where a fresh film forms and flows down. In between, liquid and gas can mix turbulently and two-film theory took no account of that. By 1950 Peter had sent several theoretical essays on the subject to publishers. He laid out his thoughts on a more convincing, practical approach to design of industrial equipment in a June 1951 contribution to the journal *Industrial & Engineering Chemistry*. Much later, in a review of Peter's lifetime contribution in this field, Neal Amundson—US chemical engineering's equivalent of Danckwerts—noted: 'There was never any question in his mind about the importance of the problem to the chemical industry'. When Peter eventually published a book, *Gas-liquid reactions*, summarizing work on the subject, readers found a very large number of gas-liquid absorption processes used in industry listed right at the start.

Miles Kennedy, a young New Zealander, arrived in Cambridge in October 1951 hoping to do post-graduate research and earn himself a doctorate. By coincidence Kennedy's professor at home in Canterbury, Stan Siemon, was on a short visit to Cambridge at the time. He advised Kennedy to go for Peter as his supervisor because Siemon thought Peter was 'the bright boy of the department and the chap with all the ideas'. At their first meeting, Peter suggested Kennedy should read the June 1951 paper and come up with ideas for experiments that could test Peter's theories. Kennedy got reading; he became the only graduate research student Peter supervised in his six years as a Cambridge lecturer.

Peter's theory of gas absorption drew on an idea of 'surface renewal' first suggested by the American Ralph Higbie. An employee of the Eagle-Picher Lead Company in Joplin, Missouri, Higbie had worked on extraction processes. He developed his surface renewal concept in 1935 while preparing for a University of Michigan doctorate. Peter took up Higbie's idea that, in contrast to the two-film theory, if the mobile surface of a turbulent liquid continuously renewed itself, it would always be presenting a fresh surface to the gas. On the other hand, and in a distinctive advance on Higbie, Peter realized that:

It was clearly improbable that each element of liquid brought to the surface would be exposed to the gas for the same length of time. It occurred to me that if surface renewal did in fact occur (in turbulent flow or in film flow over a randomly arranged packing) then any part of the surface was equally likely to be replaced, regardless of the length of time for which it had already been exposed.

From there it was but a mathematical hop, skip and a jump to analysis of all sorts of absorption situations, involving simple diffusion or reactions of several kinds, with and without catalysts. In the case of the simplest absorption, there seemed to be a useful way of distinguishing between the two theories. Peter's extension of the surface-renewal concept led to the prediction that the rate of absorption of a gas depended on the square root of the diffusivity, D, of the gas (that is, D to the power 0.5); where diffusivity is a measure of how fast the molecules of gas move into and mingle with the molecules of the liquid, as it were, of their own accord. The old stagnant film model predicted a dependence on D itself (that is, D to the power 1).

At the time, while Peter acknowledged the difficulty of providing conclusive proof, he felt his theory had more secure foundations. Some people took issue with him, and he defended himself robustly. In the vanguard amongst the sceptics was one of the department's first students: Sydney Andrew, by then back at Billingham with ICI. He felt that Peter was asking too much of random packings. As liquid cascaded down a packed column, the surface renewal which took place as it moved from one piece of packing to the next could not possibly be complete, Andrew reasoned. Partial renewal seemed to him acceptable, so he christened his variation 'surface rejuvenation'. Argument raged. At one well-attended technical meeting which addressed the matter a speaker spent some time attempting comprehensive demolition of the Danckwerts theory. Finally he invited questions. A tall figure stood up in the back row. He began: 'My name is Danckwerts…' And he was more than a match for the opposition.

To verify Peter's ideas, more precise measurement was necessary of the rate at which gas was absorbed into a stagnant film of liquid over short periods of time. Peter had already done some experiments using carbon dioxide as the gas. Why carbon dioxide? At the time absorption of this gas was a significant procedure in the domestic gas industry. Town's gas, generated from coal and made up of several gases which burned to give heat—hydrogen, carbon monoxide, methane and small amounts of other hydrocarbons—also contained some carbon dioxide and nitrogen which did not burn. Absorption of carbon dioxide ('scrubbing') therefore improved the 'calorific value' of domestic gas. When methane from gas wells, like North Sea gas, replaced gas from coal for domestic use, there was still a need to scrub carbon dioxide out. But Peter probably chose carbon dioxide to test his ideas because of its convenience and safety.

After all, it is a lot more pleasant and less toxic than hydrogen sulphide gas, a possible alternative. As liquid he used water; it too is safe and carbon dioxide dissolves in it to some extent. As equipment he built himself a small drum, immersed it to its axle in a bath of water and covered the bath so that the space above the water line could be filled with carbon dioxide. He could rotate the drum at various speeds. As it turned, the surface of the drum emerging from the liquid carried a film of liquid on its surface which was then exposed to the gas for a short period. This made possible the measurement of absorption rates of gas into the film of water for different exposure times.

Peter was not the world's greatest experimentalist and his drum hadn't worked too well, so Miles Kennedy set about improving it. He also built a small packed column, four inches in diameter and a foot high, packed with half-inch ceramic rings which could model the industrial process. The development of cunning ways to measure absorption rates simplified mathematical analysis of the results. Kennedy changed the operating variables like the density and the viscosity of the liquid, as well as finding out what happened when the liquid contained something—for instance, caustic soda—which reacted with carbon dioxide. This was 'absorption with reaction', in which Sherwood felt progress had been so slow. Throughout Kennedy's research project (and many others in the programme which Peter organised on gas absorption in the following decade) the objectives were clear. To develop a laboratory model, the use of which would allow direct design of production-scale industrial gas absorption equipment for any absorption system, without reverting to tedious and costly small-scale trials. Writing in 1982 Peter reflected that:

These objectives were not attained completely. However, the work throws a great deal of light on the subject—sufficient, at least, to discredit the older and less sophisticated design procedures.

He lamented some unavoidable limitations—having to use laboratory-scale equipment with small packings and thin columns—and strove for large-scale work on equipment of production size.

Experiments on packings of a scale approaching those used in industry generally require to be made on industrial premises. The limited number of experimental campaigns I was able to mount on this scale [columns half a metre wide, packing measuring 25mm across] indicated that the methodology could be applied more successfully than on the laboratory scale. I conclude that the work goes a long way towards providing rational methods of design.

Others contributed original ideas for the measurement of absorption rates. Peter's colleague John Davidson got his first research student John Cullen (a future head of the UK Health and Safety Executive) to try a system pouring gas-free water onto the surface of a ping-pong ball. Water formed a transient thin film absorbing gas within a box which surrounded the ball, and the geometry made calculating the area of the thin film of liquid straightforward. Davidson and Cullen extended the arrangement to a series of up to 20 ping-pong balls strung out one above another, imitating 20 chunks of packing in a column. The results suggested to some that there was full surface renewal as the water passed down from ping-pong ball to ping-pong ball. On the other hand, others felt that the geometry of ping-pong balls differed too greatly from that of industrial packings for the results to have meaning. Sydney Andrew was scathing.

In his unique position as Peter's only graduate research student, Miles Kennedy had unusual access to him. At first Kennedy, like many others, was 'scared stiff'. People were acutely conscious of Peter's war record and he didn't do small talk. But things warmed up. Kennedy was on a scholarship from the New Zealand army and, when King George VI died, he had acquired tickets for the funeral procession on 15 February 1952 on Horse Guards' Parade in London. He dressed up in his lieutenant's uniform sporting a black armband and, with his wife Betsy, took the train to London. On the way back they were grabbing their bicycles at Cambridge station when they ran into Peter. In view of Peter's war record, Kennedy had kept his army connection secret; at tea-time next day Peter made a crack about being knocked down with a feather when he saw Kennedy in what he at first thought was a brigadier's uniform.

In March the Kennedys invited Peter to supper; he gave them the impression that he rather envied them their cosy domesticity. Then, in November 1952, Peter rang the Kennedys and invited them to tea. They enjoyed roasting chestnuts over a laboratory Bunsen burner, while Betsy warmed her feet by the resulting glow. Peter's colleague John Kay and his sister looked in briefly, and when they left Peter suddenly offered the Kennedys some gin; he looked at Betsy and said ruefully: 'Sorry, I've only got water to go with it'. After that Kennedy found Peter much easier to chat to, both socially and in the laboratory. And the contact brought another benefit: 'He taught me more about writing English than any of my English teachers at school had ever done. And once the time came for me to write up my research, he refused to look at anything I wrote until the

THE DEPARTMENT OF CHEMICAL ENGINEERING
TENNIS COURT ROAD,
CAMBRIDGE.
TELEPHONE 58231.

Sept. 24th

Dear Miles,

Can you turn up, shaved & showered, for your oral examination at 1100 hrs. on Sept. 30th ?

Yours sincerely

M Danckwerts

Peter's note to Miles Kennedy summoning him to defend his doctorial thesis and (l) 'Brigadier' Kennedy in 1952

thesis was bound and submitted.' His external examiner Anderson Storrow from Manchester seemed to like the result. It all stood Kennedy in good stead later when, back in New Zealand, he pursued an academic career culminating in a professorship at the University of Canterbury.

Another topic that Peter addressed in these productive few years in Cambridge was mixing. The problem faced by industry was how to decide when they had mixed things up enough and could stop. In 1952 Peter published some thoughts on this in a journal called *Applied Science Research*. He was looking for a way to define and measure 'goodness of mixing'. In the view of many, including Anthony Pearson who later worked with Peter as his assistant director of research, it contained ideas and proposals which were seminal at the time. Peter's aim was to help industrial engineers design and select apparatus which would achieve good mixing. He wrote that the study of mixing processes suffers from the lack of any way of putting numbers to 'goodness of mixing'. What might be easy in the kitchen, where you can see what is going on, is much more challenging when the mixture is in a closed vessel with metal walls. With practical application in mind, Peter argued that putting numbers to goodness of mixing must be related as closely as possible to the visual impression of whether the components of the mixture were well or badly mixed. Four entirely theoretical papers Peter wrote in the 1950s provided, in his words, 'a valuable disciplinary background to mixing'.

The chosen method of measuring goodness of mixing had to be convenient to carry out, of course, and applicable—without changes—to as many different types of mixture as possible. Peter warned against relying on purely arbitrary tests, each one chosen to suit the mixture under consideration. These, he considered, would lead to measurements of doubtful significance. Pearson explains that Peter then proposed two measurable statistical parameters which he termed 'scale of segregation' and 'intensity of segregation'. Peter described 'scale of segregation' as a measure of the size of clumps of, as yet, imperfectly-mixed components, and 'intensity of segregation' as a measure of the extent to which the clumps differ from each other in composition. He concluded that a knowledge of these two parameters would 'provide a good deal of information about texture, in a quantitative form that would allow different degrees of mixing to be compared'. He emphasised that 'the two quantities are virtually independent and represent aspects of goodness of mixing which cannot be defined by a single quantity.' Scale of segregation, he pointed out, might be important in determining the performance of a reactor: the better the molecules of two reacting materials can get to grips with each other, the better the yield of product.

'Sadly he did not follow up his own ideas,' says Pearson. 'Had he done so it might easily have led to him becoming an even more revered figure in the science of chemical engineering than he is now seen to be.' Peter considered that making the necessary measurements would be both difficult and tedious, and other engineers certainly seem to have felt the same way. No-one has taken up the gauntlet. US mixing expert Julio Ottino pointed out in 1990 that whereas the objective of other investigators of mixing has been to relate the mixed output to the fluid mechanics of the mixing, Peter 'focused primarily on the characterisation of the mixed state'. But in practice no widely understood or calculable criterion has emerged.

Whereas Peter's name is referred to in further development of all these matters, there are a couple of areas in which he had something specific

named after him, emulating his hero Tom Sherwood. One came about
from a procedure which industry uses a lot: passing a mixture—say, of
gases A and B, 'fluids' in the jargon—down a tube packed with granulated
solid, which could be a catalyst of the reaction between them. He
imagined that the colour of a fluid flowing down the packed tube was suddenly changed from colourless to red. In theory, if the body of fluid
moved like a piston, the boundary between the two colours would remain
sharp. In practice it blurs. This is because when fluid flows close to the
solid packing it is slowed by viscous forces; where it finds a free channel
through the packing it moves faster. And if it encounters turbulence its
speed changes rapidly and in an irregular way. Peter estimated, however,
that if the packing has no channelling and is quite randomly arranged, it
can be a useful approximation to say that all parts of the fluid behave in
about the same way. That means that the only thing influencing the
sharpness of the boundary between red and colourless parts is diffusion:
that movement which is due to the energy of the molecules themselves and
which makes their movement like a 'random walk'—the path followed by
a drunk leaving the pub door. Diffusion blurs the boundary. Using the
standard rules for diffusion, Peter deduced 'Danckwerts boundary
conditions' to help engineers work out what would happen—in this case,
how much reaction of A with B would take place. Peter wrote some thirty
years later about his discovery:

[It has] reached the status of being a primary reference which is seldom cited... it
has served and continues to serve working chemical engineers because it does not
blind them with mathematics... the introduction of the 'Danckwerts boundary
conditions' for the flow of fluid through a bed of packing of finite depth gave rise to
a lengthy and almost metaphysical discussion in the literature, although I thought that
I had fully justified them in [my] first publication.

There was a sobering footnote. In 1982 Peter was invited to review a
book of contributions to an American celebration of a century of
chemical engineering. It was, he claimed, 'a lot less boring than it sounds',
especially one section by Arvind Varma, an American of Indian origin
interested in the use of mathematics in chemical engineering, which spoke
directly to him. Peter responded:

I have felt for some years that chemical engineering is weighted-down with more
mathematics than it can support, but this section is of great historical interest. For
instance, I was amused to find that the treatment of the tubular reactor with
longitudinal dispersion, together with the 'Danckwerts boundary conditions' which I

put forward in 1953 and which were credited to me for at least 25 years, had been anticipated by Langmuir in 1908 (I should have been less amused in 1954).

There is a second practical tool that bears Peter's name. It has to do with gas absorption. He published the idea in 1963 and it is now called 'Danckwerts-plot technique'. It is aimed at helping work out what size of equipment is needed to cope with absorption of specific gases in a gas mixture. Suppose that carbon dioxide is amongst the waste gases from some combustion process—one generating energy, for example. It is going to be absorbed into a liquid in a scrubbing tower. There is an equation that tells chemical engineers about the rate at which carbon dioxide will be absorbed. But the equation features two items—K, which is a 'mass transfer coefficient' and a which is the surface area provided for absorption by a unit volume of the column. They tend to crop up together and it is very hard to measure each separately. Now here's Peter's idea. Put a soluble catalyst into the scrubbing liquid. It accelerates absorption but, it is assumed, does not change the surface area. Peter worked out the theory to predict exactly how the absorption rate would vary with the amount of catalyst added. Bingo! A way to find the separate values of K and a: vary the amount of catalyst and measure how much the absorption changes. Peter suggested plotting the results as a graph—the Danckwerts-plot— of the amount of carbon dioxide absorbed against the amount of catalyst added. The graph makes it easy to read off the values of K and a. But what about that assumption? Does the catalyst added really have no effect other than speeding up absorption? It is a measure of the Danckwerts-plot's usefulness (and perhaps of the faith of engineers) that this method was employed for many years without anyone checking the assumption. Until 2005, that is, when a group of Dutch engineers found it to be OK.

Peter's reputation was already beginning to grow. Miles Kennedy bumped into a group of Americans visiting the department in the early 1950s and discovered that they were well aware of Peter. 'They expressed astonishment when they found that he hadn't been in the laboratories working since the crack of dawn. They asked him how he came up with his original ideas; perhaps it was on the evaluation of mixing, or maybe the analysis of residence-time distribution. His reply was: "In the bath". He was serious but I doubt that they thought so.' On more than one occasion in the future Peter could be heard claiming that the British had more good ideas than the Americans because the British (or at least those of his generation) preferred baths to showers.

Acoustics, food and fire balloons 19

BESIDES PETER'S MAINSTREAM CONTRIBUTIONS to the theory of chemical engineering, he made excursions during his years of academic indolence into other applications of science that caught his eye and related to his experience. It was a characteristic of the breadth of his interest and the sharpness of his mind that he was able to do this so widely. In 1950, for instance, he wrote a piece on ultrasonics for the trade magazine *Power & Works Engineering*. It explained the basics of what was then a new technology and reviewed uses to which industry was, or might be, putting it. Clearing away two possible sources of confusion early on, he made the nature of ultrasonics crystal clear:

In the first place, the term 'supersonics' has sometimes been used to describe the subject under discussion; however, in modern usage 'supersonic' is generally taken to mean 'having a velocity greater than that of sound' while 'ultrasonic' means 'having a frequency greater than that of audible sound'. Secondly, the ultrasonic signals used by such instruments as Asdic [the wartime anti-submarine detection system later renamed Sonar] and echo-sounders are as different from Radar as sound is from light.

After an elegant summary of the physics of ultrasonics (and how the very short wavelength makes possible the production of sharply directional beams rather than a broadcasting of waves in all directions), Peter provided short master-classes on how to generate ultrasonics for industrial use and how they work in Sonar detection. Then he summarised what was then their current usage in materials testing. Finally he got on to how ultrasonics might be used in the future. In particular he wrote about cavitation; it happens in pumps, for example, when a piston is withdrawn too rapidly from its cylinder and—under the effect of negative pressure—the body of liquid flowing into the cylinder 'breaks', forming cavities. He wrote:

When the tension is relaxed these cavities disappear again and the opposing walls of liquid come together with great violence. The momentary pressures produced at the instant of collapse can be very high indeed… A calculation by Lord Rayleigh [the physicist who, amongst many other things, gave his name to the scattering of light by molecules of air that makes the sky appear blue] showed that [in some cases] pressure of several thousand atmospheres may arise. The effect is in some ways similar to that of the hollow conical explosive charges developed during the war for drilling holes in armour plate or concrete.

This last observation came straight from his Combined Operations experience and the visit to the island of Walcheren to see why Allied shelling failed to take out the Nazis' concrete defences. Throughout his life he retained a strong interest in the nature of explosions, the penetration of missiles and the timing of detonation.

In his article Peter illustrated how ultrasonics can be used to investigate the effects of cavitation without the need for elaborate moving models. Marine engineers know how cavitation produces a pitting of metal surfaces, and ultrasonics causes a rather similar effect. One theory is that ultrasonics tear from the metal surface the film of oxide which normally protects it; there was support, Peter suggested, from the successful development of an ultrasonics-agitated soldering iron for aluminium which cannot be soldered by conventional means because of the very rapid formation of an oxide layer. Peter quotes as his source for the whole article a 'most valuable and comprehensive' German text, originally published in 1937—*Der Ultraschall und seine Anwendung in Wissenschaft und Technik* by Ludwig Bergmann (translated in 1938 as *Ultrasonics and its scientific and technological applications*). The Americans had published a version of it and Peter may well have picked up a copy while in the States a few years earlier; then again, his German was quite good enough to make use of the original text. In his final cadenza on ultrasonics, Peter speculated on all sorts of things ultrasonics might achieve in the future, from de-aerating boiler water, through the prevention of 'bumping' of boiling liquids to the driving of mice from vermin-infested houses.

In the early 1950s world production of fat was estimated to be four or five million tonnes a year less than was needed to feed the world's population. In a couple of articles in *Food* magazine Peter looked, with Stan Sellers, at the case for producing protein and fat from yeasts. They described how *Torulopsis utilis* might be used to produce fat in a semi-continuous plant. The work was not based on practical experimentation but pulled together data provided by others; waste sulphite liquor from paper manufacture—which Peter and Ray Baddour had run into in

Bangor, Maine, during their MIT course—appears as a possible nutrient. After a cost analysis, the articles concluded that the process could hardly be competitive to other alternatives, but suggested that 'a small commercial-scale plant based on molasses from sugar beet refining' was not beyond consideration.

While Peter was in Cambridge, his brother Micky was developing a career in journalism with a technical flavour. He later wrote speeches for that charismatic ICI chairman of the 1980s, Sir John Harvey Jones. Together the brothers conceived a plan to write a book of simple experiments attractive to children: *100 things a boy could do*. The idea was that children—and in the 1950s that meant boys—might carry them out and perhaps get excited about science. Peter encouraged his Cambridge colleagues to suggest to him 'things to do'. Determined only to include 'things' he could make work himself, he would disappear to his office at Tennis Court Road to try out something new for the book, leaving his colleagues, John Kay particularly, in fits of laughter. Some of the experiments drew on Peter's childhood experimentation with relatively harmless explosives, backed by his wartime experience of more lethal ones. Others were either spectacular or just fun. One involved Marmite, but the detail is lost in the mists of time.

Some of the stimulating things had to do with triboluminescence, the fluorescent glow created in a darkened room when two strips of adhesive plaster stuck firmly together are ripped apart. The ripping plays havoc with chemical bonds on the two surfaces, with the result that electrical charges are formed and discharged. Many crystals give off light when they are cut, and for the same reason. Peter used a roll of adhesive plaster in the 1950s; sticky tapes weren't generally available until later. In more recent times grown-up boys have experimented further and found that if the tapes are pulled apart in a vacuum, the process can produce X-rays. Jetex engines for model aircraft were in their heyday in the 1950s. Brought to Britain from the United States, they gave boys more range to their model aircraft and had attracted Peter's attention as presents for his nephews. Crepitating explosives were part of Peter's own childhood: nitrogen triiodide crackled nicely when unsuspecting siblings walked on it.

Fire balloons were inevitably another of the topics, for Peter was a lifelong enthusiast. He made his own, and could be found on a winter's evening on Cambridge's Midsummer Common releasing them with the help of some enthusiastic assistant. His brother-in-law John Garson, based on the Wirral and married to Peter's younger sister Diana, caught the bug from him. Later his son Ian Garson inherited the tradition, not to say a certain competitive edge in balloon design. Fire balloons are essentially made of tissue paper. Cross wires across the open base of the balloon support a blob of cotton wool at their centre. The cotton wool is soaked in methylated spirits and set alight. Hot gases rise inside the tissue paper envelope and are caught there, just as they are in a hot-air balloon. Assuming the canopy has been made air-tight and the whole thing doesn't catch fire, the hot gases take the balloon high into the night-

And up she goes
Courtesy Ian Garson

sky. There it glows eerily. Many a naïve and credulous observer has mistaken a fire balloon for an extra-terrestrial spacecraft.

The making of a fire balloon requires care and creates a great feeling of achievement. Peter would not have understood how buying a Chinese mass-produced article off the shelf could compare with it. Glueing the panels of tissue paper together is tricky, but holding the whole thing vertical while lighting the fuel is even more so. There is always a risk of fire, especially in cross-winds. But once it is launched the balloon should stay aloft until the fuel is exhausted. Alternatively, if the tissue paper catches, the whole thing can come down in a mass of flames. Peter's balloons, using meths as fuel, burned out quite quickly and so were relatively safe; using a candle for heat means that the ballon flies for longer but increases the risk of setting fire to some unfortunate farmer's haystack. Today's health and safety rules may take a dim view of the whole scene but even fifty years ago the Garsons always made sure that the wind took their balloons out in the direction of the Irish Sea. Those who watched Peter launch fire balloons on Midsummer Common in the 1950s felt that the dangers to neighbouring houses or farmers' crops wouldn't have even crossed his mind.

In the event no publisher would touch *100 things*. Health and safety considerations made it a liability; its time had passed. Looking back on this doomed venture, nostalgic nephews and nieces feel Peter was showing characteristics typical of a Danckwerts: a fascination for mechanical things, finding out how they worked and making them himself.

BACKGROUNDER:

HOW TO MAKE A FIRE BALLOON

Raw materials: Twelve sheets of tissue paper 50cm by 75cm; a stick of glue (Pritt stick, for instance); pencil, ruler and sharp scissors; 250cm of thick wire (2-3mm diameter), bendable but able to retain its shape; a length of thin binding wire (0.5mm diameter); insulating tape; masking tape; pliers; cotton wool— enough to fill a clenched fist; methylated spirits; a box of matches; a large table at least 2m x 1.50m; a plastic tablecloth to cover it; an enthusiastic friend or friends. Allow two hours for construction.

Construction (a): Panels. Lay four sheets of tissue paper precisely on top of each other on the tablecloth: draw pencil lines as in diagram 1 and cut off the shaded areas. These four

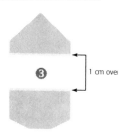

sheets will form the top part of the balloon envelope. Keep off-cuts as patching material in case the envelope gets torn later. Lay out four more sheets and trim as in diagram 2 to form the bottom of the envelope. Now take one cut sheet from each group and lay them on either side of a single uncut sheet, as in diagram 3. Glue them together— one at a time, as the glue dries quickly. Make

sure that there are no holes for air to escape from later. Make three more panels in the same way: all four measure 150cm down the centre line.

Construction (b): Envelope. Lay one side panel exactly on a second. Pull away the upper panel at the top end (section A, diagram 4) and apply a band of glue about 1cm wide to the edge of the lower sheet. Re-lay the upper sheet on the lower one and smooth down. Continue the gluing down the same edge of the lower sheet, a section at a time (B, C and D in diagram 5) avoiding glue-free gaps. Don't let the glued panels get stuck to the tablecloth! Fold back the unglued part of the upper panel at its centre line and lay a third side panel on top. Glue this one in the same way to the other side of the lower sheet. Three panels are now glued together. Fold back the unglued part of the third panel along its centre line. Place the final panel precisely on top.

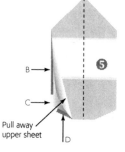

Before gluing it to the edges of panels 1 and 3, pick up the whole assembly and invert it—like tossing a pancake. This lessens the danger of gluing multiple panels together . Glue the edges of panel 4 to the free edges of panels 1 and 3. *Construction (c): Wire ring and cross wires.* Bend a 136cm length of the thick wire into a circle. Overlap the ends by 3cm and bind them together with insulating tape (see diagram 6). Mask any jagged edges. Now lay a straight piece of thick wire across the diameter of the ring so that about 3cm projects outside the frame at each end. Bend the last 3cm at each end at a right angle and bind it to the frame of the ring with insulating tape. Repeat with a second length of wire at right angles to the first. Where the wires cross, secure them to each other with thin binding wire; leave about 15cm to hold the blob of cotton wool. In what follows, make sure the end of the wire does not damage the tissue. Grip the central binding with pliers and squeeze. The binding wire bites into the cross wires, giving extra stability.

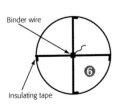

Binder wire

Insulating tape

6

Construction (d): Assembly. Open up the aperture at the bottom of the tissue paper envelope and present the wire ring to it. The ring should sit a centimeter or so inside the envelope all the way round. If the ring is too large, cut away some tissue paper. If the ring is too small, make a bigger one from scratch. Now, using 3cm lengths of masking tape, attach the wire ring evenly inside the tissue envelope (diagram 7). Add more pieces as required; at tissue joints use one piece of masking tape on each side of the joints. The balloon is ready and can be gently folded away until needed.

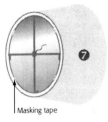

7

Masking tape

Launch: Wind is the crucial element. It should be very light or dead calm. Night launches are the most spectacular and cold nights are best for lift. While an assistant holds the envelope vertical by its uppermost point and opens the panels as far as is possible, soak the blob of cotton wool with meths and, lying under the balloon envelope, secure it firmly with the binding wire at the centre of the cross wires. Ignite the meths and let hot air fill the envelope. If the envelope catches fire it will be the assistant's fault, of course. As the envelope begins to rise, the assistant can let go. Stand up, holding the balloon by its wire ring. As the envelope fills, it tugs upwards: release the balloon, stand back and admire.

Flight time: about 20 minutes.

© Peter's nephew, Ian Garson.

Note: Peter used ten sheets of tissue paper in five panels of two, arranged vertically. The 'improved' Garson model 'fairly shoots up, even on a warm night' while Peter's can 'hang around in the air wondering whether to go up or not'.

Origins of a bestseller 20

LTHOUGH PETER WAS UNABLE TO interest a publisher in his compendium of things a boy could do, it was a different story in his professional field. For decades American journals had ruled the chemical engineering roost and European chemical engineers looking for a truly international platform for their work had to look across the Atlantic. There was an element of 'cap in hand'. After WW2 Europeans were keen to create a level playing field. In 1950, a slight, immaculately-dressed and highly articulate central European was visiting centres of European excellence with a proposition. And naturally he visited Cambridge.

Paul Rosbaud was by then a very experienced publishing consultant in science and engineering. He had learnt his trade in Berlin in the 1930s as an employee of Springer-Verlag, and was on first-name terms with a formidable list of top scientists throughout Europe who could contribute to Springer journals. He remained in Berlin during WW2 and continued the business while living a dangerous life providing Britain with information on German progress in technology. Now based in London, he could see that the time was right for a journal focused on the application of science to chemical engineering, precisely Peter's field. It would be aimed at chemical engineers who were able and willing to provide a scientific backbone for their subject and to show by experiment that the backbone stood up to reality. Although Rosbaud intended that Europeans would be well represented, giving Europe's scientific engineers a natural home for their research papers, he also understood that the readership (and potential contributor base) would be fully international.

Every journal needs a good cast: first an individual like Rosbaud with vision and the energy to drive it on, then a board of editors who know their stuff and are able to assess and turn round contributors' papers reasonably quickly and—when the writing deteriorates into action—a publisher who can print quickly and promote worldwide. As the instigator, Rosbaud, like Peter, had an extensive address book of acquaintances. He also had a good understanding of the subject matter and so knew whom to ask. And there was a number of characteristics he shared with Peter: both had been extraordinarily brave in adversity during WW2, both were instinctively inclined to keep their own counsel—especially about their wartime activities—and both made friends and contacts effortlessly.

By a series of coincidences, Rosbaud found himself working in the early 1950s for Robert Maxwell. After the war Maxwell had been casting around for openings in the world of science and technology publishing which, with considerable vision, he saw as having great potential for making money. As 'Mr Fixit' in Berlin from 1945 he had helped the Springer-Verlag publishing house in its efforts to regain the dominance in science and engineering that it had held in Europe before the war. His efforts had allowed him to acquire a share in the booksellers Lange, Maxwell and Springer. But Springer management did not entirely trust Maxwell and protected its journals from him—those were produced by Butterworth-Springer, a UK-German collaboration dating from an agreement in 1949. That agreement was a difficult one and by 1951 it fell apart. Rosbaud ran Butterworth-Springer; he wanted to buy the business and so keep his relationship with Springer, but couldn't raise the cash. Maxwell saw his

---

BACKGROUNDER:
COINING A COLOPHON

The photograph of the head of the goddess Athena on a gold coin which served as the first Pergamon Press logo
Courtesy WP Jaspert

Looking for a suitable logo for the new publishing company Robert Maxwell had acquired from Butterworth and Springer, Paul Rosbaud showed WP Jaspert— at the time an assistant production manager with Butterworth-Springer and the son of old friends of Rosbaud in 1930s Berlin—some photographs of gold Greek coins from the British Museum's collection. Rosbaud knew a thing or two about Greek coinage; he and Jaspert chose a head of Athena minted in Pergamon and so provided the new company with its name.

opportunity. He bought the fledgling publishing company with five book titles and three journals: Pergamon Press was born. Rosbaud had misgivings about working under Maxwell, but he stayed on as scientific director and at first Maxwell, who was busy with other matters, left him to get on with it.

With two elements now in place, Rosbaud set about assembling the third vital ingredient: the board of editors. He was the prime agent in persuading a group of top European chemical engineers that the time had come to launch a new journal emphasising the scientific basis of their discipline. He named it *Chemical Engineering Science*. His schedule included Cambridge and he spoke to both Peter and his colleague Kenneth Denbigh, two leading lights of chemical engineering as science. Denbigh remembered much later that Rosbaud 'offered us something quite special' and proposed 'a journal which emphasised European achievements. He was the driving force... and it has remained one of the most successful journals in the field'. When news of Pergamon Press's intention to launch *Chemical Engineering Science* reached New York, it caused consternation. The American Institute of Chemical Engineers was planning a similar product with an identical title. At the time virtually all 'learned journals' were sponsored and published by professional and academic bodies, not by upstart independents. The American Institute's committees did the only thing their traditions suggested was possible in the circumstances: they delayed their project for a couple of years.

Chemical Engineering Science began life in October 1951 at a frequency of four issues a year. At first it was slim: volume 1 covered the years 1951 and 1952. A group of seven engineers had agreed to serve on the first year's editorial board, one each from Britain, France, Italy, Switzerland, the Netherlands, Norway and Belgium. They were supported by an advisory board of 21 Europeans and one Australian. The journal's subtitle *Génie Chimique* showed serious pretentions to a wide audience: in the 1950s French aspired to equal standing with English in international communication. The famous Toulouse scientist Joseph Cathala penned the first article of the first issue and he wrote in French. The whole venture was a clarion call to continental European engineers. Although they were not called chemical engineers (the profession did not exist in continental Europe), they did things which in the USA had long been called chemical engineering.

The British anchorman on the launch editorial board was Maxwell

Bruce Donald. He had just been made Ramsay professor of chemical engineering at University College London, one of the few British establishments to take the subject seriously. Donald—with American qualifications from Massachusetts Institute of Technology and real experience of production in the recovery of nitrates for fertilisers from Chilean guano—was much admired. Peter, having taken part in initial discussions with Rosbaud in Cambridge, was enthusiastic about the venture. His name first appeared on the editorial board listing of *Chemical Engineering Science* in volume 2, dated 1953, and his seminal article on residence-time distributions led the first issue of that year. By then the board had swelled to ten. Peter published several other bits of his own innovative thinking in *Chemical Engineering Science* over the next few issues. Denbigh was equally enthusiastic but didn't join the editors until 1955, by which time he had moved from Cambridge to Edinburgh; after a further move to Imperial College London, he continued as an editor, until 1965.

In 1955 Rosbaud made an inspired addition to the team. The American Neal Amundson was on leave from the University of Minnesota and enjoying a sabbatical year with Terence Fox's Cambridge department. Rosbaud met him and offered him the American editorship of the journal from 1956. Amundson's professional reputation was second to none and he played a major part in establishing the journal in the USA before stepping down in 1969.

By the mid-1950s Robert Maxwell was turning his attention to the publishing of journals. That meant Pergamon Press. Relations between Maxwell and Rosbaud were uneasy at the best of times and, once Maxwell began to take more interest in the day-to-day running of Pergamon, the inevitable happened: after many niggling differences of opinion, they fell out decisively. A disagreement over a US contract in 1956 led Maxwell to sack Rosbaud, whose publishing instincts had served Pergamon so well. In those days Maxwell's reputation before a wider public was apparently untarnished; it darkened as time went on, and Peter became well aware of the potential dangers of their association. But the value of *Chemical Engineering Science* to his chosen profession outweighed all that. Here is Peter recalling Maxwell in a *New Scientist* piece dated 1983: generous with just a hint of irony and a sting in the tail:

> A Czech by birth, he was awarded the Military Cross for his action in the German Ardennes offensive of 1944, during which (I believe) he formed a unit of the Pioneer

Corps into a fighting force. This was a splendid entrepreneurial venture and he has been on the attack ever since – as a one-man information explosion and the biggest influence on British printing since Caxton. I first met him during the early 1950s and I became editor of what I believe was the second journal in the now enormous Pergamon Press stable. (The first was edited from the Vatican.) He is remembered for his period as a Labour MP (1964-70) partly because he introduced dehydrated mashed potatoes into the Commons kitchens.

The throwaway mention of the Vatican is vintage Danckwerts. Had a holy father really been editing Pergamon Press's first journal? Or was it simply that a previous editor had been a staunch Roman Catholic? The background is fairly exotic. The journal in question was one of three that Pergamon Press inherited from Butterworth-Springer in 1951. It was called *Spectrochimica Acta*. Publishing papers on spectrochemistry and astronomy, it had been founded by a professor called Alois Gatterer in 1939. Springer Berlin published the first issue in May of that year; three issues a year followed until mid-1944. Then publication ceased abruptly as the Russians closed in on Berlin and Germany was falling apart. Gatterer was an academic of the old school: a Jesuit theologian at Innsbruck University as well as a scientist. He directed the astrophysical laboratory of the Vatican Observatory, which operated on the first floor of the Pope's summer home in Castel Gandolfo, and was a member of the Pontifical Academy of Sciences. As a means of keeping his journal going, Gatterer suggested to Pope Pius XII that the Vatican Observatory might take up its publication. The Pope was keen, and four issues appeared between 1947 and 1949, effectively edited from the Vatican.

The 1950 issue of *Spectrochimica Acta* was published in London. The catalyst for this change was a Dutch intermediary called Frederick van den Heuvel. A count of the Holy Roman Empire—and, according to an authoritative version of Butterworth's history, 'a shadowy and influential person' known as 'Fanny the Fixer' to intimates—van den Heuvel had excellent relations with the Vatican. He had spent WW2 as MI6 station chief in Berne and the British decorated him for it. Van den Heuvel cut an impressive figure with imperial whiskers and a black homburg. Late in 1946 he had attended a meeting in London of top British scientists, Butterworth representatives and Robert Maxwell as part of the British government's efforts to kick-start post-war science publishing. Working on behalf of Butterworth-Springer, he had bought *Spectrochimica Acta* from the Vatican by the end of 1949. When that company passed to Maxwell

BACKGROUNDER:
PAUL ROSBAUD, SCIENCE,
ESPIONAGE AND VISION

Born in 1896, Paul Rosbaud was the third illegitimate son of an occasional liaison between his mother Anna and the cathedral choirmaster of her home town of Graz in Austria. Educated as a physicist, Rosbaud took a doctorate in the physics of metals, doing exciting things in X-ray cinematography. But he decided research wasn't for him. He joined the staff of the German magazine *Metallwirtschaft*, halfway between a learned journal and a trade magazine. As talent scout for interesting articles, he ranged far and wide. The Springer publishing house, a major world force in scientific journals, recruited him in 1932 for its flagship magazine *Naturwissenschaften*, Germany's equivalent of *Nature*. Suave, assured and absolutely trustworthy, Rosbaud made extensive contacts and friends amongst scientists and engineers at the cutting edge of their fields in pre-war Germany. But Nazism was anathema to Rosbaud. In 1938 he persuaded Lise Meitner, who worked on nuclear fission with Otto Hahn, to avoid arrest by leaving Germany, and then sent his Jewish wife Hilde and their daughter to safety in England. He himself stayed on to fight what his biographer Arnold Kramish calls a deep, personal, silent war against Adolf Hitler. He speeded up the publication of research of German nuclear advances in the late 1930s, so that the implications could be seen internationally. When war broke out

Rosbaud, as *Der Greif* or 'the Griffin', sent coded messages on the state of German war science and technology— the atomic bomb, the V1 and the V2— via contacts in Norway to his London spymaster Eric Welsh. Ten years after WW2, the US Justice Department conceded that 'his activities on behalf of the Allied cause were successful and of such importance that even today they cannot be disclosed'. In 1945, as the Russians closed in on Berlin, Welsh spirited Rosbaud out to a new life with the embryonic Butterworth Scientific in London. In 1949, when Butterworth and Springer agreed to form a joint UK company, Rosbaud became scientific editor. But Butterworth felt Springer got the better of the deal and in 1951 wanted out. Rosbaud tried to buy the Butterworth share, but didn't know about raising money. Robert Maxwell did; failing to sell on to Springer, he then had to buy them out too. The actual amounts involved, as in much of Maxwell's life, differ from account to account. Rosbaud decided to stay on in the new company. Fired by Maxwell in 1956, Rosbaud survived on consulting jobs until his death from leukaemia in 1963. He was buried at sea, reputedly leaving behind £500, a gold watch, a medal from the American Institute of Physics and an enduring mystery about both his wartime activities and his finances. In 1991, when his wife Hilde died in Switzerland, her estate was worth more than a million pounds— possibly a gift to Rosbaud from the Allies for his wartime contribution.

Paul Rosbaud at 65
Photographer: Lotte Meitner-Graf, London,
courtesy AIP Emilio Segre Visual Archives,
Physics Today Collection

and became Pergamon Press, *Spectrochimica Acta* was part of the deal.

The dismissal of Rosbaud had deprived *Chemical Engineering Science* of its driving force. But another dynamic personality was on hand. Peter was not a Maxwell employee but just the kind of person Maxwell needed and would hold in esteem: a fine war record (like my own, he would have thought) and that quiet reticence which suggests either an aristocratic background or an unwillingness to suffer fools gladly... or perhaps both. And Peter's peers held him in high regard. They showed it when, before the end of the decade, they gave him overall executive editorial command of the journal and he held the position until 1982. The journal was turning out to be a best-seller in the somewhat rarefied arena of chemical engineering science.

Brushes with romance

21

ESIDES HIS BIT PART in the world of Cambridge bohemia, there were other aspects to Peter's social life in Cambridge in the early 1950s. For one thing, he got to know Philip Bowden and his family. Bowden was a Tasmanian who had lived in Cambridge—with a break during WW2—since arriving to take a higher degree in 1927. In the process he had become an international authority on the science of friction. Since 1936 he and his wife Margot had leased half of a remarkable Cambridge house: Finella, originally a Victorian mansion on Queen's Road. When Bowden's niece Naomi Wolfhagen (her mother and Margot were sisters) arrived in Cambridge from Tasmania in the summer of 1949, the Bowdens invited her to live in. It was, she found, a very welcoming household with four young cousins, not to mention a continual flow of interesting family friends in and out of Finella West. Bowden offered his niece a place in his department which went by the intriguing title of 'The physics and chemistry of rubbing solids', or PCRS. She worked as a secretary, typing the PhD theses of young men studying in the laboratory.

A good-looking young woman, Naomi was much in demand amongst the university personnel. In those days almost all the colleges accepted only male students; the rare and rather exotic female student was either

Internal decorations in Finella
Courtesy Morley von Sternberg/RIBA Library
Photographs Collection

a bluestocking cycling in from Girton or a serious-minded maiden sequestered in Newnham. Peter may well have bumped into the Bowdens through Philip Bowden's acquaintance with Elizabeth Vellacott and the wider Mitchell circle. He, in common with quite a few other Cambridge men, found he got on well with Naomi. She admired him; she knew about his George Cross, although he never mentioned it. They saw something

---

BACKGROUNDER:
FINELLA
The Victorian villa originally called The Yews sits in attractive grounds on Queen's Road, Cambridge, facing 'the backs' of colleges across the river. The exterior is quite conventional but the interior is very remarkable, thanks to the attentions of Mansfield 'Manny' Forbes, an eccentric English don and minor scion of a Scottish clan, who rented the property in the mid-1920s. Forbes rebelled against the philistinism and lack of imagination of his English public school education. In the period 1927-29 he and the Australian architect Raymond McGrath turned The Yews into a modernist hymn to Finella, legendary 'queen' of a Pictish tribe, who escaped retribution for regicide at the end of the tenth century by flinging herself into a waterfall. It became a showcase for the use of modern materials in interior decoration, with light, copper-clad multi-panel folding doors, an aluminium-walled bathroom, mirrored ceilings and a rubberised floor with Pictish motifs reflected in the decoration. For a decade Finella was a meeting-place for young artists. In 1931 Forbes organized the exhibition of Jacob Epstein's controversial sculpture 'Genesis'. It portrayed a pregnant woman. On delivery, the heavy sculpture went straight through the floor; once reinstated at floor level, Forbes charged a shilling (10p) a head to view it and collected quite a large sum. Visitors walked into a room containing two figures covered by white sheets. When all was ready, Forbes emerged dramatically from under one sheet, spoke briefly and then removed the other sheet covering Genesis. When Forbes died young in 1935, Finella's owners—Gonville & Caius College—employed the architect Peter Bicknell to split the house into two flats. (After viewing the original Forbes/McGrath interiors architectural historian Nikolaus Pevsner reputedly commented: "A milestone in the coming of what might be called expressionism in decoration'; when he saw the Caius conversion he allegedly groaned: 'Vot a disaster'.) In 1936 Caius leased one half—Finella West—to the Bowden family. When Naomi Wolfhagen arrived she was amazed by the modernist interiors and its myths: Forbes had allegedly painted a Grenadier Guard on one of the bathroom ceilings (or was it on the bath itself?) but no trace remained. Caius took the property back in 1980 and converted it again, this time into rooms for its Fellows.

of each other—she helped out at Peter's annual party for department students—but she had plenty of invitations from other potential suitors.

In the summer of 1950, Naomi had arranged to meet an Australian friend in Verona with a view to spending a fortnight exploring Italy. To get there, she and Peter hatched a plan for a joint trip to Italy. They would travel with John Buckatzsch, a good friend of Peter from his Balliol days, and his wife Wilma. The party travelled in Buckatzsch's Alvis tourer, which excited attention wherever they went. For old times' sake, their route to Verona took them through Peter's pre-war stamping ground of Salzburg. They found bed and breakfast on *Hellbrunner Allee*, surrounded by the beautiful linen and embroidered flim-flam of impoverished Austrian gentry; it was the same street in which Peter had lodged in 1935.

Continuing on their way over the Alps and into Italy, the party paused for refreshment on the Brenner Pass and Wilma took some photographs. Naomi busied herself preparing the snack while Peter, in tie, blazer and handkerchief in the top pocket, sat characteristically aloof. Buckatzsch had observed Peter's somewhat distant manner with Naomi and impishly joked with him that he needed a love philtre: something to make him embrace life more avidly, rather than just enjoying it from the sidelines.

Naomi's parents and brother, the Wolfhagens, arrived in Cambridge for a visit in the summer of 1951, and she went with them on their six-week

Brenner Pass, 1950: (l to r) John Buckatzsch, his magnificent Alvis tourer,
Naomi Wolfhagen, Peter Danckwerts

trip back to Tasmania on the P&O liner *SS Himalaya*. She spent 1952 in Tasmania but in mid-1953 booked to return on the *SS Orion*. This time Philip Bowden provided her with a self-contained flat at the rear of Finella West. She renewed her friendship with Peter.

In 1953 the Tube Investments (TI) Group, then one of the thirty largest companies in the UK, had asked Bowden to set up a research laboratory for them. The Group brought together more than 50 manufacturing companies producing steel and steel tube for products like bicycles, gas cookers, vehicle exhaust systems and rings for jet engines. It was fashionable for blue-chip companies to operate some sort of research facility; TI wanted it to be close to a major university yet far enough from its own operating sites to avoid researchers getting dragged into production problems. Hinxton Hall, nine miles outside Cambridge, had been on the market for a while and filled the bill. Budgets were generous, and Bowden hired Elisabeth Vellacott, a good friend of his wife Margot, to decorate and furnish Hinxton Hall. His new position brought with it use

---

BACKGROUNDER:
JOHN BUCKATZSCH,
FRED DAINTON,
AND WILMA

Peter's Balliol friend John Buckatzsch had been at secondary school in Leeds with Fred Dainton; the two made local history by getting places at Oxford. It was while they were all together at Balliol that Peter first met Dainton. The two were to meet again in the 1960s when Dainton was a member of the syndicate in nominal charge of the Cambridge department of chemical engineering. Dainton became a major force in British science administration; amongst his long list of achievements, two of the best were the vice chancellorship of Nottingham University and as the driving force behind the new British Library in St Pancras, London, opened just before his death. John Buckatzsch, an economist, statistician and very likeable person had contracted tuberculosis in his youth; it blighted his short life with long illnesses and several operations. Buckatzsch had met his future wife Wilma when her friend Barbara Hazlett had married Fred Dainton and Buckatzsch was best man. He and Wilma married in 1947. Already missing one lung, he became seriously ill in early 1954. Peter drove Naomi to Oxford to see the Buckatzsches in January of that year, and later spent a number of weekends with his friend while he was dying, lifting his spirits with champagne. Wilma had already lost one husband: John Alexander, a wartime airman. A couple of years after Buckatzsch's death she married a third—writer and traveller George Crowther. With him she built a career as field zoologist and Oxford don. Peter kept contact with her to the end of his life.

of a car, but neither he nor Margot could drive. So Naomi took on chauffeur's duties, and while Bowden attended meetings in Aldermaston, Waltham Abbey or London, she and her aunt explored the hinterland. During this second stay in Cambridge Naomi grew fonder of Peter; he often came to dinner and they dined out together; he invited her to the Emmanuel College May Ball and they disappeared off at weekends touring East Anglia by car.

The Bowdens loved skiing and were in the habit of taking a holiday each spring in the Austrian mountains. They planned a trip in April 1954 and Naomi invited Peter to go along too, knowing how he loved Austria and mountains. To her surprise she had to work hard to get him to join in. 'It

---

BACKGROUNDER:
THE BOWDEN FAMILY

Philip Bowden came to Cambridge from his native Tasmania in 1927 to take a higher degree in electrochemistry. A young Tasmanian friend of his, Margot Hutchison, followed him in 1931 and they were married. As a researcher Bowden was a contemporary of the novelist C P Snow, who later drew on Bowden in his series *Strangers and brothers* for the character of Francis Getliffe, the 'gifted, wise and sensitive scientist'. Lecturing in Cambridge in the 1930s, Bowden became interested in friction and the lubrication of surfaces when they rub together. A series of brilliant experiments led to a theory of friction that brought him an international reputation and, in 1948, election to the Royal Society. He realized that when one solid surface lies on another the area of real contact is quite small: when pushed towards each other, those contact spots deform and stick together; sliding causes high temperatures at the contact points, or hot spots. In 1936 when the Bowdens

moved into Finella West, the other half was leased by Paul Sinker, later head of the British civil service and later still director general of the British Council. In 1939 Peter Bicknell, architect of the Finella conversion, and his family moved in. When war broke out Bowden, his wife and their young son Piers were on holiday in Australia. Feeling that he would benefit the war effort more in Australia than in Britain, Bowden persuaded the Australian government to let him set up a research unit in Melbourne on lubricants and bearings: tribophysics. There Bowden applied his ideas on hot spots to explosives in which tiny air bubbles can act to facilitate detonation. The family returned to Cambridge and Finella West in September 1944 bringing two more sons, Humphrey and Jonathan; a daughter Sophie arrived soon after. In the later 1950s the Bowdens released fire balloons in the gardens of Finella (perhaps Peter had seeded the idea) and in the 1960s the venue became famous for spectacular firework and fire balloon parties.

was like trying to prize a limpet off a rock,' she wrote home. 'All sorts of imaginary difficulties flock to his mind.' He liked skiing, though, and was very attracted to Austria. An extra incentive may have been Philip Bowden's research interest in the rubbing of skis on snow and ice which offered the possibility of skiing faster. Finally he succumbed. They descended on St Anton and a variety of friends came too.

Lord (Theo) Chorley was an old friend of the Bowdens. At 60, he seemed fragile but took skiing lessons without untoward effects. 'In the evening we foregathered with Philip and Margot to brew tea,' Naomi recalls, 'and from this hour Theo flowered, entertaining us with his conversation. There seemed to be hardly anyone he had not known, and from his wonderful memory anecdote after anecdote came pouring forth.' In Zürs, on the way to start a favourite run, Peter and Naomi shared a T-bar ski lift. 'On the first trip I was nearly thrown off by Peter's greater weight. But we worked out a way I could balance.'

While Naomi had been away in Tasmania there had been a development in Peter's Cambridge social life. The fallout from the Mitchells' outing with companions to Monaco in 1951 had been dramatic. Peter Mitchell had deserted his wife in favour of Helen Robertson, and Eileen Mitchell was extremely upset. Friends rallied round to cheer her up: Bryan Robertson took her to Paris and introduced her to Alice B Toklas, Picasso and the Dior dress show. Peter Danckwerts was also sympathetic and, although rather on the edge of the group, eager to express his sympathy.

Eileen was a lovely young woman and widely admired; Mitchell had described her twelve years earlier, when they had first met, as resembling the Mona Lisa. Gradually a strong mutual attraction developed between Peter and Eileen. They became lovers. There may have been some similarity between the two Peters: a similar hairline, the big brow, the slightly predatory look. But they were totally different in style. While Peter Mitchell was self-confident and outgoing in almost any company, Peter Danckwerts appeared to the Mitchell circle to be reserved and a bit detached. Where Mitchell's life style was bohemian, Peter's was much more conventional; he preferred Lady Taylor's tea and cake parties to Peter Mitchell's rackety Grange Road evenings.

The affair between Peter and Eileen continued for several years and it says something for Peter's ability to organize his private life that, even though Elisabeth Vellacott was a close friend of both Eileen Mitchell and Naomi Wolfhagen's aunt Margot, Naomi knew nothing of Eileen and, in all probability, *vice versa*. Things came to a head early in 1954: Eileen realised that she was pregnant.

This development must have been behind Peter's indecision about the invitation to join the Bowdens' skiing trip that April. Eileen took a considered decision to continue the pregnancy and Peter, like any good Wykehamist in that situation, proposed marriage. Eileen refused him. Perhaps it was a gut feeling that it wouldn't work, or perhaps she accepted too willingly the advice of her trusted friend Vellacott, whose opinion she often sought. Vellacott had her eye on another candidate for Eileen's attentions who had also been a participant in the fateful 1950 trip to Monaco: the medic John McNeil. And later she did marry him. But in the meantime Eileen gave birth to a son in August 1954—Peter Jonathan Mitchell, known to the family as Jonnie. Over the next few years Peter didn't give up his attempts to 'land' Eileen, but to no avail. They remained friends; he contributed to Jonnie's upbringing and played an occasional part as a hands-on father. But the rejection as a formal father may have encouraged him to consider leaving Cambridge. During 1954 he applied for and accepted a job at the UK Atomic Energy Authority based in Risley, just north of Manchester. In October 1954, and two months after Eileen

Mitchell had given birth, Peter's new job took him away from Cambridge.

Naomi was most upset to see Peter go. They went out for a last lunch in Clayhithe, a mile or two down the river Cam, where Naomi recalls they ate oysters 'like at a funeral wake'. In November Peter called on Naomi when he was passing through Cambridge and she found out how he was getting on. He seemed exhilarated both by the job and the practical opportunities that it brought him to apply his engineering skills. Then in February 1955 Peter wrote and invited Naomi to go on another

Eileen Mitchell with her son Jonnie by
Peter Danckwerts in 1956
Courtesy Julia Mitchell

St Anton, April 1954 (l to r): Margot Bowden, John Graham (a friend of the Bowdens),
Naomi Wolfhagen, Peter Danckwerts
Photo: Philip Bowden, courtesy Naomi Canning

Austrian skiing trip, again to St Anton. The fortnight passed off without
great incident. They met again in April and finally in June. By this time
Peter was travelling incessantly and missing the relative ease he had
enjoyed in Cambridge. Meanwhile a candidate for Naomi's hand was
closing in. Peter Canning, a Tasmanian engineer, travelled halfway around
the world to find her and persuade her to return home with him. In July
1955, ten months after Peter had left Cambridge, Naomi Wolfhagen
boarded SS *Stratheden* bound for Hobart.

Once back in Tasmania she wrote Peter a considered letter, 'putting her
cards on the table'. He replied in what she felt was a cavalier fashion and
the prospect of a further relationship disappeared. About a year later Philip
Bowden returned to Tasmania for a short holiday and took Naomi aside.
He had a message to deliver which seemed to cause him some embarrassment. Peter had asked Bowden to tell Naomi that the reason he had been
diffident towards her was that his wartime accident in Sicily had left him
unable to father children...

Halcyon days in atomic energy 22

WHY DID PETER JUMP SHIP in 1954 from his Cambridge lectureship to accept the position of deputy director of research and development for the United Kingdom Atomic Energy Authority (UKAEA) industrial group? One strong influence was his admiration for its leader, Christopher Hinton. Another was that the prospect of promotion within the Cambridge department was no more than a remote possibility. Fox was young and energetic and his team were almost as much in awe of him as the students, so there seemed little prospect of the professorship becoming vacant. Sellers, Kay, Denbigh, Gray and Peter himself—all had moved on by the mid-1950s. Of the seven-strong team built up from 1948 to 1954, only John Davidson and Denys Armstrong stayed for the long haul. Kay, who had moved to UKAEA in Risley from Cambridge in 1952 and by 1954 was taking up a professorship in nuclear engineering at Imperial College London, certainly bent Peter's ear on the attractions of a job in atomic energy.

The romance of the technology undoubtedly moved Peter. Atomic energy had an entirely positive image in those days; enthusiasts viewed it as the future of power generation. Some, who may have been asleep during economics lessons, even saw it as likely to be too cheap to meter. In 1946 the Americans had suddenly stopped all communication on matters nuclear. The British, in spite of having provided a great deal of the expertise for the Manhattan project which produced the atomic bomb, suddenly found themselves out in the cold. So the Attlee government in Britain launched Tube Alloys, the cover name for a British independent nuclear project. British scientists already knew how to make bombs, either from plutonium or enriched uranium, and quickly three men (the first later knighted and the other two knighted and then ennobled) took charge. John Cockroft looked after nuclear research in Harwell, William Penny headed weapons research at Aldermaston and Christopher Hinton undertook the nuclear engineering work. Hinton's efforts resulted in a uranium purification plant at Springfields, a uranium 235 (U-235) enrichment plant at Capenhurst and two reactors plus accessories for the extraction of plutonium and separation of unchanged uranium and the products of fission reactions at Windscale, today known as Sellafield.

It was also undoubtedly true that Peter felt a strong urge to get some industrial experience with engineers actually doing the job. As an academic he had been teaching undergraduates how engineers might do things and how they actually managed to do them. MIT had shown him the benefits of getting his hands dirty, and that had always been on his mind. Then there was money. He reflected on several of these issues in print:

Having spent six years as a lecturer in an engineering subject, I felt the need for industrial experience. Atomic energy had, in those days, a romantic appeal and the salary lifted me from poverty to relative affluence. The Authority was able to pluck me from a lectureship at Cambridge and offer me double my university salary (I got another 50% increase in the next two years). It went to my head; I bought the first new car in my life—the notoriously overpowered and underweight Ford Zephyr—and courted death daily with my optimistic driving.

Until then he had been making do with a second-hand Morris Oxford. The salary which turned Peter's head was the consequence of a policy adopted by the Ministry of Supply when it established UKAEA earlier that same year. The idea was to free

A 1954 Ford Zephyr

atomic energy development from the 'dead hand' of the Civil Service. Length of service was a major influence in the Civil Service on pay and promotion; the Ministry felt that merit rather than longevity should be the deciding factor and Peter was a beneficiary. He prefaced any discussion of the matter with the modest suggestion that UKAEA may not have got a particularly good bargain in his case. But the organisation managed to attract him and a good number of other highly-qualified people who would otherwise never have touched the Scientific Civil Service with a bargepole.

The glamour of the UKAEA project moved Peter and Hinton's brilliance as an engineer dazzled him. He frequently drew on what he learned and experienced at UKAEA and wrote in *New Scientist* 25 years later:

Hinton is, I believe, one of the great British engineers. Having had to learn the science and technique from scratch, he succeeded by sheer brilliance and unremitting driving force to get his plants designed, built and commissioned on time. It was a reign of terror, and the degree of hyperactivity was comparable with that associated with the Manhattan project.

As Deputy Director of Research and Development of the Industrial Group, Peter was getting close to Hinton, at least in theory.

Peter arrived in Risley on 1 October 1954. He immediately became acutely aware of the sacrifices involved in leaving his Cambridge environment; Risley is north of the river Mersey and on the edge of

---

BACKGROUNDER:
CHRISTOPHER HINTON

Perhaps the most powerful British engineer in the two decades following WW2, Christopher Hinton began his career as an engineering apprentice. He left grammar school at 16 and was taken on by the Great Western Railway in Swindon. Bored after six 'unnecessarily long and wearisome years', he secured a place at Trinity College Cambridge and read mechanical sciences. Hinton joined ICI and by 1930, at the age of 29, was the company's chief engineer. The experience gave him a firm grasp of the technical, financial and organizational challenges of large projects and the ability to bring them home on time and on budget. He put all this to good use during WW2 with the Ministry of Supply overseeing the efforts of up to 30,000 workers. Size alone, he wrote, is not a problem provided that management is not afraid and knows how to create appropriate structures to cope. Post-war, Hinton joined the British atomic energy project Tube Alloys and, working out of Risley, drove industrial production of plutonium: an entirely new technical challenge. He chose the production sites at Springfields and Windscale and, lacking time and resources for pilot plants, his teams built full-scale plants which he later described as 'monuments to ignorance'. As the military requirement for plutonium increased, newer plants offered the chance to use the heat produced to generate electricity—the peaceful use of atomic energy; Hinton took Harwell's design studies and turned them into the Calder Hall nuclear power station. Doubts about the price and availability of natural uranium led him to press for the fast breeder programme at Dounreay, even though he recognized the immense engineering problems that would be involved. Hinton went on to chair the Central Electricity Generating Board. He was rewarded with a knighthood and then a peerage. A man who brought professional competence, energy and enthusiasm to all he did, Lord Hinton died in 1983.

Warrington—'that queen of cities', Peter noted acerbically. His daily drive to work took him past a pub called the Cemetery Arms. But there was family close by. While he sought out a base from which he could easily get to Risley (he eventually settled for a flat in the southern suburbs of Manchester) he lodged with his sister Diana Garson and her family at Bromborough on the Wirral. Diana's husband John, a medical doctor, and his young son Ian were both converts to the delights of fire balloons, and 5 November 1954 would have been an early opportunity to share their mutual enthusiasm. In fact, no matter how busy he may have been 'Unc', as he was known by the Garsons, always seemed to find time to drop in for Guy Fawkes' Night and a spot of fire ballooning.

Peter's job, which was essentially administrative and for which he claimed to feel unqualified, took him all over the place—to UKAEA's many sites and to conferences keeping up with research. UKAEA had laboratories in Culcheth (metallurgy) near Risley, Springfields (uranium extraction, fuel-element research and manufacture) near Preston, Windscale (hot chemistry, plutonium extraction, reactor problems and heat transfer) in Cumberland, Capenhurst (isotope separation in general) in Cheshire and Dounreay (where the fast breeder reactor and associated laboratories were to be built) isolated on the extreme north coast of Scotland. If Peter's career at UKAEA had lasted more than two years, his beat would also have included Winfrith Heath, developed in the late 1950s as a site for research into nuclear power. Then his activities would have...

...spanned the island from the rocks of Caithness to the sands of Dorset.

The Industrial Group was working hard at developing civil nuclear power. The guiding concept was Pippa (pressurised atomic piles for producing power and plutonium) which manifested itself as the Calder Hall power station, the Magnox reactors and, later on, the fast-neutron breeder reactor at Dounreay. It seemed to Peter that the heroic days of the atomic energy effort were already largely past. All those involved in the initial thrust at the end of the war, achieving the impossible by stipulated dates, had suffered frightful stress. Peter thought it well captured by Margaret Gowing in her book *Independence and Deterrence 1945-52, Volume 2: Policy Execution*. But at Risley the tradition was still strong.

It was possibly the last industrial organisation in Britain where the uniquely British wartime achievement of productivity, dedication and slave-driving lingered on.

John Cockroft had deliberately steered Harwell towards the status of a university campus. The objective was to attract brilliant young men to the establishment as research fellows. But…

…in most of the Harwell divisions the sense of urgency had dwindled to vanishing-point (I think I should in fairness except Monty Finniston's Metallurgy Division). Harwell had something of the atmosphere of a Greek city-state whose citizens discussed general ideas round the dinner-table. Indeed, they produced a remarkable number of ideas for nuclear reactors which withered under the rude comments of engineers—they were known as 'after-dinner reactors'. Anyone with a major practical contribution to make tended to move to the Industrial Group 'where it was at'.

The Industrial Group had urgent problems to solve and deadlines to meet. So it set up its own dedicated research and development organisation to provide the information it required. Peter estimated that, at its peak, this organisation employed more scientists and engineers on research than Harwell.

But we were no citizens of Athens. We were more like the slaves building the Pyramids.

On that first day at Risley Peter found that the offices of his new employer comprised a collection of wooden huts erected on a bog outside

---

BACKGROUNDER:
MAGNOX DEVELOPMENTS
Peter recalled at the time:
One of the major tasks at Springfields was the development of 'Magnox' (magnesium non-oxidising) fuel elements for the Calder Hall reactors and their commercial successors. As these reactors were to be run at a higher temperature than the existing Windscale piles, it was necessary to clad the uranium fuel elements with a metal of higher melting point than aluminium. Magnox—an alloy of aluminium and magnesium—filled the bill. But the whole economy, even the practicability, of the design depended on producing a fuel element which could undergo a sufficient degree of 'burn-up' without warping or developing faults in the Magnox can. Such faults would release volatile fission products into the carbon dioxide coolant. The burn-up transmuted elements in the fuel—for instance, substituting light fission product elements (some of them volatile) for part of the uranium—and this, together with neutron bombardment, distorted the crystal structure. Fluctuations in the temperature of the fuel would be inevitable and would lead to phase changes and consequential dimensional changes which would be liable to distort the elements and rupture the cladding. In order to minimise the difference in temperature between the centre of the fuel and the coolant gas, an elaborate pattern of fins was imposed on the Magnox cladding; at the same time these fins increased the resistance to gas-flux and hence the pumping power required to circulate the carbon dioxide coolant. The optimum design for the fins was arrived at as much by intuition and trial as by theoretical fluid mechanics. All the same, inevitable and maddening problems arose…

Warrington. They were, he recalled, almost inaccessible by public transport; signs pointing to Risley were infrequent and obscure, and often foiled the efforts of even the most determined of taxi-drivers. He kicked his heels on arrival in a seedy waiting room and was finally admitted to the office of Leonard (Larry) Rotherham, Fellow of the Royal Society and the director to whom he was to report. But the induction was ineffectual and things did not improve with time. Peter complained that:

One could never pin down anyone senior to oneself except for half an hour between ending a committee and catching a train. [And when Peter finally caught up with Rotherham the response was] 'Ah, Peter, I can't stay long—just off to Culcheth—why don't you read some of these reports? See you later'. And he slid out of the door.

Rotherham remained an enigma to Peter, who at first likened his own position to that of a vice-president in the USA—holder of a fine title but unable to act effectively. It was, just as in his first days at MIT, unfamiliar territory. In 1954 ICI was the biggest process company in the country with a huge capital investment programme, and yet UKAEA's Industrial Group at that time may have been spending money equivalent to a quarter of ICI's budget and was even more research-intensive than ICI. After six years in Cambridge, sitting on no committees and supervising a single graduate student, Peter felt he would have needed to become a whizz-kid overnight to grasp the organisation effectively. The job he found 'fascinating, adventurous and precarious'. He absorbed the detail quickly and enjoyed the remarkably esoteric nature of his new knowledge.

For lack of any other instructions, I got hold of the latest monthly reports detailing every item on the research programme of each laboratory, and paid them each a visit, going through every item in turn. The personalities of the various laboratory directors varied as much as did the nature of their technological responsibilities. Some of them thought my procedure eccentric, but on the whole I managed to get a pretty good idea of what was going on, in all its diversity.

Amongst the diversity he was amused to find that a lot of chemical effort went into the preparation and characterisation of elements which occupied what were then little-understood parts of the periodic table. These included a few very light elements—boron, beryllium and lithium—plus some heavier ones—zirconium, vanadium, tantalum, uranium and niobium (chauvinistically called columbium in the USA except, as Peter noted, by the American satirist and song-writer Tom Lehrer). Then there were polonium, deuterium, tritium and all those radio-elements in the lighter half of the table formed by fission of U-235. Peter found the work was full of surprises. The textbook recipe for making boron, for instance, produced a horrid impure and amorphous mass. The melting point of uranium had to be investigated, as had detail of its various phase changes from solid to liquid and even to gas. The characteristics of plutonium were even worse but—fortunately, Peter acknowledged—not on the agenda. To protect the employees from radiation, much of the chemistry had to be done in 'hot cells' by remote control behind protective windows. That made it an order of magnitude more laborious.

Isotope separation was inevitably an important part of Peter's world. The separation of U-235 (which produced the energy in atomic power plants) from U-238 (the major constituent of uranium ores) required the vapours of uranium hexafluoride to be pumped at sub-atmospheric

---

BACKGROUNDER:

ATOMIC NUMBER, ISOTOPES & FISSION
Every element has a unique atomic number A, equal to the number of positively-charged protons (P) in its nucleus. For example, lithium has P=3 and so A=3. The only other constituents of the nucleus with any weight are uncharged neutrons (N), and for any element the number of neutrons can vary, producing isotopes. The total number of protons and neutrons (P+N) is the number seen frequently in nuclear energy discussions alongside the symbol of the element. So Li-6 has three protons and three neutrons (P+N=6); Li-7 has three protons plus four neutrons (P+N=7). The number of neutrons affects the behaviour of the element; for example, of the two boron isotopes, B-10 (P=5, N=5) is a wonderful absorber of neutrons while B-11 (P=5, N=6) is not. Uranium has A=92 and there are six isotopes, none of them stable. U-238 (P=92, N=146) is far and away the most common form: natural uranium is 99.3% U-238. The

isotope nuclear scientists want—U-235 (P=92, N=143)—is 'fissile'; that is, if a neutron with sufficient velocity strikes the nucleus, the atom breaks into pieces with the release of a huge amount of energy. U-235 makes up only 0.7% of natural uranium. Some nuclear power stations can use uranium as it comes while others need it slightly enriched in U-235. Nuclear weapons, on the other hand, need super-enriched uranium, with U-235 raised to between 40% and 85%. The process of purifying uranium from 0.7% U-235 to 85% is a tedious one. Fission of U-235 doesn't just release enormous energy. Several more neutrons are formed and they can go on and collide with more atoms of U-235 and liberate more energy and more neutrons. This is a chain reaction liberating energy; indeed, every collision can cause branching of the chain reaction. In nuclear power stations the chain reaction is controlled by neutron absorbers but in nuclear weapons it is designed to result in a massive explosion.

pressures through porous membranes. U-235 hexafluoride passed through the membrane preferentially but the amount of separation achieved in a single stage was infinitesimal. Satisfactory separation required the building of huge 'cascades'. Uranium hexafluoride, enriched in U-235 by passage through a membrane, passed on from one stage to the next, while what was left (depleted in U-235) returned to the inlet of the previous stage. Each stage involved a compressor, a cooler and a membrane, and it took literally thousands of them to achieve the required enrichment of U-235.

There had been almost no British chemical engineers available during the design of the membrane separation plant, so physicists had been left to cope. They rediscovered the principles of recycling, reflux, tapering and equilibration time, and wrote long treatises about them—matters, Peter noted, already long understood in the American chemical industry. The resulting plant had unhelpful characteristics—for example, its response to changes in operating conditions was extremely sluggish. Unbelievably, the response time to changes was measured in months, much worse than trying to stop a large oil tanker travelling at full ahead.

The control of a cascade of diffusion cells offered a nice problem. Generally speaking, all sophisticated controls had been abandoned in favour of the most primitive of passive devices. [As a result,] waves of pressure would travel down the cascade, raced by the plant operators who soon acquired a taste for the sport. A typical plant operator, it was often boasted, was a retrained bus driver.

There were other isotopes to be separated—for instance, lithium 6 (Li-6) from natural lithium. The more abundant form of the element is Li-7. The two lithium isotopes are unique amongst lighter elements: they are stable, yet can undergo useful energy-liberating fission reactions for kick-starting a hydrogen bomb. UKAEA knew that the Americans were buying up lithium in order to strip it of the preferred Li-6, so the first thing to do when purchasing 'natural' lithium on the open market was to check that it had not already passed through the Americans' hands, been stripped of Li-6 and was being resold as an industrial chemical.

How much Li-6 was needed? Enough for one hydrogen bomb, it seemed, and by any means except self-destruction. The research and development team at Capenhurst—under the direction of Hans Kronberger who had fled Austria as a teenager, studied in Durham and been interned during part of the war—designed, constructed and commissioned a small-scale plant. When it looked as if it was going to work,

Capenhurst's Director of Operations Kenneth (KB) Ross tried to take over the plant and get the credit for it. Rotherham was away, and so Peter had to fight a battle with KB on his own. KB was a tough opponent; he had directed the huge Anglo-Iranian refinery in Abadan and was 'the last man out' when the Iranians ejected the British in October 1951. This was industrial life, red in tooth and claw, but not a wasted experience; Peter found analogies in academic life later on. The would-be academic or corporate in-fighter need look no further, Peter advised, than Cornford's *Microcosmographia Academica*, first published in 1908.

In the end, Kronberger got the credit. He received a red-carpet visit to Aldermaston as a reward.

Hans told me that he waited in [William] Penney's drawing room until the great man arrived showing signs of physical distress and saying 'Gosh, I'm exhausted!' Hans commented sympathetically on the strain involved in the production and test of Britain's first hydrogen bomb. 'No, no,' said Penney. 'I've just played 18 holes of golf and I'm rather out of training.'

Boron 10 was another separation challenge: it is a voracious absorber of fast neutrons and was required for the control rods in the Dounreay fast reactor; the rods absorbed neutrons in order to modify and control the reaction. But B-10 makes up only 20% of natural boron and had to be separated from the majority B-11. Kronberger's team took up the challenge by distilling boron trifluoride. As the amount needed was small, the distillation column was quite narrow. On the other hand, in order to get the requisite number of separation stages (known to chemical engineers as theoretical plates) the column was so high that it had to be supported by an electricity pylon. It was filled with very expensive wire-gauze packing. It worked, but there was consternation when the heavy isotope B-11 came out at the top of the column, instead of the lighter B-10.

I have never heard physical chemists talk so fast, explaining that this was only to be expected. At an international conference in Amsterdam, Hans made an in-joke about it, referring to 'boron *leicht/schwer*' and writing BOLS.

Preparations for plutonium production at Windscale were in train, and Peter was involved. The plan was to build two graphite-moderated, air-cooled, natural uranium piles. It was a pretty dramatic business, as the engineers involved had never built such a thing before. At one stage Sir John Cockroft returned from a visit to Oak Ridge, one of the USA's main sites for the Manhattan project, with the news that a similar American pile which had already been built was fitted with a filter on the exhaust stack to trap radio-active particles. At Windscale the stack on No 1 pile had already reached 120m (400 feet), but a decision was nevertheless taken to incorporate an air-filter at this height. Peter noted how it allowed Christopher Hinton to display his virtuosity as an engineer: when the No 2 pile was built it was also provided with a filter 400 feet up in the air—for reasons of aesthetics and symmetry. If Cockcroft had not brought back that news about the filter, Peter felt at the time, Britain might have provided the first and most serious example of nuclear damage to the environment. But he needn't have worried; in 1967 Cockroft revealed that he had discovered later that the filters were unnecessary. The Americans had put them in when they discovered radio-active particles of uranium oxide on the ground near their new reactor at Oak Ridge. But it transpired that the particles had come not from the reactor pile at all but from a nearby chemical plant chimney. All the same, during the Windscale fire of October 1957 the filters probably prevented a disaster becoming a catastrophe.

Amongst the operations at Windscale was a chemical plant for dissolving the uranium fuel elements from a nuclear reactor and separating the plutonium they contained, the fission products and what was left of the original uranium. The process itself Peter described as an unquali-

The Windscale site in October 1957, with Christopher Hinton's two filters
high up on the 400 foot piles
Courtesy Louise Rawling & Ivor Nicholas

fied success. The design was based on work done at the Canadian nuclear research facility at Chalk River, Ontario, by an Englishman, Robert Spence, who had then moved to Harwell. In Canada Spence used an amount of plutonium about the size of a pin's head, yet the Windscale process was much superior to that first used by the Americans. The laboratories at Windscale had to deal with samples of highly-radioactive solutions from the plant. Delivered enclosed in thick lead containers, they were manipulated in shielded 'hot cells' with windows consisting of two sheets of plate glass with concentrated lithium bromide solution between them. Access to the hot laboratories was, of course, tightly controlled. Operators had to wear overalls, overshoes and a cap. Before leaving they had to scrub their hands until they passed scrutiny by a hand-monitor.

If they remained contaminated extremely drastic measures were taken to remove the outer layer of contaminated skin. Hence the saying 'you can always tell a radio-chemist; he washes his hands before he has a pee.'

The laboratories at Windscale, like everything else, had been designed with almost no experience. The pessimistic assumption was that superficial contamination would be widespread, and that it would be a very time-consuming business decontaminating people to the level required to cross the barrier. In view of this there were a number of internal windows where those incarcerated could hold up their notes to show those outside what they had been up to. This allowed assessment of the degree of decontamination they needed.

Disposal of very low-level radioactive wastes, such as rinsings from laboratory and laundry, was direct to the sea. A pipeline had to be towed off the beach, sunk and pegged down. Before it was laid, an elaborate series of trials code-named 'Seanuts' was staged off the coast. Large quantities of fluorescein (a fluorescent dye) dissolved in water were dumped into the sea from a trawler, and Windscale staff monitored its progress and dissipation. Geoffrey Taylor, at the time the doyen of British physicists and whom Peter had known during his time in Cambridge, was called in to discuss the results. He pointed out that when the tide was flowing parallel to the coast—from north to south, for instance—the water near the surface flowed southwards much more rapidly than the water near the bottom. This might have led most people to the prediction of a very long smear of tracer. But Taylor's insight also told him that vertical mixing (due to turbulence in the relatively shallow water) was so rapid that almost all the effluent, moving rapidly between top and bottom, moved with the mean velocity of the tidal flow. The net result was that the tracer formed an elliptical patch which moved north and south with the tide, growing gradually more elongated. Peter recalled that Taylor subsequently extended the principles of this study to dispersion in capillary blood-vessels and in petroleum pipelines.

In the end the results of the trials were not particularly relevant: the limiting factor turned out to be fixation of radio-elements by mud and sea-weed near the discharge point. But the Seanuts exercise provided atomic scientists with a lot of open-air exercise and some early warnings in public relations. When the trawler came back to harbour with its decks awash with flourescein solution—an evil-looking liquid with a yellowish-green fluorescence—the cry of 'radioactivity' immediately went up. Eventually it was found expedient to change to a colourless tracer which could be detected only by its fluorescence under ultra-violet light.

The Springfields laboratory was required to produce metallic uranium and uranium hexafluoride for the diffusion plant by an enlightened modern process. A 1940s plant for making metallic uranium already existed, but it was primitive and had simply answered an urgent need at the time; elegance and optimisation played no part in it. Laboratory chemists at ICI had worked out what to do: convert the uranium ore to yellow cake and treat that with nitric acid to turn it into the water-soluble nitrate—uranyl nitrate. Then extract the solution of uranyl nitrate with a solvent insoluble in water; impurities remained in the water. The 'plant' was essentially no more than a scaled-up laboratory bench, and procedures captured the worst aspects of letting laboratory chemists scale up for production. When a chemist wanted an organic solvent in the 1940s he reached for the bottle of ether on the shelf in front of him; amazingly, that most volatile and inflammable of liquids had been the chemists' solvent of choice for the plant. And all the process operations—like precipitation, filtration and reduction, which laboratory technicians would have achieved brandishing nickel spatulas—were undertaken by process workers wielding shovels.

The rest of the purification process was quite a long business: add ammonia to the uranyl nitrate to precipitate ammonium diuranate, heat and reduce this to uranium dioxide, bring on hydrofluoric acid to get uranium tetrafluoride and finally use chlorine trifluoride to make the hexa-fluoride. And only then could the process of enhancing the content of U-235—by diffusion of hexafluoride through a membrane—begin. For

Peter the long process was a rainbow cascade of colours. Ammonium diuranate is yellow (like its impure form yellow cake) but reduction of diuranate went through the orange of uranium trioxide to the black of the dioxide and then to green tetrafluoride. The final step in the production of uranium metal was the reduction of uranium tetrafluoride with calcium, an expensive metal of doubtful purity and availability. Strenuous efforts were going on at Springfields to develop a method using magnesium instead.

Never having been involved in extraction metallurgy before (except for iron-making [on the MIT course]) I was startled when a uranium ingot covered with slag was tipped out of its crucible and pronounced 99.9% pure.

Peter took part in the effort to put the extraction of uranium onto a modern chemical engineering footing. A much less dangerous solvent replaced ether in a continuous cascade of mixer-settlers. This produced very pure uranyl nitrate. As a means of handling solids continuously, shovels gave way to a series of fluidised beds in which reactions between gases and solids took place. As part of the technical team, Peter flew to the States in October 1955 and visited Argonne National Laboratory outside Chicago. The Argonne establishment, a spin-off from the wartime Manhattan project, was the centre for research and development of US nuclear energy generation. In 1951 it had been the first to use nuclear-generated electricity to light up a string of electric bulbs. Peter's role was to find out what he could about American experience which might help process activities back home. His report dealt with aspects of uranium enrichment ranging from alternative technologies for producing uranium trioxide to the optimum size and shape of holes in the base support of a fluidised bed for denitration of ammonium diuranate. But some of the problems were persistent:

When concentrated uranyl nitrate was sprayed onto a fluidised bed of uranium trioxide particles [the nitrate] was supposed to decompose into uranium trioxide particles. While it may do so without problems in a 2-inch column, in an 8-inch column it produces golf balls or cannon balls rather than particles [if the conditions are not carefully controlled]...

Such were the excitements which engineers had to cope with in order to get a full-scale plant operating successfully.

Grazing in Geneva 23

A CHANGE IS AS GOOD AS A REST, and towards the end of his first year with UKAEA a unique international event gave Peter a break from the difficult technical challenges of nuclear fuel production. 'Atoms for Peace' was a ground-breaking get-together in Geneva of those of the world's nuclear scientists who were apparently working on 'peaceful applications'. Russian and Western physicists had not assembled *en masse* for discussion since 1937, and the major post-war nuclear powers—the Soviets and Americans—had never met like this before. In the event they exchanged pleasantries and a certain amount of technical information. Peter was a UKAEA delegate. Alongside the large delegations from the two big powers were, Peter observed, bit-part players like the British, Canadian and French delegations and also a huge assembly of on-looking nations:

Africans and Orientals in garish gowns and bright head-dresses, Latin Americans in quasi-military uniforms including gold-laced caps, medals and dark glasses (side arms had to be checked with the portière).

The whole show ran for twelve days, 8 to 20 August 1955—almost exactly a decade after the first use of an atomic weapon—by invitation of the United Nations organisation in Geneva's Palais des Nations. 1,400 delegates and 900 journalists from 73 countries attended. The leader of the British delegation, Sir John Cockcroft, described the event as 'a test of physical fitness unparalleled in our experience'. The impact was not just physical. International cooperation amongst scientists received a boost. Igor Kurchatov, who had been the driving scientific force behind the Russian atomic bomb and until then a shadowy figure to scientists in the west, suddenly appeared in April 1956 in front of an audience at UKAEA's Harwell site. He talked about thermonuclear fission, argued for removing the cloak of secrecy covering it and proposed international co-operation. At a follow-up conference in Geneva in 1958 he got his way.

Sitting on the podium in 1955 with UN secretary general Dag

Hammarskjold was the conference president Homi Bhabha. Peter registered that Bhabha was foremost amongst speakers offering hostages to fortune. It was strange, he felt, that an Indian physicist speaking for a country where 80% of the energy came from burning dung should be predicting that fusion power was only ten years away. Unfortunately Bhabha died in an air crash shortly after the time ran out. George (GP) Thomson—son of Nobel prize winner JJ Thomson, who was primarily responsible for discovering the electron, and a Nobel prize winner himself for showing how an electron can behave like a wave as well as like a particle—also stuck his neck out. He saw man solving the problem of controlling hydrogen energy 'within a generation'. There must have been a feeling of optimism conjured up by the euphoria of the occasion.

Sitting on the other side of Bhabha was MIT professor Walter Whitman, who had been a benevolent presence during Peter's chemical engineering practice course in 1946-47. Whitman was in charge of civilising aspects of the meeting. Sensing that delegates would be exhausted by the rough and tumble of a multi-lingual conference, he had organised a large room in the Palais where they could try to relax and recuperate. Peter preferred to recharge his batteries elsewhere, exercising his German:

Atoms for Peace: (l to r) Dag Hammerskjoeld, Homi Bhabha, Walter Whitman
Source: The Times of India Group. © Bennett, Coleman & Co Ltd. All Rights Reserved

The packed assembly hall in Geneva
Courtesy Getty Images

One of the 'friendly' delegations had a truly beautiful girl receptionist and I refreshed my spirit as I sat beside her for some happy hours while we tried to answer the multi-lingual questions of passing delegates.

There were innumerable evening parties to attend, and Peter's fine grasp of a wide range of applications of science made him a versatile member of the UKAEA squad. The Czech delegation invited two of the British delegation to dine. Peter went along with GP Thomson and felt that together they gave good value:

I started well by mentioning Aurel Stodola's classic treatise on steam turbines (which GP had certainly never heard of). During the rest of the meal we put on a pretty good double act; the talk ranged through space travel, the continuous cultivation of micro-organisms and solid-state electronic devices, all visionary at the time. None of the Czechs guessed that I was a virtual impostor.

At the Russian party Peter noticed how Slavonic gloom dissolved as the vodka flowed. A young Russian diplomat came into range and asked him what he thought was the most important political problem in Europe:

Conscious of my lack of political awareness I said 'I don't know—you tell me.' Before he was submerged beneath a sea of flushed faces he managed to say 'The reunification of Germany.' [Later] I myself clinched a rivet in the iron curtain by breaking into unseemly laughter over the drawings for a Russian nuclear power station. The building was fronted by a heavy classical façade of the Corinthian order.

The conference was largely a sales campaign, Peter judged, for those countries which claimed to have something in the nuclear power generation line to offer to those who hadn't. The Americans led the way, as might have been expected: they knew more, had a lot to say and relished their role as salesmen:

Above all, they exhibited a swimming-pool reactor. Looking down into the water one could see the beautiful blue glow of Cherenkov radiation—atomic energy made manifest. One almost expected to see MGM bathing beauties gliding through the depths.

Contributions to the technical programme were designed to tempt by showing what had been, or could be, done rather than revealing anything technically useful. The idea was to put on a good show and try and get some contracts. The Dounreay fast breeder reactor, then under construction, was presented as a mass of pipes and boilers with a little black box at the centre—the nuclear core—with no indication of what should go inside it, something the British themselves were wrestling with.

In Peter's estimation Christopher Hinton's presentation was the most dramatic and uninhibited. Hinton described Calder Hall, the world's first serious industrial-sized nuclear power station, with relative freedom and as engineer to engineer. He was able to chew over decisions affecting the economics of the plant: the operating pressure, the size of the pressure vessel and what wall thickness was practicable. But he said nothing about the crucial fuel elements—canned uranium rods—on which everything depended. Calder Hall's primary purpose—to produce weapons-grade plutonium—was not allowed to spoil the pacific atmosphere of the meeting. Indeed, Peter felt that the engineers whose first interest was electricity production might even be hailed in due course for their subtle interpretation of military needs.

Hinton's presentation, like all the others, was translated simultaneously into three other languages in the set: English-Russian-French-Spanish. Peter, with his interest in multi-lingual performance, or at least in its practitioners, was fascinated. He realised that the in-house team had done a crash course in nuclear jargon and considered that they coped excellently with prepared scripts. But they were liable to panic during question periods when unfamiliar phrases like 'xenon poisoning' might appear without warning.

Peter had the chance to catch up with a lot of friends and acquaintances.

Peter (l) oils the wheels as an enthusiast presents an unexpected
olive branch to a Russian delegate
Courtesy *New Scientist*

Dick Moore—the designer of the Calder Hall power station and based,
like Peter, at Risley—was there. They had shared the toughest of times: like
Peter, Moore had been in London as a sub-lieutenant RNVR in
September 1940 and had won a George Cross for making unexploded
parachute mines safe. He had worked alongside the man who had first
showed Peter how it was done: Dick Ryan. Now Peter discovered a little-
known fact about Moore—he had a fear of foreign water. Moore had
driven all the way to Geneva from Risley...

...in his own car with jerry cans lashed to it, and had drunk nothing but
Warrington municipal water throughout the journey.

The Geneva meeting attracted not only John Cockroft but Ernest
Walton. They had been joint winners of the Nobel prize for physics in
1951. Working with Rutherford in Cambridge in 1932, they had been the
first to engineer the artificial disintegration of an atomic nucleus: using
600,000 volts they accelerated a proton beam and smashed a lithium
nucleus. Peter had heard that Walton was the intellectual of the two and
Cockroft the engineer of the project. Now he found that while Cockroft
fronted up the British delegation in Geneva, a very demanding role,

Walton was enjoying a quiet life as a professor at Trinity College Dublin. With some difficulty, Peter managed to persuade the two of them to be photographed together. He was struck by the extraordinary contrast between the two men but delighted to find that Walton's primary ambition in life was to acquire a Wimshurst machine. Drawing on his teenage experience in the attic of Merton Lodge, Peter was able to advise. And it certainly helped him get his photograph.

Developing Dounreay 24

A FAST BREEDER REACTOR was the dream of every country research-ing nuclear power in the 1950s. The idea was that the fuel in the core would undergo fission to generate energy and in the process 'breed' more fuel, a very attractive economic package. During his second year with UKAEA, Peter was given the job of helping to plan the installations for Britain's fast breeder reactor research project at Dounreay.

The Dounreay site during construction in the 1950s
Courtesy Dounreay Site Restoration Ltd and NDA

The fast breeder reactor differs from a so-called thermal reactor. In a thermal reactor, like those at Calder Hall, the fuel is almost totally U-238 (uranium in its natural state) or U-238 very slightly enriched in U-235. Moderators like graphite slow the fast neutrons generated from fission to energies at which they would react with U-235 rather than with the large excess of U-238. In contrast, the fast breeder reactor core might be a mix-ture of plutonium (Pu) and uranium metals, or oxides of Pu-239 (20%) and U-235 (80%). In this case neutrons emanating from the core are not slowed down (hence 'fast' breeder) but could be made to bombard U-238, either by surrounding the core with tubes containing U-238 or by including U-238 in the core itself. The result of the bombardment would be the conversion of small amounts of U-238 to Pu-239 which could be extracted and used as future fuel. Studies in the 1950s showed, conveniently, that it would be surprisingly easy to control this process and avoid a low-grade nuclear explosion.

Peter's team gave some attention to the economic prospects of the plans. The cost of recycling the core and blanket materials, involving decontamination from fission products and refabrication, was enormous. And the capital investment in U-235 or plutonium required a very large heat rating (kW/kg of plutonium) in order to make the reactor an economic generator of electricity. At the time plutonium was worth about 100 times as much as gold. The studies Peter and his team carried out showed that the installation would be about the size of a modest stately home. And in good time local occupants of similar property showed

---

BACKGROUNDER:

PLUTONIUM

Plutonium has an atomic number of 94. Plutonium as Pu-238 (protons=94, neutrons=144) was first prepared in 1940. The element was named after "Pluto" as it followed close on the heels of the discovery of neptunium (Np), named for the astronomical body next to it; it was thought at the time to be the last possible element in the periodic table. The light-hearted suggestion of Pu as its symbol passed without comment into the record, a tribute either to chemists' sense of humour or their lack of awareness. Amongst more than a dozen isotopes the most significant one is Pu-239 (P=94, N=145) formed by neutron bombardment of U-238 and useful for nuclear weapons and power production. When a neutron resulting from fission of U-238 hits an identical uranium atom, it is absorbed

and the atom becomes U-239. U-239 is short-lived and breaks down into Np-239 which in turn breaks down rapidly into Pu-239. It is estimated that, together, all the power stations in the world today produce about 20,000kg of Pu-239 a year. Making enough of it for an atomic bomb was an important part of the WW2 Manhattan project. In due course an atomic bomb weighing more then four tonnes and based on 6kg of Pu-239 (Fat Man) was dropped on Nagasaki; 20% of the plutonium underwent explosive fission, equivalent to 20,000 tonnes of TNT. The Hiroshima bomb (Little Boy) dropped three days earlier was based on U-235. During the cold war atomic power stations for electricity generation produced Pu-239 as a side-product, but official sources were shy of publicity about its extraction for use in nuclear weapons.

themselves to be enthusiastic for the programme:

The Queen Mother, herself the owner of the nearby Castle of Mar, once demonstrated her well-known aptitude for small talk by telling me how well she thought the
space-age Dounreay reactor fitted in with the rugged coastline.

The rate of heat release from each square centimetre of the surface of
the central fuel elements would be of the order of 1KW—that is, about the
same as a domestic toaster but from a very tiny area. The only coolant
reckoned to be capable of removing heat fast enough was liquid sodium,
or possibly a sodium-potassium alloy which had the advantage of
remaining liquid at ambient temperatures. Such a heat transfer medium
was completely untried. So naturally it gave birth to a whole new research
and development sub-discipline, complete with novel challenges. The
Nusselt number (a dimensionless number which is the ratio of heat
transfer by convection to that by conduction) was of a new order of
magnitude. What about electromagnetic pumps with no moving parts?
How to deal with the sodium oxide which would inevitably form as air
leaked into the system? Hot liquid sodium would heat water to produce
steam for electricity generation; if liquid sodium and water were to meet
they would react violently together. What about the challenge of designing a boiler which would transfer heat from sodium to water without any
possibility of an explosive encounter between the two?

Peter reported later that the arguments were long and exhaustive, and
that prophets of doom were essential participants. One of them originated
the concept of 'thermal bowing'. Suppose two fuel rods are intended to
be parallel, but in fact one of them is slightly bowed. The spacing being
so close, bowing would restrict the flow of coolant between adjacent rods;
so they would heat up even more and eventually touch and probably melt.
There was no shortage of good ideas about how to deal with this
problem, but UKAEA's down-to-earth engineers knew the difference
between a good idea and a sound piece of engineering. The arguments
were still in progress when Peter left UKAEA.

Another scenario with science-fiction overtones of doom was a general
melt-down of the core. This might lead to a supercritical pool of fissile
material gathering at the base of the reactor. The results were hard to
predict but would have been spectacular. That problem, Peter noted
ironically, was easier to solve:

We had an official visit one day from a group of American fast-reactor experts (plus
people who had come along for the ride). The objective was to exchange ideas and

experience. Presumably they expected to get more out of the exchange than we did, but they did at least have a small experimental fast reactor operating. We had heard that they had experienced unexplained power surges during operation. During the formal meeting, in the bar and over a protracted lunch we tried to get some information out of them. They seemed strangely coy, almost embarrassed. It was not until they were on the point of leaving that they admitted that they were authorised to say that their fast reactor had just had a melt-down.

At an early stage Larry Rotherham had told Peter to work out the establishment for a research and development organisation to be attached to the Dounreay reactor. He did so as well as he could, and showed Rotherham the staff list.

'Double it' he told me, and thus we started our recruiting drive for Dounreay. Inevitably we attracted a large number of eccentrics—Scottish nationalists and bird watchers amongst them. There were also many who liked the idea of freedom from noisy committee rooms and tobacco smoke. A great attraction was that trout fishing was free; nothing less than a salmon was worth charging for.

Peter was also involved in plans for an effluent pipeline at Dounreay. It would run out into the sea to dispose of low-activity waste. The concept was just like that at Windscale, but the coastline off the north

Destination Dounreay? A December 1955 party on a fogbound runway in Aberdeen

coast of Scotland was quite different: sand in Cumberland, jagged rocks off Caithness. Peter conceived the idea of investigating the dispersal of effluent by bringing some fission products up to Scotland from Windscale. When dribbled into the sea, very low concentrations of radioactive tracer could be measured with great accuracy, and Peter thought that the dispersion of the different waste products could be followed individually.

This job, technically a great deal more difficult than it may sound, involved a team from Windscale [and] a few hundred curies of mixed

Christopher Hinton at Dounreay
Courtesy Dounreay Site Restoration Ltd and NDA

fission products. They were heavily shielded in lead, and getting the package to Dounreay was no easy business. This lethal brew then had to be fed very slowly into the sea while samples of rocks, fish, seaweed and so on were collected.

It turned out that the limiting factor in the migration of fission products was the concentration of certain radioactive elements in the foam blown ashore by northerly winds:

After damning my eyes for an impractical academic, Hinton decided to build a tunnel, about the diameter of [London's] Northern Line tube, out below sea level to deep water. His eyes sparkled with enthusiasm and I could sense the spirit of Brunel in the room as he developed the idea. It may seem funny to those who were not among my colleagues, but the only reason why the original experiment was done was because I met Hinton in the lavatory one day and was able to put the idea to him. 'Exactly the sort of thing we should be doing,' he responded.

During his employment with UKAEA Peter had found time, especially during the earlier days when he was based at Risley, to visit Cambridge and fulfil duties towards his son as far as he was able and time allowed.

His professional priorities and ambitions had changed during his time at UKAEA. In early 1956 he got a call from Dudley Newitt, who occupied the Courtaulds chair of chemical engineering at Imperial College in London. Newitt was nearing retirement and the message between the lines was that Peter might eventually take Newitt's place. Initially the position of 'reader' was offered, but money was a sticking point. Thanks to UKAEA's enlightened attitude to pay, Peter was already earning much more than any reader would command. Newitt took the point. He spoke to Robert Maxwell and Pergamon Press—who valued Peter's performance on the editorial team of their journal *Chemical Engineering Science*—and the result was that Imperial College offered to fund a new professorship 'and in other ways increase the inducement'. The title of the job—professor of chemical engineering science—showed the direction in which Newitt wanted the subject to grow. Peter looked back later on his time with UKAEA as 'two convulsive years' and found the offer a 'merciful release'. He accepted.

BACKGROUNDER:
A BABY-SITTER IN KITZBÜHEL
*An atomic-energy-fuelled skiing fantasy
by Peter Danckwerts*

One Eastertide in the mid-1950s I
was sitting transfixed by London in a
smoke-filled committee-room, doodling
in the margin of my agenda. My mind
was far away, in the alpine heights and
valleys of Austria beneath a cloudless
sky. Fate had been unkind, as usual;
I had arranged to go skiing with some
friends but my boss fell ill at the last
moment and I had to stand in for him
for ten days. Meanwhile my friends had
come home and I was going out there
all alone. I was worried about my *après-
ski*—who was I to drink and dance with
after the dangers and exertions of each
day's skiing?

Eventually I arrived in Kitzbühel after
midnight and I had to wake the porter
of the hotel, who gave me a room in the
Annexe. The next day I limbered up and
then I bathed and changed and went to
look up a girl to whom I had been given
an introduction. She was a lumpish and
governessy English girl, who asked me
whether my ski boots fitted. I went to
bed very disconsolate.

The next day I did a lot more skiing.
I came down to my solitary table in the
hotel dining-room, wondering where
I could spend the rest of the evening.
A few minutes later a blonde came
in alone and was shown to another
solitary table on the far side of the
room. She was no Marilyn Monroe, but
she looked good enough to eat. We each

spun out our meals until no-one but
ourselves was left, except for a waiter
who wanted to clear the tables. I was
a shy young man, but I felt so desolate
that I called over to her in German to
ask whether she was alone and whether
she would like to go dancing. The
answer to both questions was 'yes'
in Central-European English.

Goodness me, how that girl could
dance! I had only to put my right arm
round her waist and take her right hand
in my left, to slip out on to the floor for
a few bars and know that as dancing
partners we were sublime. It is difficult
to convey to modern-style dancers what
it used to be about—the body-language,
the electricity, the willing but splendid
submission of the woman to the man,
the fluidity and wittiness of the steps.
Was it all really about sex? Oh no, I
don't think so; I have had the greatest
of pleasure in dancing with very plain
girls with whom I would not otherwise
have passed the time of day and whom
I happily left on their Mummys'
doorsteps afterwards.

But my blonde was not like that. In
the small hours we eventually gave up
and skidded precariously back to our
hotel along the ice-rutted main street
of Kitzbühel, flushed, giggling and
exhausted. I kissed her an ardent good-
night but she put me firmly aside and
retired to her room which, curiously,
was next to mine.

Next morning I was woken up, late
and hung-over, by the chamber-maid
with a message from *die Dame* next
door. Would I take her out skiing?

Peter and babysitter in Kitzbühel's Café Praxmair

Eventually we went off together. I must say that her skiing was not as good as her dancing, but she kept up with me—she just fell down more often than I did, and I waited for her. That evening we dined at the same table—to the disapproval of the elderly Austrian head-waiter—and off we went for another evening of swooping and gliding in each other's arms. This time she didn't turn me away from her room when we returned to the hotel. In fact for the rest of our stay I was seldom out of her sight—day or night. She did, however, have some friends in another hotel whom she used to visit occasionally but whom I never met.

The weather was beautiful, with constant sunshine and a cloudless dark-blue alpine sky. There were all the ingredients of a perfect holiday of the sort always held out by the package tour dealers but so seldom realised. Eventually it came to an end, and the time finally came to say farewell. A few tears were shed, we exchanged addresses and I never heard from her again.

I was very naïve at that time. It was not until several years later that other events led me to think about that holiday. After all, I had been a fairly senior scientist in a very sensitive government enterprise. I had been warned that when I was abroad I should not disclose the nature of my employment. Well, my blonde had never showed the slightest interest in what I did. But what about the uncanny coincidence of our neighbouring rooms and our solitary tables; the mutual passion for dancing and much else? Who were the mysterious 'friends' to whom she reported?

Very rum indeed, I thought. I had obviously been 'fitted-up' but by whom? Was it Them or Us? For a time I liked to think it was Them, and only my personal charm had deflected her from her mission to worm secrets out of me. But later on common sense took over and I decided that she must have been one of Us. After all, as a hard-drinking bachelor with a roving eye, what a security risk (or target) I must have been. At that time I was supposed to know (although I never could remember it) the output of plutonium in the UK. I had been provided with a baby-sitter in Kitzbühel.

Living the life Imperial 25

L ONDON'S IMPERIAL COLLEGE OF SCIENCE AND TECHNOLOGY, into
which Peter parachuted in October 1956, is a spin-off from the Great
Exhibition of 1851. The whole complex of colleges and museums
in South Kensington, not to mention the Albert Memorial, were founded
on the handsome profits of that unique event. Peter recalled on arrival that
Prince Albert, a century earlier, had spotted the symptoms of the English
disease—a lack of interest and preparation in things technical and
scientific. The prince, Peter mused, wanted to bring British facilities for
teaching technical and scientific skills up 'to the level enjoyed by any small
German princedom of the time'. In 1956 the department of chemical technology and chemical engineering within Imperial College was a veritable
rabbit-warren of sub-departments under the overall control of a single
professor: the legendary Dudley Newitt. Presiding over this disparate
grouping, involving activities described as technology as well as
engineering, Newitt was Courtaulds professor of chemical engineering.
Peter himself became the Pergamon professor of chemical engineering
science.

Everything about this South Kensington establishment seemed complex:
Imperial College itself was both part of the University of London and had
quasi-autonomy, so there was twice the usual number of controlling
administrative committees to cope with. And Newitt certainly knew how
to cope. He may have presented a bumbling appearance to the world but:

He knew how to work the University machine, which in London requires real
finesse. [His] combination of diffidence and old-fashioned politeness sometimes led
people to underestimate his capacity for getting his way—possibly to his advantage.
The steel in his character showed when he had to deal with the clash of claims and
personalities which arose from the peculiar and ramshackle nature of his department.

Peter noticed that whenever Newitt joined a conversation, informal or
professional, he seemed to begin with a hesitant 'Er...' but what he then
had to say was to the point and well worth waiting for. Newitt was

Peter's kind of man. They had a shared interest in guns and explosions: what Peter called 'a schoolboyish enthusiasm'. Newitt's father was a ballistics engineer and had shot in the 1908 Olympics. Newitt himself had spent WW2 as director of the scientific research service which provided back-up for the Special Operations Executive (SOE). Churchill had charged the SOE to 'set Europe alight' by means of non-military skul-duggery. It operated out of an address in Baker Street, and its members were known as Baker Street Irregulars, because their business had some-thing in common with the group of street boys who carried out spying tasks for Sherlock Holmes. The small plaque on the wall outside their headquarters read opaquely: 'Inter-Services Research Office'.

The SOE position gave Newitt the opportunity to devise booby traps and demolition devices. He even got involved in ways of disabling time-bombs, something of which Peter could also claim experience. But Newitt drew the line at parachutes. Required by his superiors to take out a small refinery on the edges of the German wartime sphere of influence, he refused point blank to be parachuted in. He went instead on foot. Another of Newitt's characteristics inspired Peter's admiration:

He had what I regard as the premier qualification for a scientist, namely a distaste for administrative matters. As far as I could make out, all the papers relating to departmental finances were kept in the top right-hand drawer of his desk. If a question arose he used to fumble around in it and bring out a piece of paper—usually the relevant one.

A soldier in the Middle East during WW1, Newitt had commanded Sikh regiments in Mesopotamia and became very attached to all countries east of Suez. Not that he returned to them often: 'We had to shoot a few people who were attacking our troops, and they've got long memories in those parts,' he would say by way of explanation. A keen fisherman, he enjoyed a day's relaxation on the river bank, and is said to have landed the largest fish ever caught in the Euphrates with rod and line. When he visited India, fishing was rarely quiet relaxation: former

Dudley Newitt, the senior professor in Peter's area at Imperial College
Courtesy Malyn Newitt

students would line the banks to applaud every catch. As a consequence of this mutual affection, a great many diligent Indian students could be found working long hours in the Imperial College laboratories. It was an attraction that rubbed off on Peter.

Newitt's engineering interests were wide but he was particularly known for work on reactions at high pressure. He had been in Germany before 1914, where Fritz Haber and Carl Bosch were busy devising an innovative high-pressure process to make ammonia out of hydrogen and nitrogen. In Peter's view, Newitt's physical insight was unerring in activities as diverse as drying porous materials, conveying particles from one place to another in a flow of air, and making powders into pellets. He was also interested in the thermodynamic properties of steam and of refrigerants. Peter found that Newitt had an idea, well before its time, that insecticides should be sprayed on crops as bubbles rather than as drops. Bubbles would hit the lower surfaces of leaves as well as the upper ones and so make the insecticide more effective. For a related project, in which he wanted to picture better how fluids might flow around obstacles...

...[Newitt] sent some students with suitcases to stand still in busy passages of the Underground. The resulting disturbance of flow could be observed from above (until he was asked by the management to desist).

A great gatherer of industrial support, Newitt insisted that Peter should buy a wind tunnel. Seeing the purchase as inevitable, Peter had enough feel for the subject to draw up specifications and check the design. He wrote that finding practical uses for the tunnel was more of a challenge but it 'turned out quite successfully'. Newitt's leadership ensured that relations between his five sub-departments—fuel technology, combustion (which included process safety), thermo-dynamcs, low temperature work and chemical engineering—were, on the whole, cordial.

Another facet of Newitt with which Peter inevitably sympathised was his interest in sailing. Competition for the America's Cup was resumed in 1958 after a pause for WW2. Yachts in the 12-metre class took part; in the run-up to the main event the British challenger *Sceptre* designed by David Boyd lost to the US entry *Columbia*. Newitt proposed that together they should buy the loser. Peter felt he must have been unconsciously signalling indications of wealth. 'She'll be going for a song,' Newitt told him. 'Best materials. Shorten the mast and turn her into a cruiser.' When this came to nothing, Newitt had the idea of building a yacht in the Mediterranean and—a proposition instantly recognisable by anyone

familiar with the habits of ageing despotic professors of the 1950s—
manning it during the summer months with relays of students.

A gourmet and wine connoisseur, Newitt frequently took Peter across
Queen's Gate to the Gore Hotel, where the chef knew what he was about.
As a conversationalist at the table, Newitt could draw on his very wide
experience to cap any story. If someone were bold enough to tell a tale
from WW1 in the Middle East, Newitt might respond: 'When I was sent
to arrest Lawrence of Arabia...' Peter knew from experience that company
like this was best enjoyed over a good meal. Newitt's philosophy, as he told
Peter several times, was 'It's not money or titles that count, but privilege'.

Newitt was the senior professor and although Peter had understood
from the start that he would eventually inherit, he found that there was
competition. Belgian-born Alfred Ubbelohde, resident professor of the sub-
department of thermodynamics since 1954, had something of a reputation
as an empire-builder. His eye was certainly focused on the Courtaulds
professorship. One of his professional interests was in the philosophical

---

BACKGROUNDER:
FORMS OF ADDRESS: 1

Peter Danckwerts had been brought up
during the 1920s and 1930s in a world
of fairly stuffy formality. Even in the
1950s people who worked quite closely
together addressed each other by their
surnames. He himself was only
addressed as 'Peter' by close family.
People on more or less the same level in
society might at first address him as
'Professor Danckwerts'. As they got to
know him better, it would soften to
'Danckwerts'. Those in more junior
positions and yet in day-to-day contact
used the handle 'PVD' between
themselves, although not to his face.
Little did they know of its origin in
the WW1 paravane development and
minesweeping at sea. PVD effectively
took the place of a given name. In
1960s came a thaw. When Peter left
Imperial College in 1959 he was

replaced by his old Cambridge friend
Kenneth Denbigh. Denbigh inherited
Peter's secretary Emma Maccall.
Returning from a trip to the United
States in the mid-1960s, Denbigh told
his secretary that 'everyone over there
uses first names. What's yours?' 'Eileen',
she replied, 'but I don't like it.' Graeme
Jameson, a new young Australian
member of staff, came to the rescue.
He started calling her 'M' after Miss
Moneypenny's boss in the James Bond
films, then all the rage. It was but a
short step from 'M' to 'Emma'.
(Jameson had the Midas touch in other
ways: 300 of his eponymous flotation
cells—for recovering the finer particles
of coal and other valuable mined
minerals which might otherwise be lost
during washing—operate around the
world. Those in Australia currently
provide his country with some
$2.4 billion a year in export income.)

implications of thermodynamics. Take entropy, for example: it is a measure of order and, despite efforts by some of mankind to reverse it, disorder is growing. Is the universe therefore gradually running down and what are the implications? And so on. In his spare time Ubbelohde collected Chinese pottery and jade.

Inevitably known as 'Ubble Bubble' by the students, Ubbelohde held court at Imperial with his secretary Georgina Greene, a busty blonde with famously blue eyes. As members of staff knew only too well, she wielded considerable influence. Arriving in her office to try and catch a moment with the professor, a lecturer might find himself taken to task for addressing her with his hands in his pockets. In Peter's day Ubbelohde kept a sky-blue Rolls Royce. Department members were used to overhearing exchanges like:

Ubbelohde: 'I think I'll use the Rolls today.'

Greene: 'You can't, I need it.'

Greene sat in on her boss's conversations, even confidential ones with Newitt. So she would learn about proposed developments, like possible changes in personnel, before they had been announced to the world, and that could cause friction.

In 1958 Peter took on a new secretary. Emma Maccall was a glamorous and lively young woman; she caused quite a stir amongst the students. One of those stirred at the time admitted that most students had a crush on her. It wasn't just the looks. She had a warm and welcoming character, could always find time for people she liked—that is, most people—and knew how to get along with those she didn't like so much. And she was always good for a bit of gossip. Maccall found Peter's style attractive and assessed him as 'a delight to work for, a kindly man with a great sense of humour and considerate, although slightly impersonal'. She particularly admired his ability with language: Peter dictated letters and other short items in a single take and they were, Maccall appreciated, models of clarity.

What of Peter's other professional colleagues? Roger Sargent arrived in London in May 1958 from Paris where he had worked for Air Liquide in a seventh-floor office on Quai D'Orsay. The two men overlapped in London by only a year, but it was a valuable time for both of them. Sargent found that Peter's arrival at Imperial College was a breath of fresh air. Peter visited European departments, knew people in Europe and invited them to visit London. One European contemporary, Hans Kramers—professor at Delft University of Technology, another beneficiary of the

Shell company's investment in engineering—had been in frequent contact with Peter since 1948 and later described him as his intellectual 'sparring partner par excellence'. They had, he said, tried in a friendly way to out-do each other. As Sargent saw it, 'Peter opened up the department to the world.' For his part, Peter urged Sargent to visit industrial companies during the vacations to get a better view of what industry was up to and what its needs were. He passed on some names at ICI Billingham. The relationship blossomed.

Then there was the older guard—Jack Burgoyne, John Coulson and Jack Richardson. Burgoyne was reader in process safety. During WW2 he had been high up in civil defence and got interested in studying industrial accidents scientifically. When hostilities ceased he began consulting with organisations and companies on accidents and the identification of hazards (before they could turn into accidents). He found that he enjoyed it more than teaching and established his own consultancy, J H Burgoyne and Partners. It became the obvious consultancy firm for industry to call in on safety matters. He had left Imperial College to go full time as a consultant in 1964.

Coulson had joined the staff in the 1930s, and after the war had run the chemical engineering course virtually single-handed. Richardson had initially worked on safety with Burgoyne, but switched to work with Coulson. They taught the chemical engineering course together and in 1954 published the fruit of their collaboration: the eponymous standard textbook, still familiar to every student of the discipline. It was the perfect companion to the Imperial College chemical engineering course and in later years grew to six volumes and several editions.

Coulson had departed for Newcastle in 1954 and when Peter arrived he discovered Richardson somewhat miffed at being passed over for leadership of the sub-department. All the same, they got on well together; any discontent was directed not at Peter but at Newitt. Peter described Newitt as a 'cunning old fox'; Richardson assessed Peter as 'a funny bloke but pleasant, amenable and very able'; Peter acquired the occasional nickname of 'Hurly-burly' for his habit, when agitation blew up in the department over something non-scientific, of commenting 'I think I will disappear until the hurly-burly blows over'.

In 1958 Richardson resigned from the Imperial College staff and went to work for Boake Roberts in Stratford, east London, returning only to mentor his remaining research students. Then he accepted a professorship at the University of Swansea and spent the rest of his career there.

One of Peter's objectives at Imperial was to bring the chemical engineering course up to date. He spelt out his aims in the new professor's traditional 'inaugural lecture' presented more than a year after arriving, presumably to let him get his feet under the table, on 12 November 1957. To start with, no ivory tower approach:

The university department has no reason for existence at all except in relation to industry. Our job is to serve industry, but without losing our independence. Whether in teaching or research, we are bound to be influenced by the views of our opposite numbers in industry—if it were not so, it would mean that we had sealed ourselves up in an ivory tower; but it is we who must finally decide how we can best make our contribution.

And he was unapologetic about teaching science as but one element of engineering; students, he felt, could only develop the other two elements—experience and judgement—later on. The undergraduate course should do what it could to instil the engineering outlook and emphasise that no engineering problem can be solved by just applying science:

We underline the importance of using intelligent approximations and of making the best possible estimate from incomplete information. Satisfactory plant and processes are often arrived at by intelligent empiricism which outpaces scientific development.

He was happy to acknowledge the role of practical common-sense and initiative in keeping the wheels of industry turning, but...

...any British industry which is not content to remain technically stationary, or to import all its new ideas from foreign countries, is going to need engineers with original minds and scientific educations. This means that we must try to attract more of the first-class minds which at present turn to the study of pure science or the humanities, and we must offer them a course worthy of their metal.

Peter had an audience for this major lecture from a wide range of disciplines and it was a chance to describe to them what kind of animal a chemical engineer was. He explained the defining characteristics as against those of any other kind of engineer. In industry a chemical engineer might be designing a plant, in charge of running it or beavering away at research and development. The job is about processes which achieve change in the properties of matter on a large scale.

It would, in fact, be more appropriate to call ourselves 'process engineers', as is done in Germany, but it is too late for that now. Many of the processes are quite simple in principle. Some are commonplace household operations, and distillation is one of the oldest and most congenial of technological processes. The chemical engineer's task, however, is to do for a penny what any fool can do for tuppence.

The time-honoured British way of designing a chemical plant was by collaboration of chemists with plant design engineers—a recipe for a scaled-up beaker:

[This] is adequate for the design of plants that handle small amounts of material or produce a small number of different materials such as dyes and pharmaceuticals; it was quite inappropriate for designing the high-output, continuously-operating plants typical of the modern chemical and petroleum industries.

The design of a process must get things done as cheaply as possible. Ignorance usually has the opposite effect; it results in large safety factors, leading to a bigger plant with unnecessarily inflated capital costs.

Inevitably there was advice to both students and staff, straight out of the MIT practice school textbook, on getting practical experience of industry during the vacations. Industrial staff simply didn't have the time to supervise, so university staff had to be there and act as tutors, selecting projects which tested what the student had learned at university, illustrated its usefulness and showed a real plant in operation.

BACKGROUNDER:
RESEARCH IN A TECHNOLOGY COURSE
Peter's November 1957 inaugural lecture as professor included an open-eyed review of what research offered in a technology course. As devil's advocate, and recalling Terence Fox, he saw that excellent teaching could be done without any research programme at all. And he certainly didn't want to see all the top engineering brains 'gravitating automatically to the back room in industry' when they were needed throughout the business.

Still, choosing a research topic presented a challenge. It was common to develop a model for an industrial challenge, test it laboriously over a range of conditions and arrive at a formula.

Only too often the model in fact behaves quite differently from the full-scale equipment because the principles of scaling up have not been observed; under these circumstances the formula is of little value and the research only serves, in the words of Lucky Jim, to throw pseudo-light on non-problems.

Peter felt that university research, without any asphyxiating commercial demands, should strive for a deeper insight into the mechanism of industrial processes 'by looking at them with a scientist's eye'.

One of the broad objectives of research ought to be to enable process and plant designers to achieve more in their design on paper with greater confidence. A bridge designer can do all his work by calculation with reasonable confidence in the results. The designer of a novel type of chemical or process plant simply does not have at his disposal sufficient data or design procedures.

Peter had other ambitions not touched on in his inaugural address. They included adding to the undergraduate course a module on materials science, then a fashionable subject and something Peter had experienced at first hand in Dounreay. He taught the course himself with enthusiasm. It gave students more insight into the choice of materials available for process equipment, like vessels and pipework, when they designed chemical plants. He also oversaw the upgrading of that part of the course which dealt with the analysis of a process, so that students might choose with more insight the operations that made it up. And after his experience in UKAEA, he wanted to make sure that students understood the nuclear industry. He himself kept in touch with industrial visits and he oversaw a wide variety of research topics.

During his time at UKAEA Peter had maintained his contact with the Pergamon Press journal *Chemical Engineering Science*. No longer an academic, he had been relegated from the editorial board to its advisory board. Now back in the university world, he could be reinstated as an editor. One major advantage of Robert Maxwell's journal was the speed with which it published submitted work. Its ability to outstrip the journals of the professional associations proved very popular with

Peter in Yugoslavia (August 1959) visiting the Boris Kidrič Institute of Nuclear Sciences, now the Vinča Institute in Serbia

contributors who invariably wanted to see their efforts in front of their peers as soon as possible. The journal was providing publicity for the research work of chemical engineering scientists worldwide. By 1957, volume 7 of the journal had changed its subtitle appropriately to *Le Jounal International de Génie Chemique* (an addition which persisted until the end of 1983 by which time French had lost the battle to match English as the international language of science). The editorial board had by now become 'editors', suggesting they had rolled their sleeves up.

Then at a 1958 meeting at Pergamon's headquarters in Fitzroy Square, London, chaired by Maxwell, a vote of all the editors present made Peter executive editor from volume 9 onwards. Over the next 24 years he was to drive the content of the publication so that it became one of Pergamon's most successful journals. Each geographical area was equipped with its own editor and Peter encouraged them all to operate in a sovereign fashion, taking full control of decision-making. He himself certainly retained a role as the final arbiter on standards, but all his editors knew what was expected of them. Peter's style meant that the editorial board met irregularly and was often incomplete; his approach to minutes and filing meant that records for his period at the helm are scarce. But the main objectives were achieved: quality and relevance in what was published. Meetings of the board were often held at Maxwell's Oxford headquarters, Headington Hill Hall, where the host reputedly provided sumptuous lunches. But Peter's view remained that any meeting lasting longer than two hours was not productive.

Over the years Robert Maxwell's reputation deteriorated. Peter began to worry whether he could trust Maxwell. Perhaps *Chemical Engineering Science* would be sold off, or Maxwell might desert it in some way. But in spite of his doubts, Peter was generous. He wrote in 1982:

This success story has depended in part on the vigilance of successive editors in various regions excluding rubbish, but to a large extent on the efficiency of the publication and production process, for which Bob [Maxwell] must take the credit.

Playing with fire 26

WHEN HE FIRST ARRIVED AT IMPERIAL COLLEGE, Peter rented a flat in Ennismore Gardens and by February 1957 he was already in residence. This neighbourhood of elegant mansion blocks and mews was known as a very good address. 'Of course', Peter wrote, 'I lived in the basement.' In contrast contemporaries remember Peter's flat as being above ground level and rather elegant. Just to the east of Exhibition Road and only five minutes from the office, it offered an opportunity for the life of a 'bachelor gay' in the old-fashioned sense. At the same time, Peter kept up his supportive visits to Eileen Mitchell, mother of his son Jonnie. In May 1957 Peter turned up in Cambridge in his Ford Zodiac. Eileen, her elder son Jeremy and little Jonnie jumped aboard and Peter took them to the RAF Waterbeach airshow, just north of Cambridge. The star of the show that year was the prototype English Electric P1 (Lightning) fighter, then under development. It had two engines which functioned like rockets and could propel the aircraft vertically up-wards at amazing speed. As they watched, Peter explained how it worked.

From his base at Ennismore Gardens Peter was able, in his spare time, to renew his interests in the arts and in ballroom dancing. It was on an evening out to the ballroom in early 1957 that, through mutual friends, he made the acquaintance of a glamorous new lady. Peter undoubtedly sensed that Lavinia Harrison was a genuine member of the aristocracy. Lavinia's mother, Edyth Lavinia Bingham, the daughter of a rear-admiral, was the niece of the fourth earl of Lucan (the Lord Lucan who disappeared in 1974 was the seventh). Lavinia's father, Duncan Alwyn Macfarlane, a military man, had risen to the rank of Brigadier-General with the King's Own Scottish Borderers. He retired in 1913, married Edyth who was 28 years his junior and the couple lived at Dunain Park House, a mansion two miles south-west of Inverness on the road to Loch Ness. The Macfarlanes were significant players in local society: the Brigadier became Colonel-in-chief of his old regiment from 1928 to 1938, and read the lesson of a

Sunday in Inverness cathedral. From his characteristic military tread, the congregation could tell with their eyes shut who was approaching the lectern. If he made a mistake, his training would kick in: he would rap out 'As you were' and start the sentence again.

Duncan Macfarlane
Courtesy Charlie Macfarlane and
Glenfinnan House Hotel

Edyth gave birth to their first and only child, Lavinia Anne on 4 February 1923. Her husband's diary entry reads: 'At 11.25pm a baby daughter arrived. A fine child with dark hair and good lungs. A dear wee thing with well-marked eyebrows. Poor Edie exhausted but well. And so to bed with great thankfulness.' A gentle and kindly old man, he doted on his daughter and she loved him. In contrast, Edyth seems to have regarded Lavinia as an inconvenience. She had enjoyed a Victorian upbringing with a stern governess and assigned her daughter to the same woman. As she grew up Lavinia found some solace in local friends: Marjorie Close, her cousin, and Anne Wallace from the nearby village of Corriemoney, became lifelong friends. Lavinia attended a local school in Dochgarroch but her mother found the local accent undesirable and sent her instead into Inverness to a 'ladies school'. Finally Edyth

Edyth Macfarlane
Courtesy Charlie Macfarlane and
Glenfinnan House Hotel

banished her daughter to St George's in Ascot, Surrey. Once St George's had been a mixed school with Winston Churchill amongst its less happy pupils. By the 1920s it was a finishing school for girls. Scottish surnames were quite rare and she found herself nicknamed 'Biscuits' after the Glasgow manufacturer Macfarlane Lang. Lavinia's time in Surrey left a legacy: a reasonable all-round education and wonderful manners. She had an aristocratic drawl and used it to effortless effect, while feeling privately that it could alienate her from those around her.

As WW2 approached Duncan Macfarlane was failing fast; he died on 9 April 1941. Lavinia, now a strikingly good-looking young woman, oozing laid-back charm and good breeding, was now free to play a part in the war effort. Lavinia wanted to join the ATS (the women's branch of the army) but her mother found it socially unacceptable. As the daughter of a rear-admiral, she suggested the Wrens (WRNS) to her daughter. Lavinia went to sign up one morning but found the staff never seemed to be available. So she joined the Women's Auxiliary Air Force (WAAF) instead, qualifying as a driver. The servicing and reassembling of engines was not her forte, but her examiner helped her through—'don't you think that bit might go here' and so on. She drove wonderfully—foot off the gas approaching a roundabout then strong acceleration into and out of it. Post-war, bad drivers were an irritation and she was inclined to snap: 'Get to hell out of the way'.

Her first posting was to RAF Valley on Anglesey in 1942 as a driving instructor. This was a part of the world popular with rich Liverpudlians and although snobbery was intense, socially-levelling influences—the Cheshire regiment and American servicemen—had arrived in the area. Rationing tended to collapse in the face of ready money, and it was a giddy environment for an unattached girl. Lavinia made friends with the daughter of a local family, Jane Plunkett from Trearddur Bay just up the coast from RAF Valley. Plunkett was divorced and often returned home to Trearddur Bay. She used to sail with Tony Harrison, son of another local family. The two of them frequented a dance hall just out of town: a Nissan hut, curved in shape and known locally as 'The Slug'. Lavinia went along too.

Scarcely twenty, the glamorous Lavinia made a big impact on Tony Harrison. His family had money; Harrison's grandfather and his father Frederick were Liverpool merchants. On his mother's side Harrison came from a branch of the Blessig family, once Alsatian bankers. Tony Harrison, urged on by his family and struck by Lavinia's charms, used all his considerable powers of persuasion to secure her. He was a Wykehamist two years younger than Peter. An army captain, he won a Military Cross in September 1944 for his outstanding conduct during the airborne landings at Arnhem.

That same year he and Lavinia married; Lavinia's aunt Emmie Close in Oxford organised the wedding—Lavinia only got married in the best university cities. Three sons appeared in short order: Michael in 1945, Brian in 1947 and Charles in 1951. But with Lavinia landed, Harrison seemed to lose interest. The relationship foundered. By all accounts he was obsessive and given to gambling. Back in civvy street Harrison would leave home in Hampstead, dressed for a day at Lloyd's, but spend it and a good

deal of money at Crockford's gaming house. He and Lavinia quarrelled
bitterly from the start. Lavinia's Scottish roots were important to her; her
friend Jane Plunkett's daughter Patsy recalls the three Harrison children
as 'enchanting small boys with minute kilts which they wore a lot'. But
Harrison was dismissive of 'all that Scottish mumbo jumbo'. His company
became agonising for Lavinia; she invited a faithful circle of girl friends
to spend time with her.

By the mid-fifties, Lavinia was in the throes of an agonising separation
and divorce. Tony Harrison accused her of desertion and tried to show
that she was financially and physically incapable of looking after their
children. She lived in the family home, a Kensington flat on the corner of
Sloane Street with Hans Crescent. Her husband had other interests and
retreated to his cottage in Christmas Common, Oxfordshire. Lavinia,
whom Patsy Plunkett judged to have 'model girl proportions' and to
'radiate sex appeal from every pore without the slightest effort', made
Sloane Street builders whistle with pleasure as she went past. She took me-
nial jobs to make family ends meet and, in spite of efforts by friends to
keep her spirits up, on that fateful day in 1957 Lavinia was feeling a bit
down. 'But we've found just the man for you', her friends assured her.
They were ready to take her off for an evening on the dance floor. 'Oh
God, the last thing I need is another man', Lavinia groaned. But she
enjoyed dancing and didn't resist for long. And so it was that she met a
fellow ballroom dancing enthusiast—Peter Danckwerts.

After that first evening Peter began to drop in on Lavinia at the
Kensington flat. It was a dangerous game; Tony Harrison was a formidable opponent and had he found out that Lavinia was interested in
another male, he would undoubtedly have exploited it. Peter involved himself in resolving the situation, helping Lavinia deal with legal aspects of
divorce. On his travels from Imperial College, mostly to industry and
academia in mainland Europe, he kept up a flow of laconic and enourag-
ing postcards. And in that summer of 1957 he even took her away for a
clandestine holiday on the Costa Brava while her boys were safely at
boarding school. Peter's sister Hilary Gatehouse agreed to go along with
her two children Jane and Michael as 'gooseberries'.

Peter and Lavinia arrived in Peter's recently acquired Humber Hawk to
collect them. There was a short hiatus when Hilary's husband came to
terms with the fact that the Hawk's tyres were bald and gave Peter a
lecture on a driver's responsibilities. But the party got away, crossed to

BACKGROUNDER:
COURTSHIP BY POSTCARD

Peter found postcards fun. He sent Lavinia quite a lot, usually laconically annotated, and especially during the first year of their acquaintance. On 17 April 1957 he was in Amsterdam, probably visiting Shell, and sent her a postcard reproducing an advertising poster from the 1930s promoting Wynand Fockink, purveyors of 'Liqueurs, Advocaat, Genever & Dry Gin'. The card features 'n Glaasje Half & Half, one of the company's most popular liqueurs. (In 1842 a junior member of staff accidentally mixed orange liqueur with bitters. It almost cost him his job but he was allowed to stay on when his boss sampled the resulting mixture and declared the taste 'perfect'. Fifty percent orange with cinnamon, cloves, a pinch of aniseed and the bitter herbs galangal and gentian root, Half & Half is half sweet and half bitter.) On the reverse Peter wrote: 'From your wandering boy'. Another card, again from Amsterdam and dated only three weeks later, was a bit more direct. It shows two charming children, a heart-shaped apple and the commentary:

Wat een mooie appel is dat,
daarvan geel ik jou ook wat
(What a lovely apple this is,
I can let you have some of it.)

Peter's message on the back seems to have been an attempt at diffidence: 'Can't think what this means. P'

'n Glaasje Half & Half bij Wynand Fockink, Amsterdam

Courting Lavinia 'from your wandering boy',
April 1957

Wat een mooie appel is dat,
daarvan geel ik jou ook wat.

Courting Lavinia and getting more serious,
May 1957

Dunkirk and progressed south. At this stage Peter decided that his quintessentially British car really did need new tyres. He eventually found some in the less-than-ideal surroundings of France. The party made stops overnight in Châteauroux and Carcassonne before reaching Caldes d'Etrac on the Costa Brava, midway between Barcelona and Tossa de Mar. Michael Gatehouse, then 11, recalls how Lavinia played the 'helpless woman' on the trip, getting Peter to do everything for her. In France, as the two sat flirting over Pernod, the ceremony of letting liquor drain through sugar fascinated the chidren. Lavinia seemed, to the pre-pubescent

Ulric Huggins in WW2
Courtesy Janie Huggins

Michael, to be elegant, long-legged and languid. On the beach he saw the vivid purple scars on Peter's lower legs left by his Sicilian accident. They didn't seem to slow him down, though: one day when Michael unexpectedly found the Costa Brava currents too strong, Peter dived in and pulled him out to safety. It wasn't their first encounter in these roles: years earlier at Emsworth sailing club Peter had rescued Michael from self-immolation in the deep end of the club's sea-water swimming pool.

Peter kept his private life just that, but his secretary Emma Maccall got some insight from time to time. She recalls occasional visits to the office by his friend Ulric Huggins who had been an RNVR gunnery officer during WW2. Huggins had become a business efficiency consultant and worked with his colleague Lyn Urwick in management consultancy. In future years Peter and Lavinia dropped in at the Huggins' Sussex home when the opportunity arose. Meanwhile, Peter's wartime chum Tom Gaskell had become senior physicist at BP Sunbury and was making a name for himself appearing on TV and contributing to *New Scientist*. They enjoyed chatting about old times and on Christmas Eve 1957 Peter broke new ground by taking Lavinia to Gaskell's flat for drinks. Emma Maccall met Peter's new girlfriend for the first time at a drinks party in his Ennismore Gardens flat in 1958. Her impression was also of 'an elegant creature', although she couldn't help noticing signs of the wear and tear of domestic life and of the jobs Lavinia had been forced to take to keep herself and her three boys fed. Peter carefully avoided introducing Lavinia to any of the guests, as she was still 'Mrs Harrison' and in the late 1950s it was socially awkward for them both.

In 1958, with Lavinia's divorce coming through, light appeared at the end of the social tunnel. But Lavinia realised that she wasn't the only woman in Peter's sights. She told her friend Anne Wallace, now Crole, that she was absolutely determined to 'see the others off and get her man'. She had fallen for Peter and he, despite his eclectic interest in women, for her. Emma Maccall caught him at his desk doodling 'Lavinia' with his Waterman's fountain pen on the blotting pad. Then, when her divorce came through, Lavinia bought a house at 83 Earl's Court Road, just down the road from Holland Park. Peter had his own room there.

In early 1959 news reached Imperial College that failing health had forced Terence Fox to resign the Shell professorship in Cambridge. The Cambridge chemical engineering syndicate wrote to Peter, inviting him to take Fox's place. In deciding what to do, not for the first time Peter felt the hand of Sir Harold Hartley, the *éminence grise* behind British science and engineering. Beginning as an academic, Hartley had moved into industry and business, and by the early 1950s he had taken up the cause

---

BACKGROUNDER:

ULRIC HUGGINS: LUCKY ESCAPE I

Peter's friend Ulric Huggins was a lucky man. Three years older than Peter, he first cheated death as an eleven-year-old. He and his parents were on a social trip crossing Malta harbour to visit the light cruiser *HMS Calypso* when their launch was hit by the destroyer *HMS Venomous*. A quick-thinking officer saw the collision coming, grabbed Huggins junior and threw him overboard. Carried clear by the bow wave of the *Venomous* and buoyed up by his Sunday best—a kilt—he was plucked from the waters by rescuers. His father, dragged under by the sinking launch,

managed to struggle to the surface. His mother went down with the launch to the harbour floor at 12 fathoms (80 feet). She had been sitting under a sun canopy on the rear deck and found that she was able to breathe in an air bubble caught underneath it. After ten minutes carefully disentangling herself from debris, she took a deep breath and shot to the surface to join the rest of the family. Back in London and looking for a suitable thanksgiving, she noticed many ex-service people and their families homeless and begging in the streets. So she set up the organization now known as Veterans Aid to help them.

of chemical engineering. He proclaimed it the 'fourth primary technology' alongside civil, mechanical and electrical engineering. Hartley was, like Peter, a decorated war hero, albeit from WW1. His views inspired many and not least, it seemed, the entire British establishment; Hartley took the Duke of Edinburgh in hand and taught him the importance of science and technology. In his nineties and in failing health, he still received groups seeking enlightenment and advice at his hospital bedside. And, of course, he was a Balliol man. Peter accepted the Cambridge post.

BACKGROUNDER:
ULRIC HUGGINS: LUCKY ESCAPE 2
During WW2 Huggins was a tall, fair-haired RNVR gunnery specialist. On 8 March 1941 with the Blitz still in full swing, Lieutenant Huggins and his first wife Pat (a Wren) were dining with two friends—a Belgian doctor and an Austrian nurse—at that smartest of London venues: the Café de Paris.

Housed in a basement under the Rialto cinema between Piccadilly and Leicester Square, it was advertised as particularly safe, perhaps even impregnable. The Café de Paris dance floor had been designed, omin-ously, as a replica in miniature of the ballroom of the Titanic. That night

just after 9.30pm, as Ken 'Snakehips' Johnson and his band The West Indian Orchestra were getting stuck into a rendering of 'Oh Johnnie', two 50kg bombs fell through the cinema roof and ended up on the Café de Paris dance floor. One exploded at head height, killing Johnson and perhaps as many as 80 others. The second bomb burst open, spilling its smelly contents. A waiter standing behind Huggins and serving him champagne as the bomb exploded was killed by a shaft of glass in the back; miraculously Huggins and his party survived unscathed. The Belgian doctor provided medical aid while the others brought survivors to him for attention, saving lives in the process. Returning home afterwards by Underground, Huggins and his wife looked at each other: her best frock, hair, face and arms covered in grime; his naval monkey-jacket (the standard-issue jacket with gold rings and two vertical rows of buttons) soaked to the elbows in blood. The author Constantine FitzGibbon interviewed Huggins and recorded his story both in a chapter of his book *London's Burning* and as part of an hour-long BBC radio broadcast *The winter of the bombs*. Huggins retired early to deepest Sussex, devoting himself to fund-raising for muscular dystrophy, hunting and organising horse trials.

The return to Cambridge 27

PETER SUCCEEDED TERENCE FOX as Cambridge's Shell professor in September 1959. He still had research commitments at Imperial College and honoured them by returning once a week for a while. Only one post-graduate student came to Cambridge with him from Imperial College: Ming Leung. Ming had graduated top of his class at Imperial in the summer of 1958. At Peter's invitation he stayed on to work for a doctorate; once in Cambridge, Peter passed on supervision of the project to his staff colleague David Harrison. Ming was kept busy in Cambridge. Besides his 'day job', a project on fluidisation, he shared Peter's enthusiasm for pyrotechnics and helped launch fire balloons of an evening. He also served as a rather specialised babysitter. In spite of his not-entirely-substantiated relationship with Lavinia Harrison, Peter was still a

---

BACKGROUNDER:
MING LEUNG
By the time Ming moved from Imperial College to Cambridge in 1959 he was engaged to a girl called Theresa whom he had met socially; she worked as a nurse at Mount Vernon hospital in Northwood, Middlesex. They found that they both came from the same village—Tai Po Market—in Hong Kong's new territories. The first wife of Theresa's father (he had three) was midwife to the whole village and by an extraordinary coincidence she not only delivered Theresa by her husband's second wife but Ming as well. In the summer of 1959 Theresa's father sent her a ticket to visit home on a ship leaving from Genoa; Ming, who had

just passed his driving test, suggested he drive his fiancée to Italy. But first he had to get the OK from Peter. Ming asked for leave. 'Impossible,' Peter replied. Then he asked: 'What do you want leave for?' Ming explained and Peter, perhaps remembering his own excursion to Verona in 1950, responded: 'Take three weeks.' The couple set off in an elderly Austin A30 belonging to Ming's brother, then studying in Leeds. Immediately they had crossed the Channel the engine failed, but Theresa had her hospital superannuation in her pocket—£60, just enough to pay for repairs. When she got back to Cambridge in the summer of 1960, the couple saved up and married before the end of the year.

magnet for women. In the first
few weeks his social calendar
occasionally got over-booked
and he relied on Ming to take
any surplus females punting on
the river Cam while he sorted
things out. Ming also played a
part in departmental public
relations. When Lord Godber,
then chairman of Shell
Transport and Trading, paid a
visit to the new Cambridge
chemical engineering facilities
in the spring of 1960, Ming
was one of the front-men.

Ming Leung (r) describes his work to Lord Godber (l)
while his assistant and Peter (second r) listen in

Lavinia and her three boys soon joined Peter in Cambridge. At first they
all rented a flat at number 6 Chaucer Road on the south side of the city.
It wasn't the ideal environment for three youngsters, not to mention
Florin, their new Alsatian who found linoleum flooring fickle. Lavinia was
used to having dogs around; there is a story that, in the early 1950s, she
arrived one day at Oxford railway station with a young Boxer. At the
ticket office she asked whether the dog needed a ticket. 'A box of puppies?'
queried the official. 'No, a Boxer puppy,' clarified Lavinia. 'Well then,
you'll need a ticket; a box of puppies would go free.' As it grew larger, the
Boxer would bump into the children and knock them over. So, on her then
husband's insistence, she had to get rid of it.

Florin was a gift to both Lavinia and Peter from Lavinia's close friend
Jane Plunkett and her second husband Chris Drake. Lavinia was hopeless
about canine discipline, but Drake had already trained Florin to a high
standard. At the time that Peter and Lavinia were about to take off for
Cambridge, the Drakes were about to move into London and wondering
what to do with their two Alsatians. They felt they couldn't cope with both
in an unremittingly urban environment, and so found Florin a new home.
Florin looked fierce but, according to Lavinia's second son Brian, was 'soft,
beautiful and very obedient'. Lavinia's boys were delighted to find, for
instance, that Florin would pee to order. And, under protest, he could be
made to do other things not usually associated with Alsatians: like sitting
outside a shop guarding a basket. Peter and Lavinia kept a series of dogs

thereafter, but Florin was the only manageable one.

Meanwhile Pembroke College had offered Peter a fellowship. The proposal was in stark contrast to the isolation Peter had suffered as a lecturer only six years earlier; in the interim, largely due to Terence Fox's efforts, chemical engineering had become intellectually respectable. Even so, according to Peter, Pembroke College was one of the few colleges that took chemical engineering seriously. Professorial fellows had no college teaching duties, so Peter did not become a major figure in college life, but quite soon he settled down to service as chairman of the college wine committee.

Peter's new job came with further help: a secretary already in post. He inherited Margaret Sansom from Terence Fox. This was a huge advantage; Fox was a shy bachelor, and Margaret had acted as his female complement in the department. She became 'the glue that held the department together'. Ideally suited for the job, she was always cheerful and generous with her time. Margaret remained at Peter's side throughout his time at the helm in Cambridge. She helped him communicate with staff and students, got on very well with Lavinia and undertook all manner of extra activities for them. Peter's Imperial College secretary Emma Maccall had daily telephone conversations with Margaret as she adjusted to life with her new boss. The two women found they were on the same wavelength. And one day at Imperial College a member of staff, showing a newcomer from Cambridge around, came into Maccall's office and introduced her as "Imperial's Margaret". Maccall felt wonderfully flattered.

Margaret Sansom in 1953 (l) and in retirement
Photographers: Edward Leigh (l) and Miles Kennedy

Peter kept his marriage plans, like everything else in his life, close to his chest. The date was Friday, 10 June 1960. Several stories have him casually remarking to Margaret Sansom on the day before: 'Don't make me any appointments for tomorrow morning; I'm getting married'. To everyone else, it was a bolt from the blue. The civil ceremony took place in Cambridge registry office; at the time Lavinia, as a divorcée, was not allowed to get married in church, and she deeply regretted it. Peter returned to his office in the afternoon with a couple of bags of buns from a local cake shop and asked Margaret to distribute them in celebration. He wrote to his former student Miles Kennedy on the Monday after:

I have just acquired a wife, three sons and two dogs.

With the formalisation of their relationship, Peter and Lavinia moved into more appropriate surroundings. They had negotiated the purchase of

---

BACKGROUNDER:
MARGARET SANSOM

Peter's secretary throughout his second Cambridge career, Margaret Sansom was a huge blessing to him. Although nursing an ailing mother and attending from time to time to an elderly father in neighbouring Huntingdon, she took a lively part in departmental social events and never complained of being stretched. The students really appreciated and respected her. Like them she was short of cash and she chose to holiday where they did—in Yugoslavia, a popular holiday destination for the less-well-off in the 1960s. She took the precaution of bringing a bottle of plum brandy back with her and she kept it in the bottom drawer of her office desk, available to students in moments of stress—often during examinations. The academic staff also took to her: although she had no formal education beyond school, she was highly intelligent, well-read and could hold her own with any of them. As a teenager during WW2 she and her life-long friend Angela Pegg had considered joining the motor corps in order to learn to drive, but their resolution wavered at the thought of dealing with a puncture. So Margaret volunteered for secretarial duties and ended up at Bletchley Park, home of the Enigma code breakers. From there it was but a small step to Cambridge. Possessor of a fine sense of humour, Margaret did not lack for friends. In fact, it was departmental folklore that in retirement she would be able to travel round the world visiting her god-children. Staff recall it was extremely rare that something made her angry, and even then it wasn't obvious. The cause might have been shabby treatment of a student: at the end of term one member of the staff was due to drive to his family home and Margaret knew of a penniless student with family in that direction who could have done with a lift, so she approached the car-owner on the student's behalf; he refused. That made her cross. In contrast, most students remember golden moments of Margaret's thoughtful help: when UK currency changed to decimal in February 1971, researchers from other countries—already struggling to master the duodecimal system—had a new challenge but Margaret produced a standard pack of the new coinage to help them through.

an elegant Georgian property at 83 Newmarket Road—Yorke House—from John Kay, Peter's erstwhile colleague in Cambridge. Kay had moved from Cambridge to a job at Risley with the UK Atomic Energy Authority in the summer of 1952 and, rather than sell, he let the property. In 1954 when he became a professor at Imperial College, he lived in London during the week and visited Yorke House at the weekends. Bill Wilkinson, who had stayed on to do research in Cambridge after graduating, rented

Yorke House, 1962

a room at Yorke House for a while in the mid-1950s at a peppercorn rent, on the understanding that he would keep the whole place warm and Kay's grand piano in working order. But by the end of the decade Kay was looking to sell and it was just the kind of place to suit Lavinia and her boys.

The frontage of Yorke House was elegant; at the back it wasn't far to Midsummer Common and the banks of the river Cam. The boys loved it. Peter was less keen on their teenage preoccupations and the birds, bottles and bands that went with them, so Lavinia took action. She sold a black opal ring to fund the purchase of a large shed which was erected in the back garden. There the boys could do their thing, entertain friends and not disturb the rest of the house too much.

Lavinia had a weakness for large dogs and besides Florin had soon acquired a really substantial one: Beausie, a Pyrenean mountain dog. Visitors to Yorke House would knock at the door—up a few, sharp steps—and be greeted

A Yorke House welcome (l to r): Florin, Lavinia and a still diminutive Beausie

by the substantial Beausie rearing before them, apparently threatening to put his paws on their shoulders and force them back into the street. He didn't last long, though; during a spell in boarding kennels while Peter and Lavinia were travelling, he became frantic, escaped and was run over. As a companion for Florin, the family acquired another Alsatian 'Roddie' from the local dog pound. He tolerated sharing with Florin but was, by comparison, quite uncontrollable. Peter would take them both for exercise on Midsummer Common between Yorke House and the river Cam. He was also in the habit of occasionally bringing them to his office in the department. When a young Nigel Kenney arrived for interview as a potential staff member in 1961, he was pointed in the direction of the professor's office. Tapping on the door, he opened it cautiously and was greeted by menacing growls. Inside, facing the door, was Peter at his desk flanked by Florin and Roddy.

Lavinia warmed quickly to her new role as the professor's wife. She had not learned to cook at school or during holidays at home, and it was her first mother-in-law, Theodora St George, who noticed that Lavinia could

not even boil an egg. She addressed the situation at once: 'Would it amuse you, dear,' she enquired, 'to learn to cook?' Once she had got things going, she insisted that her son send Lavinia on a *Cordon Bleu* course. The results had been quite satisfactory. Now Lavinia certainly needed those skills and she rose admirably to the challenge. Her visitors' book—the record of guests invited, menus prepared and wines served—started tentatively; a November 1960 dinner party for six at Yorke House has notes about where to find the

Yorke House steps: Peter, Beausie, Florin and Lavinia's son Charles recipes for roast pheasant and Vinnoy Krem (cream, brandy, nutmeg, sugar). But her confidence grew quickly and references to recipes soon went by the board. Stews—beef Stroganoff, duck en daube—were a great stand-by; chicken en cocotte and coq au vin featured frequently. And Lavinia's magnificent home-made meringues soon became a Cambridge legend. They were an almost indispensable dessert: in season with strawberries and out of season with raspberry ice cream.

Peter exercising Florin and Roddie on Cambridge's Midsummer Common

Yorke House was too small to allow huge dinner parties, but Lavinia could always count on Margaret Sansom to help with the organisation. Besides her everyday department duties she was expected to deal with the administrative side of Peter's executive editorship of *Chemical Engineering Science*, and over the years this and her day-to-day contact with students provided her with a large range of contacts worldwide. She commuted in to the office every

BACKGROUNDER:
FORMS OF ADDRESS 2

Putative contributors to *Chemical Engineering Science* may have been shocked to receive a one-line letter from the executive editor dismissing their efforts as lacking in some unspecified regard, but just as daunting was the 1930s style in which it was couched. Even if you knew someone quite well, it was not common practice before WW2 to call them by their first name, and especially in writing. Don Scott of the University of Waterloo, Canada, spent the academic year 1963-4 as a Shell visiting professor in Cambridge. He

had enjoyed the morning tea breaks at which he had access to all members of staff including the professor. As was their custom with all visiting professors, the Danckwerts had invited Scott and his wife to dine with them at least once at their home. On his return to Canada, and judging that he and the Shell professor had got on quite well, Scott wrote to Peter to thank him, addressing him by his first name. He still remembers how shocked he was by the reply. It began in the way Canadians would only expect to be addressed if they were about to get the sack: 'Dear Scott...'

morning on her bicycle from a modest house in the misleadingly-named Longview Terrace, some way down the Histon road from Castle Hill. The front window was only a few feet from the traffic which, even in the 1960s, was formidable. Lavinia first saw the building when Margaret was negotiating to rent it and exclaimed: 'Margaret, you can't live there'. But she did.

Unlike Lavinia's first husband, Peter took no exception to things Scottish; in fact he positively relished them. The new family regularly motored to Scotland for holidays, visiting Macfarlane territory and sprigs of the family tree. On these trips Peter had the chance to sail. At first the family was based at Dunain Park House where Lavinia's parents had lived; when her mother died in 1960 Lavinia had sold the house to her cousin, Charlie Macfarlane, who had decided, with the support of his wife Isobel, to give up trying to make a living out of farming and have a stab at the hotel trade. (When they bought their first beds, Charlie's brother commented that 'those are the first things with four legs you have bought that won't die'.) The Macfarlanes ran Dunain Park as a guest house at first and so were able to offer the Danckwerts accommodation. On one holiday the family sailed to the Black Isle, not far from Inverness, for a picnic on its rocky shore. Peter wore a Macfarlane tartan plaid around his shoulders against the wind. Subsequently Charlie and Isobel moved on to Kilcamb Lodge in Strontian on Loch Sunart, where Peter memorably found a perfectly-balanced, clinker-built, Swallows and Amazons-style dinghy to sail. Sailing continued later on Loch Shiel when the Macfarlanes moved on to Glenfinnan House Hotel.

Guests at Yorke House spanned the spectrum of Peter's interests. They included his new colleagues in the department and from Pembroke College, University luminaries, visiting professors and students from all over the world. He also invited old friends made during his time in Cambridge in the early 1950s: Lady Taylor, George Porter, Fay Sykes Davies, Tom Bacon and his family. A tradition of regular parties for departmental staff and for research students was soon established.

Peter and Lavinia also got out to see old friends. Rosemary Murray, first president of the recently-founded college New Hall (now Murray Edwards College), invited them to dine. She and Peter went back a long way: they had grown up together in Emsworth. Both their fathers were admirals.

Since the move to Cambridge Lavinia had rather regretfully given up ballroom dancing in favour of her new career. But she got a break now and then. In June 1961 Peter bought tickets to the Pembroke College May Ball. He and Lavinia assembled a party of eight at Yorke House; Lavinia's childhood friend Annie Crole and her husband Gerry were invited, along with Ulric Huggins and his second wife, also called Pat. After a dinner of turtle soup and cold salmon with salad washed down with champagne, they decamped to the dean's rooms in college, which Peter had negotiated the use of. Lance Percival, talent-spotted before he made his name in the TV satirical review *That was the week that was (TW3)*, was billed to perform. Outside the dean's window on the courtyard grass was a large marquee with a dance floor. Annie Crole takes up the story: 'There was a steel band and it was the first time we had ever heard one. The weather was warm, the windows were wide open and the band, getting its amplification via electric cables entering through our windows and plugged into sockets in our room, was very loud. Eventually Lavinia lost her cool. She pulled out the plugs, threw the cables into the courtyard and shut the windows. Shortly afterwards a young man appeared at the door looking puzzled. He insisted that we should reconnect the band as "lack of power would do irreparable damage to the organ". All those present studiously avoided eye contact.'

Social relations at work were trickier. Peter was held in awe by staff and students, and rumours about his war record swept the department. People felt he could be knighted at any time. But that same awe cut Peter off from normal social intercourse; in his department office Peter was becoming isolated. Octave Levenspiel, a Shanghai-born American who shared many of Peter's interests, visited in the late 1960s and noticed what was happening. An outgoing personality himself, he suggested to staff: 'You guys should go in there and say "Hi, Pete, come and have a drink".' But Brits shrank from the idea.

Gardening at Yorke House

The professor's projects 28

I N THE FIVE YEARS Peter had been away from Cambridge—first grappling with the peaceful uses of atomic energy and then reviving spirits at Imperial College with international fresh air—things had changed. The chemical engineering department of which he took charge in 1959 had at last acquired a permanent building. It was all made possible because the departments of chemistry had got the go-ahead at the beginning of the decade for a new joint building in Lensfield Road, allowing them to pull out of the gloomy Pembroke Street premises they had occupied since the beginning of the twentieth century. The design of their new building on Lensfield Road was controlled by the two professors of chemistry, Ronald Norrish and Alexander Todd. In those days professors of traditional departments operated like tsars, and Todd had the added status of the 1957 Nobel prize, something Norrish—his senior as a professor—would not

December 1956: Demolition of the Oil Companies' building in full swing
Courtesy Cambridge University Department of Chemical Engineering and Biotechnology

Pembroke Street in 1957: building activities are visible behind the parked cars
Courtesy Cambridge University Department of Chemical Engineering and Biotechnology

equal until ten years later. Norrish and Todd couldn't stand each other and saw to it that their new laboratories, although in a single building, were designed to keep the two of them, and their staff and students, apart. Even the joint library had a dividing line across the floor. The new building was occupied progressively from 1956, and opened formally in 1958. Those present, as *New Scientist* reported at the time, must have felt that Cinderella had truly entered her golden coach.

Meanwhile, the site of the old chemistry departments on Pembroke Street had been allocated to chemical engineering. But before anyone could move out of the temporary buildings in Tennis Court Road, the building facing on to Pembroke Street had to be modernised and the 'Oil Companies building' behind it demolished and rebuilt for its new purpose. Fox had known what he wanted and he ran the architects ragged. His attention extended to the design of benches and filing cabinets, and the effort involved was huge. With his colleague Denys Armstrong he went over the plans with a fine toothcomb. A specific characteristic of a chemical engineering experimental area is a roof high enough to cope with columns and a floor area uncluttered enough to make them accessible.

Fox envisaged a flexible area two and a half stories high, with distillation columns accommodated on peninsulas made from Dexion. Architects Easton & Robertson, regarded as a progressive company, nevertheless drew up plans which featured pillars in this space, wrecking Fox's required flexibility. But Fox was ready for them. At the next review meeting he asked why the pillars were needed. 'To support the roof,' the architects answered. 'But the roof is strong enough. Have a look at these calculations,' Fox suggested, pulling some papers from his pocket. The pillars duly disappeared. Fox could be stubborn too. He had seen some very practical laboratory doors planned for the new chemistry laboratories, and wanted the same. Easton & Robertson considered themselves to be artists; they deployed arguments implying that they couldn't possibly repeat the doors, and they produced plans without them. Six weeks later doors identical to the chemistry building's new ones were agreed.

Decades of contamination of the old building on Pembroke Street presented another challenge, particularly within the small radiochemistry laboratory. Peter recalled later:

Its parquet floor (not good practice) was found to be heavily contaminated with

radioactive elements. It had to be ripped up and disposed of; the Shell professor was seen in the back of the truck which took it away, dressed in a gas mask and full protective clothing.

Once the buildings were complete and ready for use, Fox's job was—in a sense—done. His department had a permanent home, research was flourishing and, most important of all, he had won the department the intellectual respect it deserved from the cloistered minds of Cambridge. Fox had achieved much of this by

The new main entrance takes shape behind scaffolding
Courtesy Cambridge University Department of
Chemical Engineering and Biotechnology

inspiring others. His high-stress personality made him susceptible to nervous collapses and he suffered from them during the planning of the new laboratories. When the plans were signed off, he resigned the Shell chair and returned to mainstream engineering as a lecturer. Peter certainly felt that Fox's exertions with the new building contributed to his early death in 1962, aged 50.

What were the characteristics of Peter's early years in charge? 'There was wonderful academic freedom', according to David Harrison who had arrived at the department as a lecturer in 1956, during the Fox years. 'Everyone had strengths and weaknesses, and generally the one covered the other. Denys Armstrong and Ron Nedderman liked teaching; Armstrong enjoyed administration [which must have been a relief to Peter] while others preferred research.' The department was small and very friendly: every time a new person joined the staff there was a welcoming party, and there was ritual entertainment at home with the Danckwerts at least once a year. Peter's style with his staff's research was to hand the ball—for instance, a new project—to chosen colleagues and expect them to run with it. If anyone were to have dropped the ball, Peter would no doubt have been puzzled. But the quality of the department was such that it didn't happen.

In the early 1960s Peter took John Davidson (who had been in Cambridge since 1952 during Peter's first stint) and David Harrison to a seminar at the National Coal Board (NCB) Stoke Orchard laboratories. It was about fluidisation of a bed of coal with air during its combustion in power stations; the NCB wanted Cambridge to collaborate with them on the technical challenges, and both Davidson and Harrison were working in the area. Once the talking was done, Peter simply invited them to get on with it. The NCB project was an early step in building Davidson's formidable experience of the behaviour of bubbles, particularly in fluidised beds. Aside from this decisive management style, Peter was very supportive of his colleagues. Davidson became a reader in the department at a very young age in 1964 (getting senior positions 'was jolly difficult in those days,' Davidson recalls) and after Peter was elected a Fellow of the Royal Society in 1969, Davidson followed him in early 1974. He knew that these advances would not have happened so early without Peter's support.

One of Terence Fox's great achievements was to arrange that the department be controlled by a syndicate appointed by the University. The very unattractive alternative was to be swallowed either by the huge faculty board of engineering to which it would have been a tiny appendage, or by the enormous faculty board of physics and chemistry where it might have been overwhelmed. All of the syndicate's members were academics bar one from industry, there to keep the feet of syndicate members on the ground. In addition, Shell had endowed the department very generously, and this allowed the chemical engineering syndicate to pursue its own line. Peter disliked committees of any kind, and was inclined to be witty about the syndicate saying that, before he arrived in Cambridge, he associated syndicates only with stockbrokers' shooting rights or diamond marketing. But he did appreciate that the syndicate was...

...one of those rich anomalies which flourish in Cambridge to the despair of the administrators

who must have felt it was out of their control.

So, as Shell professor, Peter was subject to the lightest of administrative touches. Even so he described the syndicate privately as 'the politburo'. There was some bravura here. Peter had seen Newitt in action at Imperial College and was himself quite capable of working the system effectively. For example, he inherited a situation where students only began a two-year chemical engineering course after two years' study on a feeder course—either natural sciences or mechanical engineering. There was a natural pressure from within his department to prepare incoming students better during those first two years; natural sciences and mechanical engineering, for their part, were reluctant to lose class time in their specialist areas to give future chemical engineers a flying start. But Peter was successful in getting a second-year module on the behaviour of fluids (fluid mechanics) inserted in the natural sciences course in 1966, and new entrants to the department were then able to hit the ground running. This was a fine achievement against the other big-hitting disciplines.

There was another influence at play on the Cambridge department's activities: the profession's representative body, the Institution of Chemical Engineers. During the early part of his professorship Peter became a member of its council and worked energetically to implement its policy to include a 'design project' within the chemical engineering degree courses of British universities. It meant finding time in a jam-packed timetable for students to get a small taste of the practical problems they might face as graduates in industry. During his year as Institution president (1965-66) he helped the organisation found an 'exploratory committee' which encouraged chemical engineering staff to spend time in industry in order to familiarise themselves with industry's needs. Often this used to be done using the 'old boy network'; proper industrial fellowships made it easier for staff to take a year's leave in industry.

As far as Peter's research projects as Cambridge professor were concerned, the most significant development was the arrival of Man Mohan Sharma as a post-graduate researcher in October 1961. The Shell endowment had money earmarked to support students from the Commonwealth, and after his experience with Newitt in London Peter was keen to bring Indians into his laboratories. Recalling his arrival in Cambridge, Sharma shows a degree of self-effacement worthy of Peter himself: he claims he was invited because his application was the only one received from India.

The young Sharma was brilliant and bubbly; he fairly burst upon the scene. It seemed to his contemporaries that he had read everything in the field and could recall it all. At his first meeting with the professor the two of them talked about what Sharma might work on. Sharma had prepared by reading all Peter's work on residence time distribution. But Peter had something else in mind: gas absorption, especially of carbon dioxide. Sharma recalls: 'Fortunately I knew something about that too'. So he joined Peter's sequence of research students carrying out ingenious experimental studies aimed at achieving greater understanding of gas absorption. Sharma suggested that he should spend a couple of months studying the literature and begin experimental work in the New Year.

Just before Christmas Peter took Sharma on a trip to Oxford by car to meet his old Balliol supervisor, Ronnie Bell. It was a chance for Sharma to describe his research plans to Bell. On the way, Peter warned: 'Sharma, if Ronnie says it is nonsense, keep quiet!' In the event Bell simply commented: 'Looks very interesting'. Sharma recalls that 'afterwards PVD told me, with that quarter-smile that was typical of him: "That's the best compliment you will get from Bell".' By May Sharma had enough original work to write a scientific paper, which was duly submitted in both their names for publication. Sharma was truly prolific: within a year he had enough material for a doctoral thesis, something which took most students three years to achieve.

During that first year Peter allowed Sharma to spend eight weeks at ICI Billingham on a vacation project to let him see how research was done in industry. While he was there a postcard arrived in Cambridge saying that the referees had accepted their joint paper for publication. Peter sent the postcard on to Sharma annotated 'The referees have swallowed the matter'. Later on Peter found funds and wrote letters of introduction for Sharma to visit chemical engineers in continental Europe. Warmth and great mutual respect grew between them. David Sutherland, an Australian graduate who was Sharma's contemporary in Cambridge, worked in a laboratory right opposite Peter's office door. Sharma would drop in on him when he knew the professor was in his office. If the professor's door was open, Sharma would whisper to Sutherland: 'Quick, we must catch the ideas as they escape through the door!'

Sutherland felt that Sharma broke through Peter's famous reserve. In contrast, most of the department met Peter's shyness. Sutherland recalled an incident early in his career. He arrived at the department one morning and was daunted to find himself going up in the lift alone with the professor.

'Ah. You are er...'

'Sutherland, sir'

'Ah, yes. You're working with er...'

'Dr Ratcliff, sir'

'Ah, yes. On er...'

'Flocculation, sir'

'Ah, yes. How's it going?'

Stranded far from home in December, Sharma was invited three years in a row to the Danckwerts' Christmas Day lunch. He shared the experience with other overseas students; while they tucked in to turkey, Sharma enjoyed a cheese omelette. He found that Peter would go to great lengths for his friends and colleagues. When he arrived in Cambridge Sharma was associated with Fitzwilliam House, originally a base for students unable to afford the fees of a 'proper college'. (In 1963, during Sharma's residence there, it was upgraded to the status of a fully-fledged college.) Facilities were relatively poor. Peter considered this choice was 'due to lack of imagination at the Indian High Commission' and felt that Sharma should share the greater benefits, and increased grants, offered by his own college, Pembroke. Everything was set up for a move at the end of Sharma's first year, but Fitzwilliam House wouldn't let him go.

In the course of his work Sharma produced some results which, Peter thought, would interest Shell because of their potential for industrial application. He wrote to Lord Rothschild, Shell's research coordinator at the time who, like Peter, was a wartime bomb-disposal veteran. In 1944 Rothschild had been awarded the George Medal after defusing explosives hidden in a case of Spanish onions in a ship's hold; he is reputed to have used, very much in character, a set of jeweller's screwdrivers given to him by the firm Cartier. As a result of Peter's letter Shell sent people to Cambridge to advise on a pre-patent application and their report was favourable. In due course a patent appeared with Sharma's name alone on it; this was remarkable, almost unheard-of, modesty on the part of his professor, but Peter was precise about this sort of thing. He allowed Sharma to publish on his own where he felt he had made no contribution himself. He was, says Sharma, 'a model of honesty and fairness'. Later Shell bought out the patent for £1,000, a bit of a windfall for Sharma whose stipend for a whole year was £600. Sharma was full of gratitude for the way Peter mentored him, and Peter eventually gave Sharma total freedom of action, only murmuring from time to time: 'Sharma, keep me informed'.

Sharma's stay in Cambridge ended in 1964. During the three years his output was prodigious. The question of who should be his external PhD examiner arose. Peter approached Sydney Andrew, who had distinguished himself amongst the department's 1948 student intake. By 1964 he was a formidable engineer with ICI at Billingham. Andrew had a reputation for never being wrong and Peter thought him a very appropriate choice. There was excitement rivalling that of an important cup tie as the day of the oral examination arrived. After it was over staff member Robin Turner asked Andrew whether he had survived the experience. 'I think so', replied the normally unruffled Andrew. Sharma was, of course, single-minded; visitors to India later in his career recall him pointing out young Indians playing cricket and

Peter addresses a 1960s Cambridge experimental rig

wondering aloud why they didn't spend their time in the library instead.

In the final months of his stay in Cambridge, Sharma spent his Sundays gathering together all the available literature on 'the absorption of carbon dioxide with reaction' into what became a 36-page monograph, eventually published as a review by the Institution of Chemical Engineers in 1966. It was the first attempt to take a scientific approach to design across a large number of different but closely-related processes. Sharma, at his own suggestion, made informal visits to two large contractors of the day involved in the design of such processes—Humphreys & Glasgow and Power-Gas Corporation—to check the calculations in the monograph against actual plant data. Finally he delivered the text to Peter at the end of his last week in England.

Peter organized finance for Sharma to return briefly in 1967 as a 'distinguished visitor'. Sharma spent two weeks in London, three in Cambridge and a week touring the UK. Where students usually got no more then ten minutes with the professor, and most visitors no longer than half an hour, Sharma and Peter could talk for three or four hours at a time. Peter was generous with his support of people he felt deserved it. In 1978 he put Sharma forward for Fellowship of the Institution of Chemical Engineers with support from Roger Sargent, Peter's ex-colleague from Imperial College who regularly acted as external examiner for the Cambridge department. Peter commented ironically to Sharma: 'Don't be disappointed if it is turned down'. In March 1969 Peter himself was elected to the Royal Society, quite a rare honour for a chemical engineer. Sharma wrote warmly to congratulate him. Peter replied to him: 'The Royal Society thought industry is important.' Sharma remembered so well the benevolent influence of his friend that when, in 1990, he himself was elected a Fellow of the Royal Society (the first engineer from India to be honoured in this way—Homi Bhabha, who was elected in 1941, was considered a physicist) he commented: 'How happy PVD would have been; if he were alive now I have a sneaking suspicion that I would have been elected earlier'.

Just as Sharma was starting his final year in Cambridge the Portuguese student Armando Tavares da Silva arrived, hoping to work for a doctorate with Peter. He found that Peter was temporarily abroad, so he was shunted off into the care of a staff member—Gerry Ratcliff. Assigned a project he came to view as unpromising, and aware that in any case Ratcliff was about to move to McGill University in Montreal, Tavares da

Silva approached Peter about a transfer. He got his way and took on one of Peter's bright ideas with possible industrial benefits—purifying crude iron powder by first converting it into iron carbonyl, which could be separated from impurities and then decomposed to give pure iron. The experimental conditions were challenging, and the department had little experience of them, but Tavares da Silva struggled on for a while. Eventually, with Peter's approval, he switched to a project which did yield results: attempting to provide evidence for Peter's 1950s ideas on surface renewal during absorption. The older two-film theory predicted that the rate of absorption would depend on the diffusivity; Peter's ideas required dependence on its square root. During the 1960s Peter asked several of his better students to try and gather evidence for one theory or the other. Amongst the challenges were the difficulty of measuring diffusivities of gases accurately and the fact that differences between them seemed not really big enough to get meaningful results. But Sharma had recently made a promising suggestion. He thought that absorption of hydrogen sulphide in amine solutions (in which the two react together and absorption depends on the diffusivities of the amines) had characteristics which might allow the challenges to be met.

Tavares da Silva chose amines with diffusivities varying over a range from 1 to 4, which was just about wide enough for the purpose. All the same, it was challenging work: hydrogen sulphide is nasty stuff which cannot be allowed to escape into the open laboratory, and all the solutions he used had to have air removed from them first, a time-consuming business. But Tavares da Silva successfully carried out a study in small packed columns and found the results, swiftly reported in the literature, to be a triumph for Peter's surface renewal model. Neal Amundson, commenting on the Danckwerts-da Silva paper some 20 years later, noted that Peter seemed at his happiest when reporting on such endeavours.

In early 1960 Anthony Pearson, who was to become another key figure in Peter's department, answered an advertisement for a job in Cambridge. He had studied at Trinity College Cambridge and graduated as a 'wrangler'—the name given to those few students each year getting first class mathematics degrees. Pearson's subsequent job with ICI had given him experience of 'fluid mechanics', which meant that he knew about the way fluids move through pipes and vessels with variation of speed and geometry. It had also led him to write to Peter about some of the problems he was encountering, a correspondence which no doubt left an impression. Peter needed such an expert and appointed him assistant director of research. Pearson in his turn was immensely impressed by Peter's seminal work in the 1950s, and stayed in the department for 13 years. He became the department's 'maths brain'.

One of the things Peter had appreciated during his time in the USA in 1946-48 was that computers, then in their infancy, would play an increasing part in chemical engineering. He was ahead of his time in realising what they could do for automatic control of experimental work, simulation of the course of an experiment and the recording and computing of results. Pearson got the job of selecting and bringing in a suitable digital computer—he chose the IBM 1620. In 1961 Cambridge was the first department in the UK not specialising in computing to invest in a computer of this type. Pearson was fortunate in having 'a very capable research student', Chris Cowsley, who automated his experments using his own additions to the 1620 and then wrote software and arranged hardware so that others could use them in their experiments. Pearson rewrote the IBM manuals to make them accessible to non-specialists and gave classes to the students on how to get the best from the computer.

Peter always placed a strong emphasis on chemical engineers providing a bridge between those responsible for designing and running complex process equipment and those studying the detailed science and mechanics involved in the process. Pearson benefited greatly from this. He appreciated that Peter had been trained as a chemist and so understood the difficulty many engineers would have in digesting highly mathe-matical theories. 'But it was equally clear,' to Pearson, 'that such theory could be invaluable in understanding chemical processes and could suggest means of controlling them. What PVD recognised was the need for a common language; a degree of overlap between the two.' In this Peter was at one with his predecessor, Terence Fox, who taught basic science and mathematics to chemical engineers ahead of his time. Peter's colleague John Davidson sums up Peter' s position admirably: 'He feigned an antipathy towards mathematics, but his own discoveries depended on the imaginative combination of simple mathematics with acute insight into physical and chemical realities'. A decade later Pearson and his professor acted again to keep the department at the cutting edge: in 1971, just before Pearson moved on to a professorship at Imperial College, they replaced the IBM computer with a Digital Equipment Corporation PDP-11/40. The PDP-11 series, which had only just been introduced and was easier to program, was the best system then available for the department's purposes.

In the halcyon years of the 1960s Peter's reputation attracted a stream of visiting scientists, especially from the USA. Anthony Benis was one of them. In 1964 in Cambridge he got his first experience of programming a high-speed computer; he solved problems in fluid dynamics with it and was delighted to find that, within a year, he had his name on two papers in high-quality journals. His acquaintance Lou Shrier, a young post-doc from Exxon in New Jersey, was another. Shrier had come for a year to work on carbon dioxide absorption. Peter suggested to him that, rather than doing further testing of theory, he should try for something really useful to industry in its efforts to scrub carbon dioxide out of gases. This was just at a time when gas scrubbing was becoming increasingly widespread in oil refineries. Shrier and his mentor hatched a plan to enhance the speed of absorption by using a combination of potash (a solution of carbonate and bicarbonate of potassium which was very good at absorbing carbon dioxide) and small amounts of amines (which were very reactive with carbon dioxide). They chose surface-active amines which would congregate at the gas-rich surface of the solution, react rapidly with carbon dioxide gas and carry it into the body of the potash solution for absorption. At the same time amine conveniently regenerated and returned to the surface.

The idea worked pretty well and, when Shrier went back to the USA, he and a few colleagues developed it into a practical process and persuaded Exxon that the concept had commercial promise. Of course Shrier's part in it belonged to his company but Peter had made no such commitment. So Shrier contacted his mentor and, rather apologetically, asked Peter to sign over his rights to the work. Peter raised no objection. Shortly afterwards Exxon, perhaps grateful and certainly with Shrier's enthusiastic support, put forward Peter for the American Chemical Society's annual Murphree award. It was a reward for outstanding research which Exxon funded and which more often than not went to an American. Peter was well qualified for the honour and in early 1970 he duly got it. In the following decade Exxon developed an armoury of scrubbing processes under the name Flexsorb which were based on amine promoters.

Tony Gillham graduated at Imperial College in 1961, a couple of years after Peter had moved to Cambridge. At Imperial he had been chairman of the chemical engineering society and so had got to know the teaching staff a bit. Some time after Peter left, Gillham was surprised to get an invitation to visit the Cambridge department for the day. At the end of the visit he was—rather naïvely he admits—amazed when Peter suggested 'if you get a half-decent degree and a grant you should come and do research with us'.

In Gillham's Imperial group of 55 students that year, only one came out with a first class degree and a third of them simply failed; Gillham got a 2.1 and spent four years in Cambridge. He worked alongside Sharma whom he found helpful and supportive of his contemporaries. Peter made a huge impact on Gillham. 'There was,' he said, 'an aura about him. He had tremendous presence and one could sense him in the room without seeing him.' Part of this was Peter's war record, about which he never spoke but which staff and students were all aware of. He was accessible to his research students: 'One got close contact with him, but it rarely lasted more than ten minutes. He was a typical Wykehamist: slightly distant and if he didn't have anything to the point to say, he kept quiet.' Gillham could see that Peter wasn't a natural lecturer, but found that he wrote extremely well and made pithy comments. 'I recall an ill thought-out proposition I once put to him in writing: he returned it with a single word scribbled at the end: "Byzantine".'

As a post-graduate, Gillham first needed to be accepted by a college. Peter told him that only Trinity, Pembroke, St Catharine's and Fitzwilliam House had heard of chemical engineering: Gillham approached St Cath's. In the laboratory he got off to a rocky start. Peter wanted to find a way of pumping powdered coal down a pipeline as a slurry, an idea perhaps seeded by his wartime contact at Combined Operations with Geoffrey Pyke whose 'pumped rivers' proposal was an untried means of transporting invading troops and munitions rapidly ashore by pipeline. The Central Electricity Generating Board at Marchwood, Southampton, was interested in Gillham's project, but the problem proved too complex, even for an experimentalist of Gillham's calibre. There were so many variables: concentration, time, temperature, previous history... even an unexpected vibration of the bench upset things.

So Peter diverted Gillham to the design of carbon dioxide scrubbers. The aim was to devise a simple piece of laboratory equipment providing data which could then be used to derive the height and diameter of a suitable industrial scrubbing column for the same process. A stirred tank with a precisely-placed agitator did the job: the absorbing liquid entered through the bottom of the tank, the gas to be absorbed through the top and—Gillham discovered—the cruciform agitator merely had to 'wipe' the surface of the liquid to provide the surface renewal data he needed. (His colleague Sharma, watching what was going on, realised that the same apparatus would also provide other data about the speeds of reactions taking place, adding to the usefulness of a technique which is now widely used.)

Gillham took the data he had obtained with the stirred tank and tested it by building and running a large column at the Gas Council's facility in Fulham. He saw this project as epitomising Peter's approach to research; it must answer industrial needs and provide practical solutions: 'a lot of academics will say this but PVD really meant it—an exceptional man'. Later, working for ICI's Mond division, Gillham routinely designed columns this way.

In due course Gillham became technical director of the company Mass Transfer. In 1970 his new employer had a contract with whisky-makers William Grant to look at doing something with fermentation waste. In the fermentation of barley or maize about 90%, the mash, was waste. The company could concentrate some of it to sell as cattle feed; it wanted to extract dissolved protein from the rest—quite a challenge when the solution was so dilute. But Gillham realised that William Grant directors were also concerned about something entirely different: a proposal to build a nuclear power plant next door to their production site in Girvan, south of Ayr. How would their all-important American customers view this development? Gillham suggested to the company's directors that they call in Peter, with his UKAEA experience, to advise. Peter recalled later in *New Scientist*:

I received one of those letters which used to lighten my leaden academic days from time to time. I went up to Glasgow by train and was met by the chief chemist who immediately pressed a bottle of the product into my hands. I made a brief tour of the distillery. The main principles are well-understood and frequently practised by unlicensed amateurs... The man who operated the [distillation] column [in Girvan] did so entirely by experience—he knew more or less where the bands of wood-alcohol and fusel-oil would be, and made sure they did not get into the product.

At that time the next generation of nuclear reactors was under development at Winfrith Heath: the steam-generating heavy water reactor (SGHWR). One was planned for Girvan. The Winfrith site had opened after Peter left UKAEA, but fortunately he knew its director, Donald Fry: throughout Peter's time with UKAEA and until 1958, Fry had been chief physicist at Harwell. Looking back in 1983, Peter wrote about the day he spent at Winfrith bringing himself up to date on the new technology:

I pressed my enquiries very sharply but could find absolutely no reason to suppose that radioactivity from the plant, when running normally, could possibly get into the whisky. I formed the opinion then, which I still hold, that the SGHWR system is extremely safe and docile. The Byzantine industrial-political process has consigned it to the shadows for the moment, but another parliamentary enquiry might yet bring it back to life.

Parliament is yet to provide such a stimulus.

At home in Abbey House 29

THE DANCKWERTS HAD NOT BEEN LONG at Yorke House before Peter caught his first glimpse of Abbey House. Just off the Newmarket Road and a little further out of town, it was visible from the garden of Yorke House. Investigating further, he was smitten. It was, he thought, an oasis of peace 'far removed from Cambridge suburbia and Cambridge intellectuals'. Lavinia, in contrast, seemed ambivalent about it. Although it suited her style, she probably felt apprehensive about the upkeep. Her boys certainly preferred Yorke House—more practical. Peter went to have a chat with Margaret Hoather—an artist and enthusiast for gardens who was renting one of the three parts into which the house had been divided by its then owner, the Cambridge Folk Museum. The good news was that Hoather was about to move out.

It was just as well that alternative accommodation was in sight, for in April 1962 the Cambridge city Council sent the Danckwerts an order for the compulsory purchase of Yorke House. They were planning a new ring-road in an attempt to relieve traffic congestion in the city, and soon the modern Elizabeth Bridge would span the Cam. The plans showed that this would cut off existing access to Brunswick Junior School, which lay between Yorke House and the river Cam; new access was needed. The route chosen by planners ran from Newmarket Road right through Yorke

Demolition in progress
Yorke House in July 1967
Courtesy: Cambridge News

House. The Danckwerts' new home would have to be knocked down. (In an ironic turn of events which Peter would have appreciated, the school buildings were themselves demolished in 2010.) Peter realized immediately that the city Council would offer miserable compensation, probably no more than one third of the value of the house. The matter dragged on, and Peter delved into the legality of it all. He found that if a firm offer for Yorke House was on the table and accepted, the city Council would have to pay an equivalent amount, rather than the planned compensation. At the appropriate moment Peter produced just such an offer; the Council then paid up the market price and in the late summer of 1964 the Danckwerts moved on to Abbey House.

Abbey House has a long history and a reputation amongst the susceptible as one of the most haunted buildings in England. Six different ghosts! Not that ghosts, or the listing of the property by the National

BACKGROUNDER:
GHOST STORIES

In any competition between haunted houses, Abbey House would do well. Accounts even went as far as a ghostly squirrel in the garden. The best information on credibility came from the Lawson family who lived in Abbey House from 1903 to 1911 and were not told of the tradition of haunting before they arrived. Professor Lawson, like Peter, was a fellow of Pembroke College, and reputed not to be fanciful. The Lawsons recorded two 'durable' apparitions: the 'animal' and the 'nun'. The first, nicknamed 'Wolfie', looked like a hare with cropped ears and pattered audibly around the house. Some had seen it accompanying the ghostly squirrel in the gardens. The second was a figure dressed in nun-like robes. Peter recorded one of the Lawsons' experiences in an article for the local Antiquarian Society:

The 'nun' appeared in the first-floor room at the south end of the house which was the Lawsons' bedroom [and later used by Peter as his bedroom]. It came in through the (closed) door, paused at the foot of the bed and then disappeared through a curtained window. It would wake Mr Lawson up with its heavy tread, but Mrs Lawson heard nothing (a later resident told me that the 'heavy tread' emanated from operations at the [nearby] gas-works).

Professor Lawson reported that on one occasion the nun prodded his foot—rather a tangible sort of action for a ghost. Mrs Lawson found it all rather tiresome and one night subjected the nun to an amateur exorcism; it was never seen again. Professor Stratton, president of the Society for Psychic Research, got permission in 1955 to occupy Abbey House for a week and set up a rota of watches by University staff and students.

Nothing happened and nothing significant has happened since,

Peter wrote.

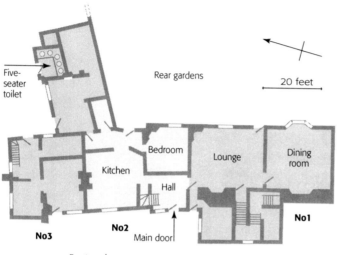

A plan of Abbey House ground floor

Trust, would put a man like Peter off. Amongst other delights, the property boasts a five-seater toilet in an outhouse (probably primitive composting units used in sequence rather than a Roman-style communal experience). Quite soon Peter, Lavinia and her boys—ghosts or no ghosts—were installed in two of the three parts of Abbey House on a long lease, number three being let separately.

Work on Abbey House began in about 1580, with bricks from neighbouring Barnwell Priory following the dissolution of the monasteries. By

Abbey House in the early 1950s: 'the gardens back and front were a wilderness'

the middle of the twentieth century there had been many extensions and the whole building was in need of considerable attention. The previous tenant, Margaret Hoather, had found the gardens a wilderness when she arrived but had taken matters horticultural in hand. The fabric of the house itself was, however, in a poor

state and this may have
accounted for the prolifer-
ation of ghost stories: the
whole   place   creaked
alarmingly in a high wind.
The Danckwerts struck a
deal with the owners, the
local Folk Museum. The
Folk Museum (and later
the city Council when the
Museum   handed   over
ownership) agreed to cope
with external repairs while
the Danckwerts would
take care of internal maintenance in return for a
reduced rent. Peter and
Lavinia immediately began
to pour cash into internal
renovation and by the
summer of 1965 the house
was in reasonable shape.

The new residents of Abbey House, 1965

Hospitality took on a
new dimension at Abbey
House. In June 1965
Peter and Lavinia held a house-warming party and 130 guests attended. In
November of the same year they laid on a buffet supper for the Shell
department. Patsy, daughter of Lavinia's friend Jane Plunkett, recalls 'the
faculty being riveted by Lavinia's glamour and style, by the splendour of
Abbey House and by the generosity and quality of the food and drink.'
They were struck too by Lavinia's simple elegance with a hint of sophis-
tication. Lily of the valley was her favourite scent: the one marketed by
the Floris company and, of course, by appointment to Her Majesty the
Queen. It wafted around her side of the house: up the stairs from the main
entrance and into her bedroom and bathroom, papered with huge
flowers which contrasted with the dark wooden panels of the rest of the
house. Parties for large groups—academics, the department, visiting
researchers and students—continued apace. On such occasions, as mid-

night approached, Peter might feel that he had had enough and would hide the bottles in an attempt to bring things to a close. At one such party in late 1965 the American Jim Carberry, on sabbatical from Notre Dame University in Illinois, was amongst the guests. A large figure with an equally big reputation in American catalysis following a six-year stint with Du Pont, Carberry was a formidable trencherman. As midnight approached and the bottles mysteriously disappeared, Carberry assembled a hunting party to seek them out. It wasn't long before triumphant shouts announced they had achieved their objective. The party continued.

Entertaining required human resources, and inevitably Peter's secretary Margaret Sansom was roped in to help organise and prepare. She had baby-sat Lavinia's youngest son Charles before he was old enough to be left alone in Yorke House and could be called upon in dire emergency for almost any other duties. If more back-up was required, the Wisbeys at number three Abbey House were very helpful. Betty Wisbey worked as a secretary at Emmanuel College; her daughter Jane acted as a kind of Girl Friday for the Danckwerts, helping with gardening, cooking for parties, driving, keeping an eye on the Danckwerts' part of the house when they were travelling, and calling in other members of the family if necessary. Jane found the Danckwerts very inclusive in their attitudes to neighbours and Peter kind, gentle and reflective.

One of Lavinia's concerns had been how to furnish her new home. There was so much space. She didn't have a lot of furniture and couldn't afford to buy it new. Here Peter's friendship with Ray Baddour helped out; Lavinia and Anne Baddour got on very well and during one of the Baddours' visits they discovered a shared interest in what Lavinia called 'old furniture'. They motored together round East Anglia looking for suitable items. Anne recognised this 'old furniture' as 'antiques' to Americans, and their UK prices seemed to her to be well below what they might command in the States. After a number of successful outings acquiring the basics for Abbey House, Lavinia and Anne were sitting over a couple of gins in a village pub when Anne asked Lavinia, 'Do you have any money of your own?' When it appeared that she did (inherited from her mother in 1960), Anne responded 'So do I: let's go into the antiques business. We can ship them from here to the States.' They needed a business name; their husbands instructed them in no uncertain terms to 'keep our names out of it'. So they chose a combination of their maiden names: Bridge Macfarlane. They assembled enough items to cram into a crate holding some 216 cubic feet, and Lavinia had it despatched across the Atlantic. Bridge Macfarlane Antiques enjoyed a short but eventful life until the two proprietors, encouraged by their husbands, were persuaded to call it a day.

To family and friends with whom he could relax, Peter was fun. His sense of humour, as closer professional friends knew, deployed irony to the borders of cynicism. In a family setting he softened it a little. Lavinia's childhood friend Anne Crole recalls how he would play to them the 1958 recording of Gerard Hoffnung at the Oxford Union reciting *The Brick-layer's Lament*: a tale of accident-prone adventures with a beam, a line, a pulley and a barrel of bricks. They laughed a lot. Peter's step-sons recall Peter as the absent-minded professor. At mealtimes a hand would reach out from behind a prop-up stand supporting the daily paper in search of, depending on the time of day, a cup of soup or a glass of scotch and soda. The occasional grunt-like laugh might follow. Brian's young daughter

---

BACKGROUNDER:
BRIDGE MACFARLANE ANTIQUES
By the time Lavinia's big box of East Anglian antiques arrived, Anne Baddour had struck a deal with an elegant interior decoration shop in Boston. The new stock seemed to have local appeal; a burglar broke in and stole all their smaller pieces. Anne found a more secure venue in Concord, Massachusetts, some ten miles from her Belmont home. The shop belonged to a local school and Anne rented it for the summer vacation. She supplemented Lavinia's stock with locally-sourced items. The launch of Bridge Macfarlane Antiques was quite a professional affair—selected guests, iced tea and lemon, cookies and an article in the local newspaper. 'One day,' Anne recalls, 'a man came in who seemed to like everything. He bought 13 tables.' Her husband pointed out that their prices were too low; they had forgotten things like transport and time. When school term began, Bridge Macfarlane

Antiques moved across the road to a first floor site. Anne hired a saleslady on a percentage deal and Anne had the impression that she did very well. Not so the proprietors. Shipping costs were inflating and Lavinia's buying criteria were also a problem. She liked ormolu, for example, but Bostonians liked simpler things. There was another factor—the better pieces, somewhat to Peter's aggravation, were finding their way into corners of Abbey House rather than into a shipping crate. The proprietors decided to call it a day. Peter wrote to Ray Baddour: It may be possible for Lavinia to visit Anne before too long. Our accountant seems to think the fare would count as a business expense. I don't know how the matter can best be resolved—I think we shall have to leave this to the principals, as you say. On her visit to Boston, Lavinia and Anne closed the shop, sold the remaining stock at auction and shared the proceeds.

Catlin Harrison recalls similar scenes later in the 1970s: 'I felt copycat pleasure with my grown-up tumbler of lime and soda. The odd comment or guffaw would rustle the pages of *The Times* and it was a benevolent presence that I sensed behind the newspaper'. She was also the happy beneficiary of Peter's love of interaction with children. 'He would respond to positive declarations such as "I like chocolate ice-cream!" With a smile he would discuss with me the merits of chocolate ice-cream, the combination of it with Grandma Vinny's home-made meringues and the melty versus the hard properties, as well as the difficulties of getting enough into my Beatrix Potter bowl.'

Part of the success of Peter and Lavinia's relationship was their complementary natures. Peter seemed aloof, apparently inaccessible and imperturbable. Lavinia, in contrast, was much more approachable but could be brittle; every now and then a plate or a saucepan would hit the kitchen wall. Peter, like his parents, was not easily moved. The boys never heard him get cross. At worst his eyes might rise towards the ceiling and he'd exclaim 'Oh, really' with a slight shrug. Lavinia kept an eye on what he wore; one day as he left for the department sporting cufflinks that did not match, Lavinia drew his attention to it. Peter replied that it didn't matter: 'The students don't look at my cufflinks,' he said, 'they look at my face and what I am teaching'. Danckwerts family slang involved addressing a wife to her face as 'old thing' and referring to her to others as 'the old woman' or OW. Together, Peter and the OW were generally easy and content with each other; Catlin Harrison again, on a tour around Britain for a few weeks visiting relatives: 'In the back of their estate car, with miles of landscape streaming by and quiet conversation murmuring from the front, I'm sure that they regularly forgot my existence and I got to see the calm and comfortable warmth that they had with each other.'

Abbey House offered more space for animals. Following the departure of Roddy from Yorke House, the Danckwerts had acquired a corgi bitch called Ferdinanda, or Ferdi. An irritating, manic little dog, she snapped at anything that moved. Move a chair at the dinner table and Ferdi would attack, growling and biting. Visitors were an obvious target and a friend of Lavinia's boys perfected a skilful technique for dealing with the situation. Calling at the front door he would stand on one leg when Lavinia answered, using the other to kick Ferdi away while greeting Lavinia politely and asking after Mike and Brian. Lavinia was none the wiser. But somehow Ferdi had never been inoculated against distemper, fell victim to it and had to be put down. The Danckwerts' short-lived Pyrenean mountain dog Beausie was followed in due course by Max, another exemplar of the breed, who proved to have more staying power. With the same long-haired coat, he too reached a formidable size. Peter was in the habit of bringing Max to work; junior staff could tell when the professor was 'in' from the distinctive odour of damp Pyrenean mountain dog about the department corridors.

One weekend when the Danckwerts were unexpectedly called away, Margaret Sansom was summoned at short notice to look after Max—by now a gigantic mountain of long, white hair. The Danckwerts took off, leaving Margaret in charge and, as time wore on, she gave Max his evening feed. Then it occurred to her that he should have a walk. She attached the lead and opened the door. Like Gerard Hoffnung's bricklayer, she realised a bit too late that Max's weight far exceeded her own. Rather than she taking Max for a walk, he was going to take her. Down Abbey Road they hurried, along Riverside and up River Lane beside the gas-works. At the corner with Newmarket Road Max applied the brakes and, with the conviction of a regular, turned into The Corner House pub and straight up to the bar.

In March 1966, Peter wrote to Ray Baddour:

We have bought a parrot. It doesn't say much yet, but I've told Lavinia to watch her tongue.

His concern was premature. Frankie the parrot defied all efforts to get it to speak English. His native language was the shriek, which he employed at ear-piercing volume whenever he identified a noisy rival: the vacuum cleaner, Charles on his tin whistle, the TV... Frankie's home was one part of a double-aviary in the kitchen; the other half was used from time to time by a white cockatoo called Zak. On occasional release into the kitchen, Frankie showed his appreciation by chewing away at the furniture or the window frames. He was particularly fond of Peter and swooped around him, sat on his shoulder and expressed devotion by taking small bites out of his ear or pulling out hairs. Just in case of unforeseen accidents—like the spillage of food, blood or avian body fluids—Peter wore an old grey jacket when indulging Frankie. It was reserved strictly for the purpose. He took to showing Frankie off to visitors. American post-graduate Mike Lockett recalls being one amongst research students attending *en masse* for a buffet supper. Peter mingled, with Frankie on his

Peter in the kitchen of Abbey House with Frankie the parrot

shoulder. The parrot had defecated down the back of Peter's jacket but neither Lockett nor any of the others had the courage to point it out.

Frankie's relations with Lavinia were not so cordial. He dive-bombed her when the opportunity presented itself and on occasion is reputed to have homed in on a plate, picked it up in his beak and let it fall onto the stone floor. Parrots discard feathers from time to time and Peter collected Frankie's, sending them as gifts to those members and friends of the family, particularly the Garsons, who were fly-fishing fanatics. In June 1966 Frankie accidentally escaped; teatimes at the department were enlivened for a fortnight by Peter's updates on his possible whereabouts. The Danckwerts advertised with a small reward for his return and, against all the odds, he was found: at The Plough public house in Fen Ditton just down river from Cambridge. Eventually Frankie became too much of a good thing and the Wisbeys next door, who kept quite a menagerie, took him in.

Lavinia's interest in animals was nothing if not eclectic. For a while she kept a goldfish in a bowl. Playfully, she named it 'Peter'. One morning when she found it floating motionless on the surface in its bowl, telegrams went out to friends advising: 'Peter is dead'.

Peter and Lavinia were particularly renowned for the sympathy and consideration they showed towards many of the visitors to Peter's department from abroad. Anthony Benis the post-doctoral researcher of Polish extraction and a graduate of Princeton and MIT, arrived in Cambridge in 1964 almost by accident. Preparing to go to Delft in Holland, he bumped into Tom Sherwood one day in an MIT corridor. Sherwood, through his friendship with Peter, suggested Cambridge as an option, and Benis changed his plans; the offer of an ICI fellowship arrived on Benis's desk from Peter within a week. Benis spent two years in Cambridge with his wife Agnes. A few days after they arrived the

Danckwerts, then still at Yorke House, invited them to dine to help them feel at home. The meal was briefly interrupted when the Danckwerts' pet hedgehog scurried into the room and curled up on the floor; Benis felt they had stumbled into a scene from 'Alice in Wonderland'. Later the women retired leaving the men to cigars and port—another surreal order of things which, for an American, belonged to an earlier world.

A year later Agnes Benis gave birth to a baby girl. Christmas was approaching and she felt she had to visit home to see her parents. On the other hand, she didn't want to expose her new-born daughter to the cold of a US winter, so Benis volunteered to look after baby Alison single-handed while she was away. Peter and Lavinia responded by inviting him plus baby for a traditional Christmas Day lunch at Abbey House. Lavinia's three boys were there too, plus an acquaintance of Benis's—Lou Shrier, on a year's leave from Exxon, New Jersey, with his wife and their son Jonathan. They dealt with some Russian champagne left by Soviet visitors earlier that month and, after demolishing the festive meal, everyone enjoyed pieces of a cake Lavinia had made. Peter enquired which liquor she had laced it with, and her innocent reply produced uncharacteristic spluttering; it was his favourite, and very old, brandy. Baby Alison slept throughout.

Benis benefited in other ways. Peter wanted to introduce a column into Pergamon's *Chemical Engineering Science* reviewing 'Soviet papers of interest to chemical engineers'. Realizing that Benis understood Russian, Peter arranged for him to visit a research library in London once a month, all expenses paid. And, on the pretext of visiting Pergamon's offices in Oxford, he sent Benis on a day's sight-seeing trip.

Another beneficiary of Peter's sensitivity to fellow professionals was the Californian chemical engineer George Standart. He had emigrated to Czechoslovakia on ideological grounds in 1948 and after almost 19 years

Lavinia and Peter in Marienbad, 1965

there he wanted out. Standart had retained his American passport but, with a young family to support, needed a senior job somewhere else. A return to the USA looked potentially tricky and alternatives were scarce. The high quality of Standart's work (he was developing the difficult field

BACKGROUNDER:
THE STANDART AFFAIR

Caltech graduate George Standart and his wife Phoebe felt ideologically ill at ease in America of the late 1940s. McCarthyism was taking hold and, although not personally threatened, the Standarts decided to leave California. Standart, who was working for Shell on distillation problems, was looking for a country with a developed chemical industry to carry on with research and pass on his skills. East Germany and Czechoslovakia appealed; one of Phoebe's forebears had been a doctor at the Habsburgs' imperial court in Vienna and travelled every summer to the spa city of Marienbad in western Bohemia. Old postcards showed a beautiful country: they chose Czechoslovakia. When they arrived in Prague in 1948, the authorities valued Standart's skills and he prospered. He was quickly appointed professor at the Prague Institute of Chemical Technology. Phoebe, a more enthusiastic left-winger than he, became a broad-casting journalist and did some editing and translating. By 1959 Standart was heading his division of the Prague Institute of Chemical Process Fundamentals and received a national prize (named after the post-war Czech president Klement Gottwald) for his efforts. By now he had an international

name in his field, but was tired of the communist bureaucracy, the lack of equipment, the ever-present secret police and the mismatch between slogans and reality. He investigated a position in Cuba but returned to Prague after four months. Then Peter intervened with a term in Cambridge and a visiting professorship in London. Phoebe and their two young daughters arrived in February 1967; there was wrangling over visas but people argued successfully on their behalf. Six months later Standart secured the UMIST professorship. Like Peter, Standart was an enthusiast for testing theories on a realistic scale. He secured very substantial UK government funding to carry out ambitious continuous projects on UMIST's large pilot plant, running it 24 hours a day and making the process work. It needed a project manager, a job that Standart had to do himself. He was a driven man and the task became overwhelming. In 1978 he embarked on a tour of the USA and, to his surprise, was warmly received. On his last stop in New York, Colombia University offered him a professorial post. Back in Manchester he agonized over the decision: a job in his favourite city or keeping faith with the country which had helped him leave Czechoslovakia. Shortly afterwards he suffered a heart attack which killed him. He was 57.

of irreversible thermodynamics for non-stationary processes) had already grabbed Peter's attention when the American submitted papers to *Chemical Engineering Science*. Peter acted as sole reviewer of Standart's substantial output, and they corresponded.

In 1965 Peter travelled to an international meeting in Czechoslovakia at Mariánské Lázně (Marienbad) with Lavinia and met Standart. Hearing about his dilemma, Peter invited him for a sabbatical term to Cambridge and by late October 1966 they were dining at Abbey House. Meanwhile Peter organised a visiting professorship at Imperial College London where his old friend Kenneth Denbigh could pull strings. Six months into the visit, the University of Manchester Institute of Science and Technology (UMIST) advertised for a professor; Standart applied, no doubt with glowing references from Peter, and spent the rest of his life based in Manchester. As his doctorate was from Prague's Charles University, getting the right robes for Manchester academic fancy dress parades proved quite a struggle.

Jerzy Buzek benefitted in quite a different way. In later life, as a Solidarność politician, he would become premier of Poland and then president of the European parliament; in 1971 he was a post-doctoral researcher interested in carbon dioxide recovery and storage. Attracted by the reputation of the Cambridge department he spent a year under Peter's supervision surviving, like other young Poles abroad at the time, on a shoestring. His memory of Peter is of a man who would sit, drink beer and chat with research students in the evening—something he never saw other professors doing. Buzek developed a great affection for him.

As Christmas approached Peter asked Buzek to carry out a procedure in the laboratory which required extremely pure ethyl alcohol as a starting material. Buzek ordered a litre and did as he was asked. The attempt seemed unsuccessful. He approached Peter who suggested he try again. Another litre was supplied but Buzek's efforts were still unsuccessful. This time Peter explained that he could provide a litre of alcohol every time Buzek needed it. Not really understanding the point of the exercise, Buzek busied himself with other work. At Christmas he and a Polish friend found themselves stranded far from home. Worst of all there was no money for vodka with which to celebrate. Then Buzek remembered the ethyl alcohol. He diluted it appropriately and the two friends got rather drunk together: 'excellent quality' was the verdict. Finally the penny dropped. Peter, aware that every Pole had a need for vodka now and then, had set

Buzek the laboratory task simply to provide it.

Lavinia's boys would often find well-known names in the house. One evening in the late 1960s the astronomer Fred Hoyle came to dinner. He and Peter had been made honorary members of the American Association for the Advancement of Science (AAAS) on the same day in 1964. Hoyle had an international reputation as a controversial thinker; his view of the origin of the universe was of continuous creation, which in those days was a hypothesis competing hammer and tongs with the 'big bang' concept. Amongst other strongly held views he believed that life must be a frequent occurrence in the universe. In a broadcast talk in the early 1950s when Australia was dominating England at cricket, he said he would wager that somewhere in the Milky Way there was a cricket team which could beat the Australians. On the evening of Hoyle's visit Peter's teenage step-son Charles was eating in. His imagination was caught by the conversation and he surprised Peter by being sufficiently up-to-date with the subject matter to join in.

The reaction of Lavinia's three boys to their stepfather was not without challenges; they were, after all, 1960s teenagers. Brian was happy to find Peter open-minded, without preconceptions and liking to talk to young people because they said what they were thinking without inhibitions. He and Peter enjoyed tickling each other's sense of humour with tales of ridiculous situations they had come across: Brian knew that if Peter found a good story but no-one to whom he could repeat it immediately, it might reduce him to silent giggles which brought tears to his eyes as he waited to pass it on.

Charles, the youngest step-son, was the most affected by his mother suddenly providing him with a stepfather (and his father with a stepmother) and could be prickly. As a teenager he became enthusiastic about environmentalist ideas, and he argued with Peter about things like recycling glass bottles. Peter's line on this was that the beaches of the world provide an inexhaustible supply of raw material—sand—for the manufacture of glass, so it was pointless to recycle bottles. Charles was concerned about discarded glass bottles as litter and by energy expended in making new bottles rather than reusing old ones, things that Peter didn't seem to find significant. Michael, the oldest step-son, was an independent soul; he struck out on his own and joined the Navy. Both his brothers went to Winchester and on to university. Charles temporarily gave up a degree course in maths and physics to go and grow food—

socially useful, he felt. He went back and finished it later with a diploma from the Royal Agricultural College.

Peter was a sceptic about religion but Lavinia was a regular attendee at St Benet's church in Cambridge. St Benet's was manned in those days by the Society of St Francis, due to a post-war dearth of clergymen in East Anglia. Lavinia liked to keep in contact with the Franciscan brothers and invited selected individuals to Abbey House once a week to dine. It was, for them, an eagerly-anticipated pleasure in their otherwise rather austere lives and they made sure they didn't miss it. Because of their long robes and the knotted rope round their waists, Peter called them 'dressing-gown men' but never to their faces.

Amongst these visitors, Brother Michael was a particularly bright member of the order. A sometime member of the Young Communist League, he was head of Anglican Franciscans in England and later became Anglican bishop of St Germans, Cornwall, assistant to the bishop of Truro, and later still minister general of the whole Society of St Francis. Peter particularly enjoyed conversation with him, but Brother Michael was an experienced missionary and Peter gradually got the feeling that an attempt to recruit him was in train. They fell out. Other brothers came and went. One who made a deep impression on the boys was Brother Lothian, a frail old man; on one visit the mighty Max got hold of his habit as he arrived in the entrance hall and swung him helplessly round and round. Only family intervention rescued him.

Relations between Peter and Eileen Mitchell remained cordial. Lavinia's boys were of an age with Eileen's twins Julia and Jeremy, and were frequent visitors at the Mitchells. They all got on well together, and with Jonnie Mitchell who was in and out of the Danckwerts' home as part of the family. Eileen was an occasional guest at Abbey House dinners and cocktail parties, and sometimes Jonnie went too. Peter paid for Jonnie's education—he attended Oakham school—and would have stumped up for Winchester had it been required. As it was, Jonnie lived in at Abbey House for six months in 1972 while he attended the Cambridgeshire College of Arts and Technology and his mother, by then Mrs McNeil, was living in Norfolk. His impression of Peter was of a benign uncle figure.

Marriage had brought Lavinia to the attention of the wider Danck-werts clan. They were struck by her style and her aristocratic accent, and regarded her as a class act with a patrician air. Most judged her 'lovely', some as 'Junoesque'. They admired her languid approach to life. At the same time, Peter had 'favourite uncle' status with his nieces and nephews. He would take visiting family punting on the Cambridge backs where nephews admired his ability to yodel as they passed under bridges. They supposed he had picked it up in Austria before the war and no doubt honed it on Alpine hillsides during subsequent skiing trips. Nieces saw the glamorous side: Micky's daughter Ingrid adored her uncle who seemed to her the most dashing, exciting and amusing of men. Peter's love of mechanical toys came in handy with younger children in the wider family: he would buy toys that amused him, play with them himself and then pass them on as intriguing presents. It wasn't unusual, for instance, to find a set of wind-up teeth on the mantelpiece at Abbey House. Jetex engines, wind-up machine guns, model steam engines and train sets were all grist to the mill. And if all else failed he usually had a bag of liquorice in his pocket. Occasionally friends turned the tables on him: a clockwork Dalek went down well as a birthday present.

Ingrid Danckwerts was a frequent visitor at Abbey House at the end of the 1970s, when she was studying in Cambridge. She was accepted into the bosom of the family and particularly appreciated her aunt's cooking. 'Lavinia prepared a lot of soups, and when she made a trifle there was always a danger that it would turn into a soup. She would add sherry to a formula: "one glass for the trifle and one for me". And then she'd add more to the trifle in case she hadn't been generous enough.' These trifles were very popular with everyone, but Ingrid was fascinated by Lavinia's 'aristocratic disdain for food hygiene'. Lavinia's son Charles provides an example: seeking out his mother one morning, he found her extracting what he described as 'dodgy bits of meat' out of a stew. Alongside an inexplicably agitated dog fixed its gaze on the stew pot. 'I'm fishing out the dog-food, dear,' Lavinia explained. 'I put it into the wrong dish.' Charles was rushing off to college and avoided that stew, but apparently everyone else said it was delicious.

Ingrid appreciated the gardens of Abbey House, a delightful area of calm and leafiness. Margaret Hoather, the previous tenant with green fingers, had left things in pretty good shape and Lavinia took over where she had left off. She had the assistance of a gardener of Romanian extraction called Gross. A productive walnut tree was one of the more interesting specimens and Ingrid had a go at producing liquor from it. But the star of the garden was a hybrid peach tree. Created by the inspired grafting of some earlier enthusiast, it produced really wonderful fruit and

Kew Gardens was said to be interested. But Gross, always a fierce pruner, eventually damaged the walnut and was so ruthless with the peach that it died. Peter christened him 'Gross the destroyer' and had him sacked.

Isolated from wider social contact in his professorial role, Peter was finding increasing compensation with alcohol. His secretary Margaret Sansom was broad-minded and was not thrown by Peter's attachment to the bottle. In early days she told Emma Maccall how Peter would lunch on a bowl of soup and a bottle of claret. In later years she would cycle home from the office with the day's empty gin bottles in her basket, so that the cleaners wouldn't make a fuss. Lavinia frequently made attempts to stop Peter overdoing the drinking. Dining out with acquaintances he was quite capable of downing a bottle and a half of wine at dinner before going home, leaving his hosts stranded. Lavinia sought help and Peter made regular visits to a local psychiatrist, although it is not clear that alcohol as an issue was addressed head on. At times Lavinia persuaded him on to the wagon. When he was made president of the Institution of Chemical Engineers for a year in mid-1965, one of his duties was a big speech at the annual dinner. As it approached, Peter spent three months off alcohol. But it didn't last, even though he understood very well what the consequences would eventually be.

Travels, travails and cancer 30

RIGHT AT THE START of their life in Cambridge and before they were married, Peter had asked Lavinia what she would most like to do in the future. She told him she wanted to travel. So he took every opportunity he got to travel himself and made sure she went with him whenever practical. Margaret Sansom organised Peter's schedules for his many foreign trips in minute detail, even keeping track of the numbers and ages of his host's children so that the right noises could be made and presents given. Peter fitted in trips around the timetable of university terms; he was in great demand as a contributor to conferences, as a consultant, as an adviser and as a personality.

The schedule was relentless: in 1960 he was in Venice and Milan, and during the summer Vancouver and Phoenix. In the summer of 1961 he and Lavinia undertook an extensive tour of the USA, starting in the Rockies, continuing on the west coast and ending in New York. Peter himself had a side-trip to Japan in November 1961. There were visits to Amsterdam, Milan and Copenhagen in 1962, plus a summer holiday in Scotland, while Peter scurried to appointments around the UK and returned to his busy Cambridge schedule.

In 1963 he accepted an invitation from the Royal Australian Chemical Institute to tour Australian departments. In those days the way out was either six weeks in a liner or thirty hours in a Boeing 707 stopping off for fuel at Athens, Cairo, Karachi, Calcutta, Singapore and Perth before reaching Sydney. In early August Peter set off first; sensibly he broke the journey in Singapore for a few days. With Lavinia due to follow, he sent her a postcard offering some useful tips:

Very hot. I should wear something light. You'll need a drink in Cairo—you can change two half-crowns and buy a Scotch. You'll find some air-conditioners to sit by near the windows of the transit lounge. Smashing Indian air-hostesses on this line. I would like a Dayak hostess.

Once they got to Australia things began to go wrong. If there is one

thing Australians don't like it
is an English aristocrat, and
that's how they perceived
Peter. He took a shooting-
stick with him to save his legs
during lengthy tours of university facilities, and this
enhanced the 'posh' image.
His Cambridge colleague
John Davidson remembers
that 'Peter could be slightly
disparaging about academic

Peter's Singapore card: four Dayak maidens
Courtesy Borneo Postcard

work which didn't have obvious industrial application,' and he spotted
some of that kind of work in Australia. At the University of New South
Wales in Sydney he addressed staff and students and cited several locally-
produced papers, suggesting that they were just academics talking to
each other—inward-looking and ignoring industry. 'There was an element
of truth in it,' Davidson allows. 'But it didn't please the Australians, to put
it mildly.' Local boy Rolf Prince, later to follow both Peter and John
Davidson as a president of the Institution of Chemical Engineers in the UK,
was there at the time. 'My recollection is of him sitting on that shooting-
stick in the middle of our main laboratory, expounding on the pleasures
of being chairman of his college wine committee.' Just as Peter had hit it
off immediately with Americans, he struggled with Australians.

The tour overall was a public relations disaster. Davidson made visits
four years later and again in 1970 and found the Aussies still upset in a
big way. Bill McManamey was a junior member of staff in Sydney. He was
considering applying for a position at Cambridge, 'our Valhalla then,'
recalls Prince. McManamey says now that: '[Peter] had a very good
mind, and saw problems very clearly.' But in his talk Peter mentioned a
technique McManamey was using in his modelling work and said it
wasn't much good. McManamey eventually accepted an offer from
Birmingham University, where he found a more congenial approach and
permanent employment.

Peter and Lavinia spent six weeks in Australia. Peter's lecture tour
often attracted publicity and he ranged widely in his subject matter. On
the radio in Melbourne he appealed to young Australians not to be
dazzled by the false glamour of pure science as presented in films, and so

be tempted to reject engineering as dull:

Pure science is an intensely exciting occupation for the genius who makes the real break-through, but for many it turns out to be just another humdrum, rather plodding profession. Engineering offers all the opportunity for brilliance needed.

Brisbane found him suggesting to Queenslanders that dual purpose nuclear plants were worth considering in Australian conditions. Coal was cheap, he realised, so nuclear power could not compete and anyway, to be economic, a nuclear power station needed an output far in excess of Australia's needs. But it was worth looking at the economic case for a nuclear station that also converted seawater into fresh water, he claimed. Water produced in this way would be very expensive, he conceded, but it might bring a good return—perhaps through irrigation in arid areas.

The Danckwerts took time off to visit Peter's brother Dick, his wife Ann and their four children then aged from 15 to 23 in Hunters Hill, a suburb on the Parramatta river, five miles upstream from Sydney harbour bridge. There he found an earlier idiosyncratic gesture bearing fruit. He had given Laura, the youngest child and then only eight, one of his quirky Christmas presents: a biology book. Laura was 'completely captivated by it'; when she got to the University of Sydney she chose the biology course.

The Danckwerts moved on to New Zealand towards the end of September and things went better. Publicity continued. In Dunedin Peter made sure he spoke directly to local agricultural industry. He urged awareness of the potential dangers of seed dressings, quipping:

It is said that cannibals will no longer eat Americans because they have too much DDT in their bodies.

Then he turned to the value of applying chemical engineering principles in the local food industry to freezing, drying and dehydrating. He dealt with iron and steel production. And there was a plug inspired by his own long-term research aims in gas absorption: to find ways to design chemical projects on paper in the same way a bridge could be designed:

At the moment if we want to design a chemical plant we have to build a small plant to see if it works. By looking into the fundamentals we hope it will not be necessary to rely so much on trial and error.

In Otago, where his first research student Miles Kennedy was on the staff, Arthur Williamson was part of the team. 'I recall him chatting about university life in general and about the way things were changing, and in particular that there was getting to be less time for academics to indulge in what he called "productive indolence"—just putting one's feet up and pondering things. I was always impressed by this term and wonder, given that the remark was made in the 1960s, what he would feel about the way things are 50 years later.'

At the University of Canterbury, the recently-appointed Roger Keey was nervous and didn't do his research efforts on mass transfer justice. Peter, seated on his shooting stick in front of one of Keey's stirred tanks, wondered aloud why Keey was bothering. Later in his career Keey saw a different side. Peter was commissioning, on behalf of Pergamon Press, a series of monographs on distinctive chemical engineering operations. Keey wrote in saying that drying—one of his special interests—should be on the list. Peter replied encouraging him to write a monograph and when in due course Keey sent the draft in, Peter made no editorial changes.

Ten days in New Zealand flew by, and Peter and Lavinia set off for Fiji and San Francisco before crossing to the American east coast to visit old friends in Philadelphia, Wilmington, Baltimore, Ann Arbor, Syracuse and Montreal. Their final flight from New York took off on 23 October; they had been away for two and a half months. To complete 1963 Peter and Lavinia visited Paris in November for a conference. And so it went on.

Throughout Peter's professorship in Cambridge he continued to work as executive editor of *Chemical Engineering Science*. It is hard to know just how it felt to be an aspiring author in that period. Many editors of such big scientific titles consider it part of their role to advise would-be contributors whose writing, or perhaps whose experimental work, doesn't come up to scratch on how they might do better. Not so Peter. He favoured a one-liner: 'Your contribution is not acceptable for publication'. Peter's key North American editor—and his equivalent in reputation in the USA—Neal Amundson had joined the editorial board in 1957 when he was at the University of Minnesota. Like Peter he was allergic to paperwork. Known locally as 'Chief', his editing duties soon led him to handle: 'a substantial proportion of all manuscripts, and all of those from the USA. Peter would on occasion chide me for accepting too many mathematical papers. But, since I was a mathematical formalist and he was an ideas person...'.

In the Soviet bloc Peter enjoyed an enormous reputation as an innovator. A Soviet scientist came to spend a year in the Cambridge department early in the 1960s, and Peter got the chance to observe the Russian temperament at close quarters. The visitor was apparently at the level of a post-doctoral person, but his unusual background led to niggling difficulties. For instance, he was not allowed to take books out of the University library.

His explanation of these problems was always the same: 'I am Soviet man'. However, he managed to get almost all the technicians in the department working for him (on a project none of us understood very well). He did not seem to be a convert to the Western way of life, any more than Soltzhenitsyn; yet when my wife and I went to Moscow a year or so later he turned up in our room with a copy of *Anglia*, a glossy English propaganda leaflet, under his arm together with a load of goodies such as caviar, vodka and Armenian brandy. He looked back on his stay in England with nostalgia, and it had given him considerable prestige back home in the Motherland.

It was in July 1965 that Peter was invited to Moscow for an international conference. He was due to attend another in Stockholm in August and a third in September in Czechoslovakia, but he managed to squeeze the Russian trip in. Peter and Lavinia travelled via Leningrad where they were exposed to a classic Soviet characteristic: the claim that every significant technical advance in world history is the discovery of a Russian. Lavinia sent a postcard home to her son Charles: 'They've got a hovercraft that runs down the Neva river and they think it's a Russian invention'.

On arrival in Moscow Peter was delighted to meet up again with his MIT teacher Tom Sherwood who was also on the list of speakers. Lavinia and Virginia Sherwood joined them for a little sightseeing. One rainy afternoon they were touring the grounds of Arkhanghelskoye, an estate some 15 miles west of Moscow, when their academician guide paused before a statue of Pushkin.

'We need more young poets like that,' said the academician. I replied 'Well, you have Yevtushenko'. He practically spat: it was not just retired colonels at Cheltenham.

Eventually their guide suggested

Sightseeing in Moscow, July 1965 (l to r): Peter, Igor Martyushin, Virginia and Tom Sherwood, Lavinia

that they should go to his flat for 'fivaclock'. Lavinia, feeling dehydrated, naïvely hoped for a glass of tea and, with luck, a slice of lemon. But it turned out to be a party which went on until after midnight.

[I] sat between Russians who spoke German and French respectively as their second languages. As the Armenian brandy flowed I became unexpectedly fluent in both languages.

Peter rashly brought up the subject of old Russian proverbs with which, he recalled, Nikita Khrushchev had been so liberally provided by speech-writers during his visit to England in 1956. Of course everyone wanted to hear a few of them, but Peter found that he couldn't remember any.

Eventually I had to fabricate one: 'Every dog has its fleas'. It sounded genuine and got a big laugh.

When they eventually reached their hotel Peter relapsed into a trance, and Lavinia spent an anxious night ensuring that his heart was still beating. The Sherwoods suffered a different symptom of over-exposure to Russian hospitality: Virginia did not stop talking all night. Another memory Tom Sherwood took home from that trip was the terrible tyranny of Soviet hierarchy which penetrated as far as food: at lunch a senior official might get a steak while a junior one had to make do with the equivalent of baked beans.

At the conference itself Peter and Lavinia were met by dignitaries plus Igor Martyushin, a Communist party member in good standing who frequently acted as translator for English-speaking visitors. A middle-ranking engineer, Martyushin had spent his school years in London and was almost bilingual. His sensitivity to language was such that he was able to translate jokes and get laughs across the linguistic barrier. Of course, Peter found him delightful.

He told us on the way to our hotel: 'This conference is going to be a disaster. In spite of instructions, most of the speakers have brought slides of the wrong size.' He did not seem to be particularly worried.

Speakers' slides weren't the only problem. Peter was due to deliver a lecture to all the delegates, one of only a few with simultaneous translation. Towards the end of his allotted time, when he found he was going to over-run, Peter decided to omit some paragraphs. He made it clear to the translator what he was doing... or so he thought. The man just went on reading Peter's original script and was left droning on for some time after Peter had sat down. At the closing session of the conference, in the Kremlin's Hall of Congresses, the British delegation was first on with its closing remarks. Peter's friend Fred Dainton was lined up to do the job.

Lord Todd, our leader, became increasingly nervous and at literally five minutes' notice he grabbed me and said 'Dainton's not here, you will have to do'. 'What on earth am I supposed to say?' I asked. 'Oh, just blah, blah, blah,' he replied. So I got off some inanities about having a jolly time and how much the ladies' programme had been appreciated. All this was being translated into numerous languages and broadcast throughout the Soviet Union, possibly the world. Subsequent speakers roamed at large over Mankind, Peace and the blessings of Chemistry. I was past caring.

All the same, it was an experience he often recalled with nostalgia when the TV news showed pictures of Soviet top brass holding forth on the same platform where he had briefly performed.

Inevitably Peter made the acquaintance of senior Russian chemical engineers. Academician Nikolai Zhavoronkov was one of the leaders of the profession in the USSR. Like Peter, he regarded theory as only useful when both verified by experiment and related to industrial practice. As a science student in the 1930s Zhavoronkov had been impressed by the government's call for engineers and switched studies to Moscow's Mendeleev institute of chemical technology. Its director from 1948, he went on to create the discipline of chemical engineering in the Soviet Union. Following the advice of Igor Kurchatov, father of the Soviet atomic bomb, Zhavoronkov had organised a secret laboratory for isotope separation in Moscow and then a faculty at the Mendeleev institute to cover all chemical aspects of the Soviet nuclear project. The laboratory still exists as part of the Kurnakov Institute. Zhavoronkov was already a major figure in the Soviet Academy of Sciences when Peter first met him. The two men shared interest in and experience of things nuclear, and got on very well together. Peter invited Zhavoronkov to Cambridge; he turned up, with Martyushin in tow, in December 1965. During his stay Zhavoronkov questioned Peter intently about his Pergamon jour-

Igor Kurchatov (r), father of the Soviet atomic bomb and mentor to his neighbour Nikolai Zhavoronkov at a meeting in the Kremlin
Courtesy Ksenia Zhavoronkova

nal *Chemical Engineering Science,* and shortly after his return home founded a new Soviet journal, *Theoretical Foundations of Chemical Engineering.*

Peter expressed the hope, in a letter to Ray Baddour, that Zhavoronkov and Martyushin's 'amusing visit' had 'served to undermine their Marxian complacency.' Two visits to Abbey House might have achieved this, if nothing else. Cocktails to welcome them in early December were followed by a small dinner party on 18 December before they left. Lavinia pulled out all the stops: prawn cock-

Nikolai Zhavoronkov (l) with Peter in front of Newnham Cottage

tail, roast duck with orange sauce and her famous meringues, with a bonus of chocolate soufflé for those who still had the space. During the visit Peter took Zhavoronkov to Newnham Cottage, the house on Queen's Road which Ernest (Lord) Rutherford had occupied as professor of physics at Cambridge University during the last twenty years of his life. They posed for Martyushin. When the guests departed, they left behind some bottles of Russian champagne which came in useful at Christmas.

It was Zhavoronkov who suggested to the USSR academy of sciences that Peter should make a second visit in 1972 to see research in Moscow, Leningrad and Novosibirsk. Naturally, Lavinia went too. Zhavoronkov himself accompanied Peter in Moscow but delegated responsibilities for Novosibirsk and Leningrad to Nikolai Kulov, a junior member of his staff. They set off for Novosibirsk in an Antonov aircraft with four pairs of contrary-rotating propellers which made an indescribable racket and added, Peter recorded, to the misery of the hangover which developed as a result of their farewell party in Moscow. On arrival they travelled to the Akademgorodok scientific complex about twenty miles south of the city along 'an abominably potholed road'. Peter found the surrounding countryside featureless and heard that it was bitterly cold in winter and plagued with mosquitoes in the summer. His host in Akademgorodok was George Boreskov, academician and director of the institute of catalysis. The mornings were spent visiting laboratories by chauffeur-driven car;

there was little for Lavinia to do besides check out the local geological museum. On a typical day they would be reunited for a lunch party which, with numerous courses and toasts, lasted until about 4pm. After that the only option was back to the hotel to sleep it off.

Before this second visit to the USSR, Peter and Lavinia had spent an Easter in Israel. They did some sight-seeing and visited Jerusalem. The technical purpose of the trip was to visit Bill Resnick at the Technion University in Haifa. Resnick and his colleagues, knowing that Peter was due to visit the Soviet Union, asked him to raise the case of a Jewish physicist who had applied to emigrate to Israel and was being harassed by the Russian authorities.

Peter tries local transport in Israel

I undertook to raise the case with the most influential people I met, but this sort of thing is not really a matter for amateurs. On the first occasion when it came up I was visiting a scientific institute. The director and his staff seated at a long table explained in turns what they were doing. At the end the director turned to me and said 'Have you any questions?' and I answered 'Yes, what about X?' There was a sharp intake of breath and I was subjected to a long harangue on the impertinence of my remark. Proceedings then reverted to normal.

A second opportunity came at a Moscow restaurant with the director and staff of a local institute. As the party mellowed with the usual toasts in vodka, Georgian wine and Armenian brandy, Peter felt he could ask confidentially: 'What's all this about X?' The director froze immediately, stopped speaking English and the rest of the conversation was conducted through an interpreter.

Our own particular host, an academician, announced unexpectedly that he would have to withdraw as he had some leave and was going to the Black Sea. I like to think he did not want to discuss X with me. X eventually got his visa and, like all good Russian Jews, is now living in the USA.

When Lavinia and Peter arrived with Nikolai Kulov in Leningrad, their host was Peter Grigoryevich Romankov—known to all as PG—a

chemical engineer with a great regard for Peter's reputation. The menfolk
toured universities and institutes. Lavinia had time to investigate the
sights of Leningrad, something for which there had been little time
during her first visit in 1965. Kulov's research student, his first, showed
her round; together they scaled the dome of St Isaac's cathedral and,
benefiting from PG's connections, got access to otherwise restricted places
like the gold storerooms of the Hermitage. Eventually Peter and Lavinia
had to return to Moscow, and were booked on the Red Arrow, the crack
night express. Peter wrote later in *New Scientist*:

[We] were in a first-class sleeping compartment as the train began to roll southwards. Also, unfortunately, were our Leningrad hosts. With scarcely an interruption in
her conversation, our hostess yanked the communication cord and the train ground
to a halt. They took their leave politely and we started off again. No recriminations,
no thousand-rouble fine.

Peter had condensed the incident for simplicity and impact. Their hosts
were PG, his fragile wife Lyudmila (a survivor of the siege of Leningrad)
and Kulov. After a fine dinner at the Romankovs' apartment, all set off
in high spirits for the station. Peter had taken the precaution of bringing
along a bottle of Soviet champagne and it flowed, along with the
conversation, in the Danckwerts' compartment. When the train
unexpectedly began to move, PG bolted from the compartment, down the
corridor and out onto the platform, shouting something about the lecture
he was due to give the following morning. His wife attempted to follow
but the train had begun to gather speed and Kulov managed to restrain
her from jumping. The spectre of her stranded at the next stop, far down
the line, without money and all alone in the middle of the night, was too
much for him. With some misgivings about retribution, Kulov made a
quick decision: he reached for the 'glittering nickel-plated handle of the
emergency brake'. The train shuddered to a halt and Romankova got off.

Kulov returned to Peter and told him he would retire to his own
compartment to await a telling off and a possible fine. Peter would have
none of it, insisting Kulov stay with him so that he, Peter, could take part
in explanations. Finally Peter got his way by producing a bottle of whisky
from his case. Kulov relaxed: 'Waiting was pleasant, and nobody came!'
Kulov was touched when Peter described his behaviour as that of a
gentleman. Next morning Romankov phoned Kulov in Moscow to thank
him for 'saving my wife'. Nowadays Kulov tells his students the tale to
illustrate—and here Peter's sense of humour may have rubbed off—the

characteristics of a 'real' professor for whom 'the forthcoming lecture is more important than his wife'.

After his time at the 1955 Atoms for Peace conference in Geneva, Peter had some experience of the challenges which his Soviet counterparts faced in attending international and even national get-togethers. They had difficulty in persuading their authorities to let them attend in the first place and when they did arrive, Peter was amused to observe, they expected to be able to present their contributions immediately, regardless of the timetable. Peter put it down partly to the bureaucracy which, he used to say, 'makes Whitehall seem like a Citizens' Advice Bureau'. Then there was the Slav temperament... Conferences within Russia, Peter observed, were no better. They never started on time because the participants had forgotten the date, or been stranded by the unreliable internal air flights or had simply not been able to get the necessary travel permits. But he found that Russians invariably regarded the situation with philosophy and good humour. In the early 1970s, for instance:

[They] referred to the pauses in their conferences as 'waiting for Fisher', an allusion to the Fisher-Spassky [1972] world chess championship during which Fisher (the US master) invariably turned up late (if at all).

Kulov went on to spend the 1973-74 academic year with Peter in Cambridge; he later became head of Moscow's Kurnakov institute. He recalls that Peter and his professor Zhavoronkov hoped that their two departments would find ways to work more closely together. Lavinia told intimates that the Russians had offered Peter a lot of money to go and work in Russia, but that it really didn't attract him. Kulov considers such offers very unlikely but remembers that Zhavoronkov made unsuccessful attempts to have Peter elected an overseas member of the Soviet Academy of Sciences. Peter's views as a young man before WW2 may have suggested left-wing sympathies, but with maturity he had certainly become more sceptical. On the other hand he wrote later about the 'twinge of envy' that Fellows of the Royal Society—Peter had been one since 1969—might feel when they compared their £30-a-year subscription with the fringe benefits of membership of the Academy of Sciences in the USSR:

[They] include apparently an automatic rise in salary on election, access to special food shops, superior housing and the use of chauffeur-driven cars. I believe there is also an element of *droit de seigneur.* Natasha, the scientific secretary of one institute I visited, was pretty, sprightly and smartly dressed, in sharp contrast to most other scientific secretaries who tended to be stocky and peasant-like, and to wear head

shawls. I subsequently heard that she had become pregnant and been married off
to a junior scientist.

By 1966 Peter and his research teams had gathered a huge amount of
data on the absorption of gases, either in liquids with which the gases
react or in liquids which have something dissolved in them which reacts
with the gas. He decided to write a book gathering together all the
experience he had gained from this research and reviewing in context all
the published work on gas absorption. He wrote:

The only book, to my knowledge, which deals exclusively with the subject is
[Gianni] Astarita's *Mass transfer with chemical reaction* (Elsevier, 1967), an
admirably clear exposition to which I am indebted. My object has been to fill the
need, still outstanding at the time of writing, for a book which will be of value equally
to the process designer, to the research worker and to the student.

The project was huge, and the effort required equally so, both in terms
of time and energy. To make time, Peter cut down on his consultancy and
other activities. He wrote to Ray Baddour:

I have been offered some interesting jobs, which I have had the sense not to
accept. I am progressively resigning from committees and intend to write a book.

For some time Peter had been an irregular attender of the two faculty
boards on which chemical engineering had a seat—physics and chemistry,
and engineering. Others sat in for him. Now his colleagues were really
miffed when he extended his rejection of time-occupying committees by
declining an invitation to sit on the University's general board. This is the
body that decides which department should get money for capital projects
and for filling staff vacancies, and every department wants its professor
to sit on it and influence decisions. But writing was an enormous effort
for Peter. It came, after all, on top of his responsibilities in the department
and his travel schedule. In the text he acknowledged his big debt to
Sharma who read and commented on early drafts and supplied a large
number of relevant industrial examples. *Gas-liquid reactions* was
eventually published by McGraw-Hill in 1970.

Reviews appeared in the technical press, one of them part-authored by
a young man at the University of Nancy—Jean-Claude Charpentier—
whose research was precisely in the area of the book. The 500-word
review praised the 'excellent collection' of data but warned readers the
book was only for specialists who, even then, would have to go back to
detail in Peter's original papers to get the best out of it. It so happened that
Charpentier arrived in Cambridge in January 1971 for six months,

financed by the Royal Society. When they met, Peter explained to him that he was the first French student to work in the Cambridge department and grilled him about his life to date.

Next day Peter took him on a tour of the streets of Cambridge, so that he could more easily choose suitable lodgings: not, in Charpentier's experience, normal practice for a professor. Then Peter set him his first task: to read *Gas-liquid reactions* again—page by page, line by line, word by word—and make a list of mistakes. Several weeks later, Charpentier presented his list. It was very short. 'I still remember that it cost me a lot of time and work to find these few mistakes', he says, 'so extremely good was the content, style and presentation of the book.' Peter studied the list and murmured to himself 'ho hum, oh dear, oh dear'.

Charpentier remembers his stay in Cambridge as *'merveilleux et très instructif'*. Naturally he dined at Abbey House where he found Peter with Frankie the parrot on board. Perhaps in thanks for Charpentier's effort in compiling that list of errors, Peter met him quite often for discussion. Sometimes the meetings were less formal. Peter set him 'tests' to which overseas students were subjected: adequate skill at punting on the river Cam (Charpentier passed), and some understanding of the game of cricket (he failed dismally). He also introduced him to sherry in all its forms. At other times they met in the department. So much so that when Charpentier returned to Nancy he was able to set up a research facility amply based on a great number of the ideas he had encountered in those discussions and in the book.

Peter was becoming exhausted, and his attachment to alcohol was making things worse. There had already been signs of discomfort. In October 1971 Lavinia wrote to her son Charles: 'We had a departmental drinks party last Friday...and we go up to London tomorrow. Peter is to see a specialist as he is very low'. Lavinia had made earlier attempts to get Peter to reform his approach to alcohol; now she had identified an ex-naval psychiatrist based in London with wartime experience similar to Peter's; she thought he might listen to another navy man. But Peter simply wasn't interested. His hectic travel schedule continued.

Armando Tavares da Silva, who revisited Cambridge to work with Peter for a few months in the summer of 1971, invited Peter to the University of Coimbra, Portugal. He was keen to get Peter's advice on the new course he had just set up and on the laboratories being built around it. Peter visited him in September 1973. A little later Miles Kennedy arrived in Cambridge from New Zealand as a visiting professor. 'We were to have a chat about what I should do to earn my keep,' Kennedy recalls, 'so he took me off to a somewhat sleazy pub. It was 10 or 11 in the morning: I think I had a beer but he was drinking gin.' When Kennedy and his wife Betsy invited the Danckwerts for a meal, Lavinia rang them beforehand to see if they would mind not serving wine. The Kennedys guessed that Lavinia was trying to dry her husband out. An old naval friend of Peter's visiting Abbey House commented with concern about the collection of empty whisky bottles in the bathroom. But it was all to no avail.

In November 1973 Lavinia was writing to her son Charles: 'Peter went to see a specialist recommended by [his brother] Dick; he has taken Peter off the barbiturates and all the other pills, said "No alcohol!" for six months, and given him a different pill. I've joined him and am drinking grape juice and water and orange while Peter drinks tomato juice heavily laced with Worcester sauce. He looks better already and is losing weight. Me too. We both smoke instead.' It didn't last. By early 1974 Lavinia was increasingly worried and Peter was finding his responsibilities as head of department too onerous. And yet he pressed on with his other commitments. In mid-March 1974 Peter and Lavinia took a month to tour India, dropping in on Sharma in Mumbai: for Lavinia it was a lazy holiday but Peter was still working hard. Margaret Sansom sent Sharma a message: 'don't offer PVD alcohol'.

By July Peter was ailing and there were more visits to specialists. Then a crisis occurred. Peter suffered what Lavinia described as 'rigours'; she was used to seeing the impact of Peter's drinking but this seemed to be different. He went into Addenbrooke's Hospital in Cambridge on Sunday 22 September 1974 to prepare for an investigative operation two days later. After the operation Lavinia seemed quite relieved: 'It turned out to be cancer of the colon as we thought. It was a major operation but he is immensely tough and came through it well.' Fortunately there were no signs of any secondary cancers; Peter returned home for three weeks before a second visit to be finally sewn up. Surgeons prefer to do this kind of operation in one go, but there must have been doubts about healing and they chose to attach a colostomy bag first and couple things up later. Peter prided himself on this toughness: he wrote to Nikolai Kulov shortly after the second visit: 'The doctors regarded me as some kind of superman because I recovered so quickly.'

A break in North Carolina 31

THE ACADEMIC YEAR 1975–76 was Peter's sabbatical year and a chance to recover from his operation for colon cancer. He had finally agreed to resign as head of department and John Davidson took on those responsibilities. Davidson was also to take over Peter's lecture course; in order to show his successor what he had been lecturing about, Peter handed him a file marked 'Miscellaneous'; it contained a few sheets of paper covered in jottings.

For some time Peter had been hatching plans with Erdogan (Ed) Alper to spend a year in Turkey. Alper had been Peter's student in Cambridge from 1968 to 1972 and was now a junior faculty member at Hacettepe University, Ankara. Alper felt a debt of gratitude for the way Peter had mentored him during his studies; as a foreign visitor sensitive to Peter's way with words, he recalls his delight at the pithy advice he received expressed in very British English: 'If I were you I would not do that' and 'There is more than one way to skin a cat'. A Turkish invitation had already arrived, the Middle East Technical University in Ankara was expecting him as a visiting fellow and the British Council in Turkey was lined up.

But Lavinia didn't really fancy this. Peter would be better able to relax, she felt, in the United States where everything would be familiar and where he would feel more at home. It was only in May 1975 that Lavinia was finally able to persuade Peter to change his plans. He wrote to North Carolina State University in Raleigh and asked to spend a year there. The campus is well placed amongst several other universities, appeared to be well-equipped with libraries and the area was home to a wide range of retired officers of the armed forces. (Alper compensated for his disappointment by visiting Peter in Cambridge during the summer of 1977 for three months.)

As Peter waited for a response, visitors continued to arrive at Abbey House, some unannounced. On 9 July 1975, a few months before they were due to leave, Lavinia answered the door to two French women who had turned up out of the blue. One of them was an old wartime friend of Peter's: Marie-Louise Perreux. She told Lavinia that she had been the only female WW2 correspondent in North Africa and at some stage had spent nine months in the Sahara with the Camel Corps. 'A real tough cookie you would think,' Lavinia wrote to her son Charles, 'but... she got haunted in the spare room and I had to spend the night with her.'

Meanwhile in Raleigh, head of department James Ferrell agreed to invite Peter for a sabbatical year on the basis of his reputation ('not many papers but they were ground-breaking') and on condition that staff member Hal Hopfenberg looked after him. Hopfenberg and his wife Patsy found the Danckwerts a house to rent in Schaub Drive about a mile from the campus, and generally kept an eye on them. The main point of the sabbatical was medical recovery and Lavinia did her best to keep Peter relaxing, initially without much success. In November 1975 she wrote to her son Charles: 'I've got Peter at home at the moment—temperature, headache, etc. I don't think he's really got his strength back fully yet and he's snowed under with work. I swept into the department this morning, collected all his papers and told everyone he wouldn't be in this week. There's no point in doing things by halves and he needs a rest. I think he rather over-estimated his capacity.'

They had a car and, as usual, Peter knew people. So they went visiting, and there were plenty of invitations. Amongst them, North Carolina State University itself couldn't resist scheduling a lecture series for Peter; everyone knew of his reputation and a large crowd of faculty joined students the first session. But it was a serious anti-climax. By this stage Peter, never a natural lecturer, was mumbling to the blackboard. Hopfenberg recalls that: 'His style was disappointing; he didn't seem to have a sense of theatre'. Attendances dwindled.

But in close conversation Peter was as stimulating as ever. He and his host hit it off from the start; although Hopfenberg was 22 years Peter's junior, he had a very sharp mind and both of them had been through the Massachusetts Institute of Technology practice course. Peter was shocked that there were no tea-time breaks in the North Carolina department to promote lateral thinking. But he was alone in this. The halcyon days of 'academic indolence' were over. When Hopfenberg was in Cambridge the following year, he could tell that people in the Shell department attended at tea-time because they had to, not because they wanted to or were prepared to enter into the spirit of it.

During his stay in Raleigh Peter took the opportunity to have an extensive cancer check-up over several days at a well-respected local hospital, the Rex. Lavinia reported to her son Charles how Peter would ring her brightly: 'I've got a radio-active liver today' or 'now my bones are radio-active'. The conclusions were, to quote Lavinia: 'fabulous: not a trace of cancer anywhere.' If there had been secondaries after his 1974 operation in Cambridge he would not have survived as long as he did.

Lavinia had been upset that, because she was a divorcée, her 1960 marriage had been a registry office affair. Now, in the rather more relaxed atmosphere of the United States, she saw her opportunity for a church blessing. It took place at the Episcopal Christ Church on East Edenton Street, downtown Raleigh. Peter wore a white suit and Hopfenberg was dumbfounded: 'It was amazing to see him kneeling in it'. For Lavinia the blessing service was important; for Peter the important bit was the party afterwards. Hopfenberg acted as barman.

Hopfenberg himself was due for a sabbatical break in 1977 and Peter asked him what he was planning to do. He explained that he wanted to go to Italy. Peter's view of the Italians may have been informed to some extent by his experience on the beaches of Sicily. 'Italy!' he exclaimed, turning up the irony dial to full. 'Do they have chemical engineers? Oh yes,

Raleigh, North Carolina: Peter and Lavinia celebrate the blessing of their marriage

there's Gianni Astarita... But one doesn't go to Italy for a sabbatical; one goes to Italy for holidays. One comes to Cambridge for a sabbatical.' Within ten days Peter was back in Hopfenberg's office: 'It's all set. The Science Research Council has offered you a senior visiting fellowship, and the best we can do is a fellowship at Clare Hall. The Italians? We'll deal with that. How much time do you really need? Two months? Let's call it six weeks. Go there for six weeks and come back to Cambridge.'

Peter and Lavinia headed home to Abbey House in the summer of 1976. Soon after New Year 1977 Hal and Patsy Hopfenberg arrived from Naples. Peter installed them in the 'Tit Box', a small terrace house they owned facing the river Cam at number 32 Riverside. It was only a few minutes' walk from Abbey House; the name referred to two birds featured in the stained glass of the front door. It was part of a small portfolio of houses adjacent to or near Abbey House that Peter and Lavinia had acquired over the years.

Every day Hopfenberg and Peter walked together to the department and back (Peter had given up cycling to work a long time ago), come rain or shine, without hat or umbrella. Usually they took the river towpath towards the city centre. They were wonderful walks, Hopfenberg recalls. 'His conversation was so intelligent. Nothing self-aggrandising, no pretence. No side.' One day nothing was said for quite a time and then suddenly Peter burst out 'Deliquescence! Does anyone understand it?' (It's what happens when the crystals of some substances left open to the air attract moisture from the air, become damp and eventually dissolve in the water they have attracted forming a solution.) Hopfenberg thought about it for a few days and on a subsequent walk gave Peter an explanation. He threw in, for good measure, why some crystals deliquesce and others don't. 'That's it,' said Peter. 'You've got it.' With the enthusiasm of youth, Hopfenberg asked him: 'Where do you think we should publish this?' Peter calmed him down: 'Hal, we didn't do it for *them*, we did it for *us*.'

Hopfenberg formed the impression that, even allowing for Peter's brush with cancer and his continuing liver problems, Peter showed signs of manic behaviour. At work he started accepting candidates for postgraduate work in Cambridge from all over the world. His colleague, and now head of department, John Davidson, became worried that the budget wouldn't stretch to accommodate them all. Then there was the little matter of car keys: Peter was always mislaying them and decided that two sets were too few. The only car offering three sets of keys was a Simca; he simply went out and bought one. Inevitably the touch of mania extended to consumables. One sunny afternoon Peter walked into Hopfenberg's office and announced 'I'm coming off the wagon.' He took Hopfenberg round to Dolomore's (a well-known Cambridge liquor store of the time). To the sales assistant's amazement, Peter began to order spirits by the case: 'First a case of bourbon in honour of my US friend,' then a case of gin, Scotch whisky, single malts, claret and so on: the order was delivered, Hopfenberg recalls, by lorry.

Lavinia's birthday came around while the Hopfenbergs were in town. Peter decided to throw a dinner party at Abbey House. On the strength of very little apparent experience he opted to do the cooking himself, and invited the Hopfenbergs to the party. 'We'll start with pheasant terrine en croute,' he said and found the recipe in a cookery book. Then he bought a pheasant from a shop on Newmarket Road. Hopfenberg knew this was a rather ambitious menu: 'But he pulled it off.' Something about kitchen operations gave Hopfenberg the feeling that Peter must have done it before. 'It was a fine party. There were just four settings at the 15-foot mahogany table, with masses of hallmarked silverware.'

The North Carolina sabbatical has a postscript. Hal and Patsy Hopfenberg went home to Raleigh in August, 1977. North Carolina State University had been busy in the interim. As a result of Peter's visit there the previous year and on the strength of his reputation, the University had organised an annual Danckwerts award for the undergraduate carrying out the outstanding 'senior project' of the year. It offered Peter an all-

A 'Benji': the value of North Carolina State University's Peter V Danckwerts award; first given in 1977, last presented in 1992

expenses-paid trip in the spring of 1978 to present the first award in person. He turned up at the Hopfenbergs' on a wet afternoon without prior notice and announced that 'Lavinia told me to tell you, or at least she should have, that I shouldn't be drinking during this visit.' Hopfenberg was surprised that Peter looked in quite good shape; in Cambridge his impression had been that his friend might not last the decade. The Hopfenbergs put their guest up in a small wooden guest extension in their back garden and he stayed for a couple of weeks. The weather forecast was terrible but it was a wet fortnight in more ways than one: Peter renewed his acquaintance with retired senior American military types; he and the generals would convene in the late morning at the North Carolina State faculty club to exchange war stories and drink tumblers of bourbon.

The curtain falls 32

P ETER FINALLY GAVE UP THE SHELL PROFESSORSHIP in July 1977 in favour of his long-time colleague John Davidson. He wrote letters to contacts in industry suggesting that they might like to fund a personal chair for him in the Cambridge department, but nothing came of it. So he lived the final period of his life in nominal retirement. Appearances in the Shell department were restricted to fortnightly lightning visits to check Max's weight on the department's large weighing machine. He would turn up unannounced with the dog, making no eye contact with students working there, and disappear as suddenly as he had appeared. Max lived to a ripe old age. Other dogs came and went; the Danckwerts acquired an old English sheepdog cross called Ben, but it was ill-advised enough to bite Peter's hand as he relaxed in an armchair and had to be put down. His replacement was Judd, a Labrador cross.

Peter gave up consultancy (and collected his thoughts about that role, reproduced in Appendix 3). He became a foreign associate of the US National Academy of Engineering in 1978, an honour that gave him a

Peter with Max at the gate of Abbey House, 1977
Courtesy Wendy Hopfenberg

great deal of pleasure. The Americans always appreciated him; they had already made him an honorary member of the American Academy of Arts & Sciences ('triple-AS') in 1964. The British were slower off the mark, but retirement triggered several honorary degrees. In December 1978 Bradford University offered an honorary doctor of technology, which he received from the hands of its chancellor Harold Wilson, ex-prime minister and prophet of the British white-hot technology revolution—by then somewhat cooled. In 1981 Loughborough followed suit and in 1983 Bath added a third. On the health front, Peter had been pronounced fully recovered from his 1974 cancer operation, yet it was clear from his scrawny appearance when he took delivery of his honorary doctorate in Bradford that chemotherapy was still on the menu.

Relieved of the strains of department and laboratory life, there were also signs that Peter was relaxing. John Bridgwater, by then a professor of chemical engineering at Birmingham University, was also a Fellow at Peter's old Oxford college, Balliol. Looking down the list of invitees to the college 'Gaudy' (the Balliol name for a celebratory get-together of alumni to which each year different age groups are invited), he noticed that Peter, although eligible, was not coming. After some time spent screwing up courage, Bridgwater wrote a letter starting 'Dear Peter'. He suggested Peter might come. Peter accepted, and even began his reply 'Dear John'. Bridgwater discovered that his old professor had known about the event but had been too shy to sign in. On the night—in contrast to his customary, and notorious, detachment—Peter seemed totally relaxed, genuinely pleased to be there and delighted to meet contemporaries.

For Lavinia, Peter's retirement was a great relief. All those sumptuous parties had been a strain, and in spite of their success, she hated them. She relished the thought now of not having to prepare another one. All the same, she could still turn it on. Peter's brother Dick, who had spent the previous couple of decades in Australia, was not long widowed. He decided to pass on his Australian medical practice to a younger man and leave his four grown-up children to develop their own lives. He returned to England and, although over 60, found a number of part-time administrative jobs. What is more, he proceeded to fall for his new secretary, Jill Lloyd. They got engaged. Peter suggested that they might marry in the Abbey Church—St Edward the Less—over the wall from Abbey House. The Danckwerts family rallied to the cause, and on 26 September 1981 a cast of 60 assembled. They included aunt Audrey, Joyce Danckwerts' sister, who had valiantly supported Dick in the early days of his first marriage when his father cut him off. The family invited her on one condition: that she wore her 'funny hat'. So she did. Lavinia laid on the wedding buffet directing a small army of Danckwerts nieces as helpers: cold salmon, sirloin, a turkey with full stuffing, ham, baguettes and copious amounts of salad. Just like old times, but it was the last hurrah.

Peter's body may have been failing but his mind was as clear and sharp as ever; he pointed it in several different directions. One was towards the

Danckwerts, Middleton and Macfarlane family trees. He was fascinated to find that both he (through the Middletons) and Lavinia could trace their ancestry back to Robert the Bruce (1274–1329), the greatest of all Scottish heroes.

Sisters at Dick's wedding: Peter's mother (l) with Aunt Audrey in her funny hat

The future of Abbey House was also a concern. It may have risen from the ruins of Barnwell Priory but there was a danger that it might shortly become a ruin itself. The original landlord, the city Folk Museum, had been unable to deal with external repairs as it had agreed with Peter and Lavinia in 1964. In fact financial difficulties led it to pass the freehold to the city Council in 1973. The Council proved no better at meeting its obligations: essential repairs had been delayed until the 1977-78 financial year, with an estimate of £21,000. By late 1977 little had been done and Peter and Lavinia decided to try and buy the freehold. The Council's property panel was in favour, but a December meeting of the full Council rejected the deal. There was opposition in principle from an ex-mayor and president of the Folk Museum management committee to the sale of a Council property into private hands (these were the last few years before the Thatcher government made such things respectable). Another ex-mayor threatened the Council with legal action. The columns of the local paper, the *Cambridge Evening News*, hummed with letters from, amongst others, Peter and Lavinia:

[£1,000] represents almost the total amount spent by the landlords since 1964. Had it not been for the very considerable sums spent by us since then, the house would by now be derelict. The exterior has not been painted for 13 years, many of the 400-year-old oak beams are rotten at ground level and some of the towering chimney stacks are in a precarious condition. A condition of the sale would have been that the purchasers should undertake specified repairs. It seems to me that this unique house would be better looked after in private hands. There would also be a saving of ratepayers' money.

BACKGROUNDER:

A POTTED HISTORY OF ABBEY HOUSE

In 1980, after considerable research, Peter published an article in the *Proceedings* of the local antiquarian society which shed light on the previously rather confused history of Abbey House. Barnwell Priory once stood close by. Founded in 1092 near Castle Hill, it was moved 20 years later to this quieter 13-acre site beside the river Cam. The Priory became an 'intellectual powerhouse' but the prior mishandled grazing rights to the extent that local peasants attacked the building during the 1381 revolt. A thousand of them, led by the mayor, pulled down walls, felled trees and drained fish-ponds, causing damage to the tune of £400. Henry VIII's 1538 dissolution of the monasteries finished the job; materials salvaged from the ruins can be found in properties all over Cambridge. Abbey House is amongst them. Peter dated the erection of the southern-most timber-framed section of Abbey House at around 1580. In the mid-seventeenth century the Butler family almost doubled its size by adding a brick extension, carrying the date 1678 on its gable end. Around 1700 the original building was faced with brick to match the extension and the interior was extensively refitted with paneling which hides Tudor timber and decorated plaster. The eccentric and monumental Jacob 'Squire' Butler, the last of the line, lived in the house from 1714 until 1759. Then the estate still occupied most of the original Priory acreage. Butler

recorded his life history on six gravestones, a hexateuch, in the churchyard of St Andrew the Less, the Abbey Church. The text is very faint now, but Peter found the wording fascinating. In 1763 one Thomas Panton bought the house; he was keeper of the King's running horses (or trainer) to George II. His son Tommy's horse Noble won the Derby in 1786 as an outsider at 30-1. Peter noted that the horse 'never achieved much else'. Nineteenth century ownership, beginning with James Geldart, man of God and property speculator, is complex; housing development claimed much of the surrounding land. By 1899, when the Askham family bought Abbey House, its gardens had been reduced to the size of today. Arthur Askham became owner-occupier in about 1922. At first he kept the gardens in wonderful condition, but as he aged both house and gardens deteriorated. In 1945 he sold to Huttleston Broughton (Lord Fairhaven)—son of an enormously-wealthy mother, equine stud owner, art collector and resident at Anglesey Abbey just outside Cambridge—for £3,675. A year later his lordship gave the house and gardens, still in a very poor state, to the citizens of Cambridge (in trust to the local Folk Museum) as 'a thank-offering for deliverance from the perils of war'. The Folk Museum made some super-ficial repairs, divided the property into three dwellings and let them. At the end of 1964 the Danckwerts took a long lease on two of them.

The Danckwerts' position was simple: sell us the freehold or carry out your contractual obligations. Peter had taken pains to research the history of Abbey House in some detail and, to help local people understand the significance of the situation, provided the local paper with a summary of it. Still the Council would not sell, although it did carry out some renovations and had the exterior painted. The Danckwerts, having already negotiated a 25-year lease in 1975 to include the remaining part of Abbey House (number three), carried out essential maintenance and occupied the whole building until 1984, taking in lodgers in rooms of the upper floors to sustain cash-flow.

A photographer from the *Cambridge Evening News*, visiting in 1978 in the wake of the Council furore over freehold, snapped Peter reading *Chemical Engineering Science* at ease on a sofa in the Abbey House lounge; the result appeared under the headline 'Council tenant Prof Danckwerts in his study', to general amusement amongst ex-colleagues.

At the same time Peter was hatching a pet idea about what might happen to Abbey House in the longer term. In January 1978 he wrote to appropriate councillors laying out his ideas:

I would dearly like to see (or anticipate) learning returning to Barnwell. The Priory was an intellectual power-house between 1112 and 1538, with international connections. It is even suggested that it started organised teaching before the University.

The Council tenant in his study, 1978
Courtesy Cambridge News

The Danckwerts' lease on Abbey House was due to run until 2006, with the option of an extension of 21 years. Lavinia, Peter mused mischievously, would be 83 by 2006 and unlikely to want to move in her old age. He told councillors:

My own feeling is that [Abbey House] should become some kind of residential institution—a mini-campus. The 16th century part of the house contains four large and fairly 'grand' panelled rooms, which would be admirable for communal use, and there is another large room in the 17th century part. There is a good deal of residential accommodation in the house itself [on the first floor and in the attic].

Outside was a considerable area of garden surrounded by walls. In the south-west corner stood number 4 Abbey Road, which the Danckwerts already owned. They also owned 20a Beche Road, a small building in the north-east corner of the garden which had outline planning permission for replacement. Peter had sounded out the owners of number 141 New-market Road in the remaining south-east corner; he judged that it might come on to the market shortly. An idiosyncratic Australian innovator called Donn Casey rented the yard and ground floor at number 141. He lived a frugal bachelor life, with an engineering workshop in the yard where he developed a variety of innovative gadgets for population control and to help the disabled. Sharing an appreciation of practical engineering, he and Peter got on well together; the Danckwerts invited Casey now and then to their dinner parties. Thus Abbey House garden, Peter felt, was an ideal part of an educational mini-campus. In his letter to councillors he warmed to the theme:

There is room for building between the old Priory wall [within Abbey House garden on the northern side] and Beche Road. Or, alternatively that space and the front could be used for car parking. Some houses or back gardens along Newmarket Road could probably be acquired. The combination of a commercial frontage and a residential rear might be advantageous.

Peter visualised a campus with 40 residents and communal rooms, absorbing the Abbey Church abutting Abbey House garden. The problem was, as he realised, money. The Council would not, he thought, regard his proposed use as 'civic' and so deserving of public funds. So he wrote to possible benefactors loosely connected with Cambridge Massachusetts, Harvard and MIT. Councillor Donald Mackay—whose father had founded the hardware and engineering company, now a Cambridge institution, only a stone's throw from Abbey House—responded enthusi-astically. He and Peter shared an engineering bond. Mackay had been brought up nearby and as a boy, with the permission of the then owner Arthur Askham, had enjoyed the run of Abbey House gardens. He shared Peter's interest in the history of the area and supported 'the return of learning' to Barnwell. 'The addition of a few interesting and learned people, to mix in with many of the salt of the earth already living there,' he wrote back to Peter, 'would be excellent.' Others, particularly potential sources of funds, were more sceptical and nothing came of the proposal.

By the end of 1980 Donn Casey had bought the freehold of number 141 Newmarket Road and, benefiting from the good relations they enjoyed, Peter contented himself with merely negotiating a gate in the wall between his garden and Casey's yard through which he could get more quickly to the row of shops and the off-licence on Newmarket Road. Later developments made Peter's idea truly impractical: 20a Beche Road was demolished to make way for a modern house, while Newmarket Road

BACKGROUNDER:
CASEY'S YARD

Peter's neighbour in the south-east corner of the Abbey House garden was Donn Casey. He was the son of two extraordinary Australians—mother Maie, an artist, writer, patron of the arts and a founder of the Australian Women Pilots' Association, and father Richard, an engineer, diplomat and, in his time, governor of Bengal and the last Australian politician to sit in the House of Lords. Casey senior was the minister whom Philip Bowden had to convince of his usefulness to the Australian war effort in 1939. Later, he was Australia's wartime minister to Washington. Donn was very bright and he rebelled. Urged by his father towards banking, he devoted his life to the cause of population control. In 1963 he followed the influential reproductive biologist Alan Parkes to Cambridge and set up a research information service. Joining the Simon Population Trust in 1966, he later chaired it and moved its headquarters to his spartan home at 141 Newmarket Road, Cambridge, beside the Abbey Church. Fascinated by gadgets, Casey kept an immaculate yard and workshop. In it he devised inventions for birth control and for people with disabilities. Most successful was the Filshie clip, a titanium and silicone rubber device which became the 'gold standard' for reversible blockage of the Fallopian tubes; it has freed millions of women threatened by a debilitating life of excessively frequent childbirth. Casey lived a solitary life, practising what he preached. His last years, succumbing to dementia, were lightened by the company of Jane Grey-Mansfield, a local occupational therapist. Donn Casey was a modest man who could turn on an old-world charm but did not suffer fools gladly. He died in 2009; a small plaque on the boundary wall at 141 Newmarket Road remembers 'Casey's Yard'.

Peter on the patio behind Abbey House: 'An oasis of peace'

became a race track for traffic in and out of Cambridge, killing off any prospect of passing trade for the 'commercial frontage'. The Council finally bit the bullet and sold Abbey House in 2002. The body of Buddhists who bought it refurbished the living accommodation and have turned Abbey House into a residential campus of sorts.

Peter was still enjoying Abbey House enormously. The back patio continued to offer him an 'oasis of peace'. Meanwhile, he continued as executive editor of *Chemical Engineering Science* for five more years and was always on the lookout for the right top-quality independent editors. In 1973 he had written to his outstanding student Sharma inviting him to join in but Sharma replied, 'I am too young to cultivate enemies'. In 1976 Peter tried again; he wrote: 'You can afford to take on enemies now.' Sharma acquiesced. After retirement from the Shell department his ex-neighbour Betty Wisbey and Anne Turner, wife of his former department colleague Robin Turner, had been helping out with secretarial duties. But by the 1980s the work was becoming a physical burden. A hand-over of responsibilities loomed.

A volunteer was needed to fill his shoes: Birmingham University professor John Bridgwater was younger, energetic and willing. Peter had guided him as a student in the 1960s, suggesting he research into industry's requirements in the handling of solids, an almost virgin field for the application of science at the time. It had proved sound and fruitful, if long-term, advice. So Bridgwater became one of the editors in 1982 and got a year's run at the executive job. His wife Diane, an experienced administrator, agreed to take on the secretarial duties. Pergamon Press offered to fund an office for her at her husband's workplace but Birmingham University authorities wouldn't hear of it. So she organised herself at home and her husband took over formally at the beginning of 1983.

Robert Maxwell appeared anxious to mark Peter's departure after 24 years as executive editor with a Maxwell trademark: a big celebratory party. The initial proposal for a venue (in London) didn't suit Peter. By this stage he was suffering abdominal pain, perhaps associated with side effects of liver cirrhosis, and avoided travelling. He had taken to carrying a hot-water bottle around which he kept on his lap to hold against his tummy, using Lavinia's traditional Macfarlane plaid (what non-Scots call a blanket) to disguise it. Eventually Maxwell conceded that the event could be organised in Cambridge, with Churchill College as host. There was a glittering turn-out on the night, but Maxwell himself failed to show.

Freed from the academic treadmill, Peter devoted some of his time to writing. In 1982, when Pergamon Press invited him to select a collection of his most influential papers to form a monograph, he wrote a delight-ful autobiographical introduction and a short preamble to each paper. Reviewing the book, his ex-colleague Jack Richardson told a story which neatly captured Peter's impact on his profession: 'From time to time, I ask a class of students to name any distinguished chemical engineer of whom they have heard. Almost inevitably I am greeted with a stony silence; when I widen the scope to incorporate the whole of the engineering profession, I hear mutterings of "Brunel", "Stephenson"... At the end of their courses they certainly know of Danckwerts! He has probably contributed as much as any other man to the understanding of mechanisms of chemical engineering operations. His ideas are now so often a part of our basic thinking, that we forget what we owe him.'

Richardson's sentiments are shared in many parts of the world where students may turn out to be better informed. In 2010 the Technological

University in Kazan, capital of Tatarstan on the Volga river, invited Peter's Russian protégé Nikolai Kulov to referee a doctoral thesis. The area is rich in oil and so quite a hive of chemical engineers. Once business was concluded, Kulov found himself at a celebratory dinner with the usual endless Russian toasts. Finally the president of the university Sergei Dyakonov introduced Kulov and, referring to his time in Cambridge, asked him to say a few words. At once, Kulov reports, several young voices around the room piped up: 'Tell us something about Danckwerts'.

The Falklands War from April to May 1982 stimulated an old interest: press reports told of missiles aimed at ships—and designed in theory to penetrate them and explode within them—passing right through and continuing on their way. Peter recalled that it had happened frequently in WW2 and little seemed to have changed. Surprisingly for those who wrote him off as becoming increasingly conservative, he revealed himself a Falklands sceptic:

I never thought I would see the day when the British fleet would leave Portsmouth to rescue some tiny island in a corner of the Empire.

The bombing raid on Port Stanley airport stimulated him to write a letter to *New Scientist* which didn't make the cut—to murmurings in the office of 'not wanting to fight WW2 all over again'—but drew on Peter's memories of German raids on British airfields:

It sounds simple—drop a bomb on a runway and it explodes and makes a crater which renders the runway unusable; but what a difficult task it is to make good, deep craters in the right places.

He went on to explain the practical matters top brass should consider when sending out their bombers to put an airport definitely out of action; at Port Stanley, he was sorry to see, they still seemed to be getting it wrong.

Peter also warmed to a 'popular science' audience. He rallied his remaining strength, and his output in the magazine *New Scientist* from 1979 to 1984, which is listed in Appendix 1, represents a true 'Indian summer' with the pen. Dick Fifield, long-serving managing editor at the magazine, knew Peter already; in the 1960s, when Fifield was with *Nature* magazine, he had regularly consulted Peter on news from the process industries. Peter fancied becoming a regular correspondent for *New Scientist*, and he was encouraged in this by a visit in early 1981 from Tom Gaskell, his old friend from Combined Operations days. In April Peter wrote to Fifield, tongue in cheek:

[Gaskell] was a prominent Fellow of the Royal Society of Snivelling Drunks at Trinity [College, Cambridge] before the war. Some of the Fellows became quite distinguished. Gaskell tells me he used to be a frequent contributor to *New Scientist* in its early days [the mid-1950s]. I have several ideas which I should like to discuss with you. The journal seems to be getting rather worthy nowadays, or perhaps I am getting frivolous...

Dick Fifield and the *New Scientist* hierarchy resisted a lurch towards frivolity, especially when Peter suggested that the publication might

---

BACKGROUNDER:
BOMBING PORT STANLEY AIRPORT
Peter's unpublished letter to *New Scientist* looked at the problems of bombing from the air. A bomber coming in low to avoid defending radar delivers a bomb almost horizontally to the ground. Assuming that the installed fuses are designed to work when the angle of impact is so small, the bomb will skid along the ground for a distance which may be hard to estimate. Then it explodes on the surface and makes a relatively shallow crater. A bomber at a great height would make a deep crater, as the bomb would land almost vertically and penetrate to an optimum depth determined by its delay-action fuse. Peter revealed the wartime rule of thumb for calculating crater diameters:

The diameter (in feet) is four times the cube-root of the weight of explosive in pounds.

To get depth of half the diameter, the bomb must penetrate about half the crater diameter before it explodes. But there are other considerations. To avoid break-up between impact and explosion, the bomb must be sturdy. Peter gave as an example the need to penetrate several feet of reinforced concrete—he was going right back to Combined Operations days and the Anti-Concrete Committee's calculations for destroying pill-boxes on Walcheren. Half the weight of a sturdy bomb may be steel. For a 1,000-pound bomb, the cube-root of 1,000 is 10, which makes a crater 40 feet wide. But the weight of explosive in a 1,000-pounder may be only 500 pounds: in this case the cube-root is around 8 ($8 \times 8 \times 8 = 512$) and the crater diameter about 30 feet. Then there was the little matter of judging how long the fuse delay time should be to allow penetration to 15 feet.

In order to meet these requirements the bombers must come in fairly high. This increases the difficult of planting the bombs on a target which may only subtend a few degrees. It also makes it more likely that non-military installations near the target will be hit. The bombers will also be more vulnerable to radar interception, fighters and flak than if they had hedge-hopped to their targets. I wonder how much has changed in the last 40 years? I shall not be surprised to hear Argentine claims that the bombers hit schools and hospitals, and that the runway was serviceable again within 24 hours of the raid.

They did and it was.

'proclaim that its editorial function is to see that standards [of scientific and technical writing] are maintained'. Undaunted, Peter began to bombard the letters editor with succinct gems and Fifield himself with short articles drawn from his earlier experiences. Space in the publication was tight but Fifield, as editor of the Forum section and with a soft spot for Peter, managed to squeeze quite a few pieces in. Topics ranged from the origins and politics of the atom bomb through the value of alcohol as a stimulator of ideas—the artist and satirical cartoonist Ronald Searle felt similarly about the bubbles in vintage champange—to memories of schoolboy chemistry. Unpublished pieces included one on Lavinia's wartime prowess in the WAAF. Peter didn't give up on negotiating a more permanent arrangement with *New Scientist*, although by June 1983 he was interested in irregular contributions without wearying deadlines:

My qualifications are old age combined with a retentive memory, plus an unusually wide spread of scientific and technical interests. Almost every issue [of *New Scientist*] brings up a point or two worthy of further comment—very often to relate it to other topics or publications, or to put it into context historically. *The New Yorker* used to, and maybe still does, run an occasional column called 'Department of Further Amplification'. The ground-rules would be: lay off the regular contributors and no bitchiness.

But the magazine's staff failed to rise to the bait.

Another matter that exercised Peter in these last years was the English language, its development, usage and how to achieve a popular and yet correct exposition of scientific and technical matters in the press.

[Some say] 'give me a properly trained journalist every time'. But what is the journalist trained to do? Cross the 'i's and dot the 't's, impose the house style and cut the piece to fit in among the advertisements? As a generalisation one might say that the scientist knows what he wants to say and the journalist knows how to say it, and they fit together as well as an amateur carpenter's dovetail joint. Add to this the necessity to splash an eye-catching headline, however misleading ('Miracle Drugs'), and one arrives at the recipe for popular scientific journalism. I have spent some 30 years of my life wrestling with students trying to express their painfully-acquired knowledge in lucid English. Should not the role of the scientific journalist be similar—namely to wrestle with the bearer of knowledge or ideas until a mutually-agreed text has emerged? Those of us (yes, even academics) who are literate feel acute pain when our scientific views are misrepresented.

In the cut-throat commercial world of newspapers and magazines Peter's time-consuming recipe of 'wrestling' has become an impossibility.

But it still happens with striking success now and then in the world of books when each of the collaborators has patience for and sympathy with the other. Time was—as Peter might have expected a journalist to write—when scientists could look after themselves. And more recently the likes of physicist Brian Cox and evolutionary biologist Richard Dawkins show that it still happens. Thinking back, Peter recalled that...

The Origin of Species was a best-seller. And J B S Haldane was a brilliant journalist: his regular scientific column in *The Daily Worker* in the 1930s could not have been bettered (except when dialectical materialism intruded). Fred Hoyle was a gripping popular expositor of his theories, right or wrong. (I got into a train with him at London's Liverpool Street station once. A single artless query about the Cosmos led to a wide-ranging review of Space and Time which lasted all the way to Cambridge [about an hour and a half in those days]).

BACKGROUNDER:
LAVINIA AT THE WHEEL
Peter remembers a story about Lavinia as a wartime driver

In 1941 my wife was a WAAF driver on an airfield somewhere in England. One night she had finished her job of delivering hot dinners to posts around the airfield and was rolling dreamily along the tarmac back to her base. Of course there were no airfield lights and her head and rear lights were reduced to pin-points. Happening to glance in her rear-view mirror she was horrified to see the silhouette of a fighter coming up behind her. She did the only possible thing – she veered off the tarmac and ended up in the sand dunes at the verge. She couldn't get the truck out, and had to walk back to her HQ. The transport corporal in charge had a very bad reputation as a bully, and he called her every filthy name his limited imagination could summon, so she slapped him on the face. He said 'I dare you to do that again'. So she did. The corporal screamed to two aircraftsmen who were in the room 'Arrest that woman!' So for several days my wife was under arrest—when she went to the lavatory a sentry was posted outside. Fortunately this was before the Military Discipline Act was extended to cover servicewomen, so she could not be court-martialled. Instead a Court of Enquiry was held and she was sentenced to three weeks confined to barracks. In the meantime she had become the heroine of the station. The commanding officer very wisely sent her away immediately on a five-week instructional course. When she came back she found the transport corporal a changed character. He was as mild as mother's milk. She never had to carry out the daily inspection of her vehicle again; the corporal did it for her while she polished her nails and read women's magazines. My wife is not an active women's libber now, but how's that for a record, sisters?

Then there was the question of language in the slightly more rarefied atmosphere of scientific journals, Peter had faced a compulsory examination in German at university in the late 1930s in order to access half the world's chemistry which was then still published in that language. But by the early 1980s most nations had embraced the need for scientists to publish in English if their work was to become widely known. Peter accepted this:

Scientists from countries which for chauvinistic or other reasons have not 'joined the ball' are indeed handicapped. For instance, I recently saw on TV the chief administrator at Akademgorodok (Novosibirsk) explaining why Russians won so few Nobel prizes. 'It's because so few foreign scientists read Russian', he said. Quite so—and he should know what to do about it. Not to write in English is to be a provincial or a peasant in the world of today.

Contributors to Peter's own journal, *Chemical Engineering Science*, could cope, he felt. The mother tongue of authors submitting their efforts during his time in charge might have differed as much from English as Hindi or Japanese. And contributions came from 'every conceivable country from Taiwan to Peru'. It was a constant source of wonder to Peter that all these scientists could still learn English 'and learn it whole'. The result might not, he admitted, be the language of Shakespeare but neither was it 'the crippled Basic English of Ogden and Richards' (a simplified form of English with an 850-word vocabulary) which was seriously propounded as a world language in the immediate post-WW2 period, even enjoying encouragement—Peter noted—from 'the old wordsmith' Churchill himself.

One of the main features of Basic English was supposed to be its very limited vocabulary. It always seemed to me that this advantage was counter-balanced by the obscurity of English syntax. What, for instance, is a Japanese to make of the construction 'I have got to get up'?

Peter was relaxed about the evolution of the language and fascinated by the contrast between word of mouth and the written word. He had discovered during his travels that, for instance, the comedian Peter Sellers was the most popular Englishman in India because of his renderings of regional variations of Indian English. But unless an Indian had spent some time in the USA or the UK, mutual comprehension was difficult to achieve. Things were not helped, Peter noted, by the Indian custom of shaking the head sideways when signalling agreement.

Nevertheless on the typescript page what they say is usually clear enough, if one is not distracted by the random scatter, or absence, of definite and indefinite articles. As an editor my attitude has become more liberal over the years. I spend little time inserting the missing 'a' or 'the', and infrequently return a paper to the author for purely cosmetic reasons. I reckon that if I can understand an article then a specialist in the field should have no problem.

Peter likened this approach to the attitude of a Roman of the old school coming to terms with Monkish Latin during the Dark Ages. In more general matters of language, Peter showed his liberal side by calling in Dr Johnson—the great lexicographer, who 'had the good sense to undermine his whole project by some eccentric definitions'—to emphasise that usage is not sacred. And he quoted *Fowler's Modern English usage* against rigidity on matters such as the split infinitive. Fowler lists five categories of splitters and his first category—the vast majority, who neither know nor care—comprises, Peter felt, 'a happy folk, to be envied by most of the minority classes'. Notwithstanding:

Personally, I would never split an infinitive, nor would I treat 'the media' as a singular noun.

Peter extended his range with some letters to his local paper, the *Cambridge Evening News*. The Latin he learned in his childhood came flooding back in a short message to Christopher South's regular column when 'schoolday dog Latin' was under discussion. Peter recalled the question and answer: '*Nasne puer?*' (Do you swim, boy?) '*No*' (I swim). He also supplied the homonym '*Malo, malo, malo*' which demonstrates three separate meanings for the same word (I would rather be, In an apple tree, Than a bad boy be). The column also revealed (and journalists hate appearing to be off the pace) that the homonym helps its readers to remember that the ablative case doesn't always need a preposition.

Peter's enjoyment of putting people and projects gently and eruditely down found its outlets. Tam Dalyell—old Etonian and legendary Westminster correspondent of *New Scientist*, unique amongst Members of Parliament in that he could write lucidly about topical science and technology and keep it up week after week—wrote a blatantly name-dropping piece in 1982 in which he described many famous scientists he had met. They included Kendrick Wynne-Jones, pictured as 'one of the great furnace-masters of the white heat of the technological revolution of the 1960s'. Peter responded:

Wynne-Jones I first met before the Second World War in the subterranean Balliol-Trinity physical laboratory at Oxford, where all the chemical kinetics were said to be catalysed by cigarette smoke. I subsequently bumped into him in the otiose round of committees during the 1960s which did as much as anything to quench the white heat of the [technological] revolution. I don't think he made an outstanding contribution to chemistry—rather in the Thatcher mould [she worked in the early 1950s as a chemist for J Lyons & Co on emulsifiers for ice cream].

Another victim of Peter's gentle humour in the same article was the communist-sympathiser and father of operational research, Patrick Blackett. Peter recalled that Blackett had served in the Navy in WW1 but had left later.

One of his reasons... had been his dislike of the social relations between officers and ratings—take three paces backwards and shout at them.

He recalled Blackett as a pretty imperious figure; they met from time to time on committees at the Admiralty during WW2 and Peter cherished the memory of a lecture he heard Blackett give at MIT in 1947. Later...

We were both professors at Imperial College in 1956-59 but the impression I got was of a professor somewhat on the Kissinger model—a status symbol for the college, but not an approachable colleague.

The writer and Labour party politician Dick Crossman was another unfortunate who strayed inadvertently into Peter's cross wires. Both were Wykehamists, but Peter lamented that Crossman had not benefited from the outstanding facilities for scientific education available at his old school. He wrote:

In his Diaries [Crossman] refers to 'isotopes, whatever they may be'. A strange appointment as [shadow] secretary of state for science in the atomic age.

Meanwhile a well-known journal of record had been picking up on his achievements. Ray and Anne Baddour had received a letter in September 1981 enquiring whether the *Guinness Book of Records* meant anything to Americans. Peter advised them that he would feature in the 1983 edition of this self-proclaimed 'universally recognised authority on record-breaking achievement'. It classified Peter in the section on honours, decorations and awards, sub-section versatility, as 'the only George Cross holder who was also a member of the Royal Society'. The Baddours remember Peter proudly adding that he had probably created another record as the first MIT graduate to appear in the publication.

Research into neglected corners of science occupied some of Peter's retirement. He considered the French physicist and engineer Nicolas-Léonard-Sadi Carnot to have been unfairly treated. Students of thermodynamics may recall Carnot's name in connection with a theoretical engine, the most efficient possible. But Peter, after painstaking research, felt that Carnot was worthy of more credit than that. He wrote a short, unpublished treatise on the subject (reproduced in Appendix 2) aimed at the scientifically-minded non-specialist; today specialists may nit-pick on detail but the essay reflects the versatility of Peter's mind.

Just a few years of this prolific retirement and hospital investigative facilities were in action again. Lavinia reported to her son Charles in March 1983: 'Peter had his barium meal and X-rays, and sees the surgeon again next Monday when all will be revealed. Let's hope there is nothing to reveal.' It was probably an attempt to examine his oesophagus for ulcers or distension of the veins at its base, common side-effects linked to alcohol consumption. By early 1984 Peter was in quite poor shape. Nevertheless, old friends and colleagues dropping in at Abbey House found his company and conversation as sparkling as ever. Miles Kennedy, Peter's research student from the 1950s, visited from New Zealand and met him in the darkened drawing room at Abbey House. To Kennedy he looked rather overweight and much older, but his mind seemed as good as it had been thirty years previously. Man Mohan Sharma, passing through Cambridge, found the same thing. He had heard that Peter was suffering from depression after the 1974 operation, but it seemed quite the opposite. 'He was in great spirits, never depressed. Perhaps he needed the stimulus of people he liked and whose company he enjoyed.'

1984: Peter relaxes with a hot-water bottle under a Macfarlane plaid, a cup of tea and Judd

No stranger to the local hospital, Peter was taken in for the last time in October 1984. His faithful ex-secretary Margaret Sansom, now retired herself, visited him and reported to friends that, in addition to numerous tubes attached to him to ease the situation, there was invariably a large bottle of spirits on the bedside table. Peter had earlier been offered a liver transplant but turned it down, pointing out that many others needed it more than him. The end came, according to the pathologist's report, when internal bleeding in the oesophagus put huge stress on his liver already weakened by cirrhosis.

Peter Victor Danckwerts died in Addenbrooke's Hospital, Cambridge, on 25 October 1984. His death removed a great interpreter of the world of science in general and process engineering in particular, brought sadness to the many who knew and loved him and his conversation, and devastated his wife. His last words to Lavinia: 'Have fun, old thing.'

Appendix 1
Peter Danckwerts' written record

BOOKS

Gas-liquid reactions, McGraw-Hill, 1970
Insights into chemical engineering, Pergamon Press, 1982

ACADEMIC PAPERS

There is a full list in Kenneth Denbigh, *Biographical Memoirs of Fellows of the Royal Society* (Peter Victor Danckwerts), 1 December 1986, volume 32, pp98–114.

CONTRIBUTIONS TO NEW SCIENTIST (L = letter):

1. *The teaching of chemical engineering* (Profile: Professor P V Danckwerts), 21 January 1960 p146–7.
2. *Better writing*, 11 January 1979, p116 (L).
3. *Unspeakable beaches*, 30 July 1981, p307 (L).
4. *Geoffrey Pike*, 20 August 1981, p490 (L).
5. *Patents on bombs*, 24 September 1981, p823 (L).
6. *Beards*, 4 February 1982, p331 (L) in response to *On beards*, Richard Wilson, 21 January 1982, pp179–80.
7. *The perils of chemistry*, 27 May 1982, p601 and responses from W J Grant, *The experimental age*, 17 June 1982, p804 and J J Meenan, *Stoic*, 22 July 1982 p259.
8. *Static idea*, 27 May 1982, p604 (L).
9. *Cavity foam*, 3 June 1982, p669 (L).
10. *Vortices*, 17 June 1982, p804 (L).
11. *Oppenheimer, Teller and the Superbomb*, 2 September, pp641–2.
12. *Worthy fellow*, 16 September 1982, p792 (L).
13. *Arsenic and old race*, 30 September 1982 (L), p932 and response from Stephen Gould, *Fast poison*, 14 October 1982, p119 (L).
14. *Sheldrake prize*, 11 November 1982, p380 (L).
15. *The fire at Windscale*, 18 November 1982, pp430–1 and response from Huw Dorkins, *Flames at Windscale*, 23/30 December 1982, p857 (L).

16. *Scotch and heavy water*, 6 January 1983, p35.

17. *Explosive rats*, 3 February 1983, p327 (L) in response to *Ariadne*, 20 January 1983, p212.

18. *The Super stroke of genius*, 17 February 1983, pp471–2 (L) and response from Peter Goodchild, *Nuclear nuances*, 17 March 1983, p751.

19. *In the land of the giants*, 31 March 1983, pp904–5.

20. *Cruise controversy*, 14 April 1983, p102 (L).

21. *Atoms for peace—the pilgrimage of 1955*, 2 June 1983, pp647–8.

22. *Supernormal*, 30 June 1983, p971 (L) in response to Ruth Brandon, *Scientists and the supernormal*, 16 June 1983, p784.

23. *Atomic ties, cocktails, blondes*, 4 August 1983, pp361–2 and response from Walter C Patterson, *No justice for Bikini*, 18 August 1983 (L).

24. *Gift of the gods?* 29 September 1983, pp960–1 and response from David G Cook, *Thanks to the bar*, 1 December 1983, p691 (L).

25. *Biography of the bomb*, 3 November 1983, pp361–2.

26. *Vivisection for the layman?* in Bernard Dixon, 3 November 1983, p365.

27. *Windscale's pipeline*, 15 December 1983, p833.

28. *Pub science*, 15 December 1983, p840 (L).

29. *To Russia with science*, 22/29 December 1983, p943.

30. *Tritium bomb*, 1 March 1984, p44 (L) and response from John Fremlin, *Tritium production*, 15 March 1984, p51 (L).

31. *Tritium production*, 29 March 1984, p49 (L) and response from D Sowby, *ibid*.

32. *From the Cavendish to Harwell*, 5 April 1984, pp24–5.

LETTERS/ANNOUNCEMENTS (A) IN THE TIMES:

1. *University News*, 10 October 1956, Imperial College professorship (A).

2. *University News*, 19 February 1959, Cambridge professorship (A).

3. *New award for technology*, 21 May 1959, p6, Peter to sit on Board of Scientific and Industrial Studies (A).

4. *Professor's advice to schoolboys*, 2 September 1963.

5. *Britons elected to US academy* [AAAS], 14 May 1964 (A).

6. *Increased emphasis on science*, 24 April 1969, p3 (News).

7. *Tax on departing Jews*, Greville Janner (with thanks to Peter and others), 13 September 1972, p13.

8. *Burning fossil fuels*, 31 January 1978, p17.

9. *Science at Cambridge*, 20 September 1978, p15.

10. *Revolutionary perks*, 1 August 1980, p13.

11. *Price of valour*, 2 July 1981, p13.

12. *University places and industry's needs*, 9 September 1981, p11, in response to *The university cuts we cannot afford*, David Marquand, 3 September 1981.
13. *Scientific output in Britain*, 29 January 1982, p13.
14. *Manners and men*, 28 May 1982, p11.
15. *Science graduates*, 14 July 1982, p9.
16. *Explaining the bomb*, 5 October 1982, p11.
17. *Soviet job-combing*, 30 June 1983, p13.
18. *Science spending and tangible assets*, 3 August 1983, p9.
19. *Obituary*, 29 October 1983, p14 (A).

SELECTED CONTRIBUTIONS TO CAMBRIDGE EVENING NEWS:

1. *Future of the Abbey House*, 2 December 1977; 12 December 1977.
2. Barnwell Priory site and Abbey House, 31 December 1977.
3. *Uncertain future for Abbey House*, 31 January 1978.
4. *City man in row over gallantry awards cash*, 4 July 1981, p7.
5. *Gallantry awards*, 8 July 1981, p9.
6. *Nasne puer?* South on Wednesday, 25 April 1984.
7. *In distress*, 27 June 1984.

Also: *War hero professor dies at 68*, 26 October 1984, p1.

BLACKWOOD'S MAGAZINE:

1. Objects of enemy origin, December 1945, pp361–368.
2. Gibraltar 1942, May 1946, pp322–328.
3. The story of King Red and Co, November 1946, pp289–297.

PROCEEDINGS OF THE CAMBRIDGE ANTIQUARIAN SOCIETY:

1. The inheritors of Barnwell Priory, 1981, pp211–234.

Appendix 2
Sadi Carnot, precocious polymath

For Armando Tavares da Silva

Shortly before his death, Peter Danckwerts made a case for recognising Sadi Carnot, who died one hundred and fifty-two years earlier, as not only formulator of the Second Law of Thermodynamics but effectively the first to postulate the First Law. It is presented here as a historical document rather than as a challenge to modern historians or philosophers of thermodynamics.

In his celebrated lecture in 1959 on 'the two cultures', C P (later Lord) Snow suggested that anyone with pretentions to culture should understand the Second Law of Thermodynamics. I wonder whether he himself understood it except as a vague philosophical concept. Did he know that, in the real world, designers of steam turbines use tables giving the entropy of steam at various temperatures and pressures to four significant figures? I suspect he knew little more of its practical applications than those popularisers, including the 'gloomy' Dean Inge, who used to write in the 1930s about the 'heat death of the Universe' which would come about when everything had reached the same temperature and nothing more could happen.

The early history of thermodynamics was concerned mainly with steam engines. Of course it took a very parochial view of the physical world. It did not consider, for instance, the nuclear reactions in the sun which produced the light which grew the plants which turned into coal. No doubt modern cosmological theories suggest that the laws of thermodynamics as we have formulated them may be of only temporary and local validity. Like Newtonian mechanics they are likely to be our best guide to practical affairs for the foreseeable future.

Most of us (especially engineers) associate the Second Law with Sadi Carnot and his cycle. Nicolas Léonard Sadi Carnot (1796-1832) was a polymath, interested in military engineering, manufacturing processes, natural history, political economy, the theatre, painting, music and religion. An early manifestation of his quality was his rebuke to Napoleon, who splashed some ladies who were boating at Malmaison. Of course, Sadi got away with a pat on the head. The First Consul was as magnanimous to appealing youngsters as the Führer.

He graduated from the *École Polytechnique* and spent some time as an army engineer. Once his attention had been turned to steam engines, he applied to them a philosophical Gallic analysis which had been lacking in the sturdily empirical

British scene. He was no scientific nationalist and did not claim that his compatriots had been the first inventors. On the contrary, he lists Savery, Newcomen, Smeaton, 'the famous Watt', Woolf, Trevithick and other English engineers as the veritable creators of the steam engine. What is more he shows in his work the qualities of a practical engineer as well as those of a natural philosopher.

Carnot's generalisations did not extend beyond heat engines which depend on the alternating thermal expansion and contraction of the working substance. The Seebeck thermo-electric effect was only discovered in 1821 and was probably unknown to Carnot. (It is not amenable to his theory of heat engines since it involves essentially irreversible processes.)

Nevertheless in his only publication *Refléxions sur la Puissance Motrice du Feu et sur les Machines Propres à Développer Cette Puissance* (Paris, Chez Bachelier, 1824) he laid the foundations of a philosophy which ultimately reached far beyond steam engines and heat engines in general to comprehend the whole of that part of science which deals with physical and chemical equilibrium. (The best access to his work for English readers still seems to be the translation by R H Thurston, together with his biography, Carnot's private notes and an appreciation by William Thomson, London, Macmillan and Co, 1890, and republished by Kessinger Publishing, Montana USA, 2007)

At the time when Carnot wrote, the orthodox theory of heat was based on 'caloric'—that is to say, heat conceived as an impalpable fluid which could not be created or destroyed. The properties and behaviour of caloric were ill-defined. The word 'energy' was not in use in its modern sense, because the doctrine of the time did not envisage that caloric could, for instance, be transmuted into *puissance motrice* (motive power) or *vice versa*, or that energy in its protean forms rather than caloric was the entity which was conserved. The well-known empirical fact that a heat engine could only operate if heat was supplied at a high temperature and removed at a low temperature was explained by analogy with the water wheel which generated motive power by the descent of water from a high to a low level. The amount of water arriving at the low level was the same as the supply at the high level. The caloric theory stated quite explicitly that the caloric rejected to the condenser of a steam engine was equal to that supplied at the higher temperature. As late as 1849, William Thomson (later Lord Kelvin) wrote: 'So generally is Carnot's principle tacitly admitted as an axiom, that its application in this case (that is, steam engines) has never, so far as I am aware, been questioned by practical engineers'.

However, steam engineers were in no position to test the hypothesis. Although in the period 1820-1840 competition between the thermal efficiencies of different Cornish pumping engines acquired the aspect of a sporting contest, the absolute efficiencies were so low that the amount of heat converted into motive power would have been hard to detect. For instance, the average Cornish engine furnished about 5.5×10^7 ft lb per bushel of coal burned, which corresponds to an overall efficiency of 5 or 6% of the heat produced by the combustion of the coal. Given the state of the art, this loss would have been undetectable, let alone measurable. In any case, Carnot was concerned with the quantity of heat absorbed by the steam-water mixture in the boiler, which involves the latent and specific heats of water which in his day were not accurately known.

The question of the amount of heat which disappeared on being converted into work was therefore not a matter of practical importance to engineers of the time. Nowadays, steam turbines operating with very high top temperatures may convert nearly half the heat of combustion of coal into motive power. The 'duty' of the condenser is an important factor in design; provision for removing in the condenser all the heat absorbed at the hot end of the turbine would be technical and economical nonsense because nearly half of it has been converted into motive power *en route*.

So it was that Carnot, young and unknown, had to compromise with the caloric theory, although it is clear that he did not believe in it. Thus it can be said that the Second Law was formulated before the First. The caloric theory was already proving an embarrassment to a more general theory of heat. As early as 1798 Rumford had shown by his cannon-boring experiments that motive power could be converted into heat, and the experiments of Joule in the early 1840s showed the same and established the 'mechanical equivalent of heat'.

Such was the inertia of the caloric theory, however, that even in 1849 William Thomson, who was only 25 at the time, retained the caloric theory in his penetrating evaluation of Carnot's work, although he then had more evidence than Carnot that the theory was wrong. In those days, at any rate, young scientists were well-advised not to challenge too stridently the perceived wisdom of their times. Thomson's restraint was rewarded by a peerage, and a long lifetime of important work in science and engineering. He realised the First Law and became one of the Great and the Good, perhaps because of his youthful prudence. No doubt similar rewards would have come to Carnot had he lived beyond the age of 36. As it was, his work went almost unnoticed until rediscovered and expounded by Thomson in 1849. As in the case of Mendel, a seminal idea did not germinate until after its originator's death.

The most striking aspect of Carnot's work is that, in spite of his scepticism about caloric, he was able to establish his most important propositions without committing himself. He used the language of caloric throughout, using *calorique* (caloric) and *chaleur* (heat) indifferently, as he states. The use of *feu* (fire) in the title seems strange. He does not use it in the technical part of his text, and the English translation substitutes 'heat'. Despite a few grumbles from James and William Thomson and from Clerk Maxwell in the latter part of the nineteenth century, it does not appear that the abandonment of the caloric theory requires any major modification of the wording of Carnot's primary conclusions.

Carnot formalised the working of a heat engine as follows. The working substance absorbed heat isothermally (that is, at a constant temperature t_1) meanwhile expanding and exerting pressure on a piston by means of which 'motive power' was produced. Then the source of heat was cut off and the working substance expanded adiabatically (that is, without absorption of heat) and produced more motive power, its pressure and temperature falling until the temperature reached t_2. This was the temperature of the *refrigerateur* or, in the case of the steam engine, the condenser. The working substance then gave up heat to the condenser at t_2 in the course of an isothermal compression during which motive power had to be applied. Finally, adiabatic compression raised the temperature again to t_1 and the working substance regained the *état primitif* from which it had started the cycle. During the course of the cycle the expansion took place at a higher temperature and pressure, on the average, than during the contraction and so there was a net output of motive power.

In the case of the steam engine (which in Carnot's time was more sophisticated than the earlier 'atmospheric' engines) the working substance was steam/water. The water absorbed heat in the boiler and turned to steam at t_1 whence it passed to the cylinder and exerted pressure on the piston (typical temperature and pressure were 130°C and 2.7 atmospheres). At a certain point (decided by an engineering combination of intuition and calculation) the supply of steam to the cylinder was cut off and the steam in the cylinder expanded adiabatically until (ideally) its temperature dropped to t_2, the temperature of the cooling water in the condenser. (James Watt patented the principle of this 'expansive' working in 1782.) The water condensed at t_2 and was pumped back into the boiler. The mixing of water at, say, 30°C with that at 130°C is an irreversible process and caused some problems in Carnot's discussion of reversible engines.

So far Carnot had only sketched a simplified model of an engine worked by a reciprocating cylinder. His major intellectual contribution was the concept of the ideal heat engine based on *reversible* processes. He showed by very simple reasoning that if a heat engine absorbs heat at a temperature of t_1 and rejects it at a temperature t_2, there is a maximum amount of motive power which can be produced by a given amount of heat and which depends on t_1 and t_2. (His terms of reference did not allow him to say that a given fraction of the heat could be *transformed* into motive power.) He argued that since the transfer of heat from a higher to a lower temperature was capable of producing motive power, every such transfer which did not result in a change of volume which was potentially capable of producing motive power could be accounted a loss. The test for such a loss was to see whether the process could be reversed—that is, whether a reversal of the change in volume would exactly restore the difference in temperature. This test for reversibility obviously fails if, for instance, heat is transferred from one fluid to another at a finitely different temperature through the wall of a pipe, since the heat falls in temperature without other effect. In a reversible heat engine, there are no losses due to the useless transfer of heat from higher to lower temperatures, and therefore the reversible engine has the maximum efficiency compatible with the temperatures t_1 and t_2.

Carnot's criterion, to translate his own words, was that *the necessary condition of the maximum is, then, that in the bodies employed to realise the motive power of heat there should not occur any change of temperature which may not be due to a change of volume.* In spite of the corona of alternative statements of the Second Law which has since been formulated this seems to be its primitive engineering origin.

If every feature in the action of a heat engine is reversible in detail, the complete cycle of events will be reversible. The *application* of a certain amount of motive power to a reversible engine will raise a certain amount of heat from a temperature t_2 to a temperature t_1, the quantities being the same in the reverse as in the forward process.

If a more efficient engine were feasible, it would generate more motive power than the engine described. This could be used to drive the latter in reverse, to produce a continuous net flow of heat at a temperature of t_1 or higher and open the way to perpetual motion and other consequences which the natural philosophers of Carnot's day felt to be absurd. Therefore a reversible heat engine operating between temperatures t_1 and t_2 produces the maximum possible amount of motive power per unit of heat, compatible with these two temperatures. This may be called the 'ideal efficiency' because it can never be realised by actual engines.

I would prefer to leave philosophers of science to analyse Carnot's chain of reasoning and the mixture of logic and practical experience which forms its links. Of course, those who followed him in the second half of the nineteenth century put the Second Law on a much sounder (and less comprehensible) footing, but he may lay claim to the first conception of the ideas of reversibility and a thermodynamic temperature scale. It followed from his major arguments that the ideal efficiency of a heat engine was independent of the nature of the working fluid—be it steam, air, alcohol, etc—and depended only on the temperatures t_1 and t_2. Thus, in principle, an arbitrary zero can be assigned to some 'natural' temperature (such as the triple point of water) and a complete thermometric scale built up by measuring or calculating the efficiencies of reversible engines operating between various other temperatures and zero.

We are accustomed to the expression

$$W/Q = (t_1\text{-}t_2)/(t_1\text{+}273) = (T_1\text{-}T_2)/T_1$$

which expresses the ideal efficiency (motive power W produced as a fraction of the heat Q absorbed at t_1°C). The absolute temperatures T_1 and T_2 are obtained by adding 273 to the Centigrade temperatures, absolute zero being -273°C. The ideal efficiency is therefore the ratio of the fall in temperature from t_1 to t_2 to the fall from t_1 to absolute zero.

This expression is often attributed to Carnot, but in fact it cannot be derived from his notions without introducing the First Law and was first put forward, not without reservations, by William Thomson in 1851. Subsequently, of course, the wheel swung full circle and thermodynamic or absolute temperatures are actually defined by the above equation. It is a far cry from the creaky, leaking Cornish engines to the universal principles of physics.

It is interesting to follow Carnot's attempts to calculate the ideal efficiencies of engines using air, water and alcohol as their working fluids. He had very little information about the thermal properties of water and alcohol and their vapours. He knew more about air; he assumed (although he made clear that this was probably only an approximation) that it obeyed the perfect gas law. He knew its specific heats at constant pressure and constant volume (with the same caution he said these would probably vary with temperature and volume). His task was to calculate the pressure-volume-temperature relationship of a mass of air as it moved round his formal cycle.

He had no difficulty with the isothermal expansion and contraction, but was quite unable to deal with the two adiabatic changes. The temperature rise which accompanies adiabatic compression is, as we should now say, due to the conversion of work into heat and cannot be calculated without introducing the First Law and knowing the mechanical equivalent of heat. The only information which Carnot had about the adiabatic compression of air was derived from measurements of the velocity of sound, and indicated that at 1 atmosphere and 6°C compression by a factor of 1/116 produced a rise in temperature of 1°C (using modern figures I have calculated the factor to be 1/109). He had some data relating the pressure and temperature of mixtures of liquid and vapour in the cases of water and alcohol. Using this information and some dubious temperature corrections, Carnot calculated the ideal efficiencies of engines using air, water and alcohol as their working fluids—the temperature drop being 1°C in the neighbourhood of 0°C— and found them to be the same within reasonable limits, as his reasoning had led him to expect. My calculations, converting his units of heat and of motive power to a common basis and using his figures, indicate an ideal efficiency of about 0.33%, whereas 'Carnot's equation' given above predicts 1/273, or 0.37%.

When it came to making calculations of the ideal efficiency of an engine with a temperature drop of, say, 100°C (typical of the steam engines of his day) Carnot had, in effect, to substitute a cascade of engines each spanning 1°C and sum their mechanical outputs. At this point the theory of caloric becomes fundamentally misleading, as it supposes that the heat rejected from each stage of the cascade to the next remains unchanged throughout the cascade, whereas the First Law tells us that it diminishes progressively according to the motive power generated. The error depends on $(t_1-t_2)/(t_1+273)$. It is negligible when $t_1=1°C$, $t_2=0°C$, but becomes important when $t_1=130°C$, $t_2=30°C$ as in a typical steam engine of 1824.

Carnot has been credited with saying that a given temperature drop yields more motive power as the top temperature t_1 is reduced. This in undoubtedly true, as shown by the equation given above, but Carnot's reasons for his statement were based on some misapprehensions then current about the variation of the specific heat of a gas with its density. In one of the voluminous footnotes which characterise his *Refléxions* he sets out to produce a general method for calculating the ideal efficiency of an engine operating between an arbitrary temperature t°C and 0°C. He concludes that the motive power produced is simply proportional to t°C, a conclusion which is both wrong and contradictory to the proposition with which he is credited.

Denied the First Law and ignorant of the mechanical equivalent of heat, Carnot could not give numerical values of the ideal efficiency which might have served as a yardstick for the performance of actual engines, which were remarkably inefficient by thermodynamic standards. A reversible engine can only work at infinitesimal speed. Heat can only be transferred across a tube wall, for instance, at a finite rate if there is a finite temperature difference which leads to 'useless' degradation in temperature; pressures on either side of a piston can only differ infinitesimally; friction leads to inefficiency in proportion to the speed of motion. Indeed, all spontaneous processes on the macroscopic scale lead to irreversible change. A rock falls from a mountain; it can be replaced, but only by the agency of an engine (perhaps a human body) which itself produces irreversible change.

In his private notes Carnot writes:

'Heat is simply motive power, or rather motion which has changed form. It is a movement among the particles of bodies. Whereas there is destruction of motive power there is, at the same time, production of heat in quantity exactly proportional to the quantity of motive power destroyed. Reciprocally, wherever there is destruction of heat, there is production of motive power... Motive power is, in quantity, invariable in nature, never either produced or destroyed.

According to some ideas that I have formed... the production of a unit of motive power necessitates the destruction of 2.70 units of heat.'

If one converts the units of heat and motive power defined in this note to a common basis, the ratio of the latter to the former turns out to be 0.9 (rather than 1.0). In this throwaway passage Carnot formulates the First Law (although he does not consider chemical or electrical energy) and produces by some mysterious 'ideas' a good approximation to the mechanical equivalent of heat. This was at least 10 years before Joule firmly established the equivalent by his experiments, which were so varied in nature as to extend the principles of conservation from heat and motive power to other forms of energy (electrical and chemical) and justify the formulation of the First Law.

In his references to the transformation of motive power to the random motion of particles of matter, Carnot comes close to the concept of entropy. But he confesses himself puzzled to explain 'why, in the development of motive power by heat, a cold body is necessary; why, in consuming the heat of a warm body, motion cannot be produced?' I wonder how many present-day engineers or physicists could answer these questions?

Apart from his insights and achievements as a natural philosopher, Carnot had much to say about engineering practice and possibilities. He observed that in his day the efficiency of steam engines was largely determined by the temperature of the steam in the boiler. There was rank inefficiency due to the irreversible transfer of heat from the furnace at, say, 1,000°C to the water in the boiler where the temperature was usually about 130°C, the corresponding vapour pressure of water being about 2.7 atmospheres. Even so, boilers seem to have burst pretty frequently (there was no anti-steam engine lobby). Carnot pointed out that an air engine could be operated at a higher top temperature without the penalty of a higher pressure and suggested that power might be efficiently produced by using the heat first in a high temperature air engine which would exhaust its heat to a conventional steam engine. He pointed out the engineering problems associated with an air engine— for example, large pumps working at high temperatures. As a matter of fact, an air engine was built and demonstrated by the Rev Robert Stirling in 1827. It used a cycle rather different from Carnot's; generations of engineering students have had to study it as an exercise, and periodic attempts have been made to realise a commercial version.

It seems surprising that Carnot did not realise that much the same effect could be achieved more simply by superheating the steam from the boiler, a practice universally adopted for the last 125 years. The saturated steam generated by the boiler is heated to a temperature much higher than the boiling point without any increase in the working pressure. The modern steam engine is typically a multi-stage turbine in which the motive power is generated solely by adiabatic expansion, which did not play the primary role in the engines of 1824. Apart from superheat, the steam is bled off between stages to be reheated. The old concept of a single top temperature t_1 at which heat was absorbed has gone by the board, and the working cycle is best displayed on a temperature-entropy diagram.

In the 1950s and 1960s, considerable attention was paid to magneto-hydrodynamic generators which would use the high temperature of a flame of oil or pulverised coal by firing it down a duct lined with magnets. The flame was a conductor and its motion generated electric power. The formidable technological problems encountered seem never to have been overcome.

Carnot realised that when a steam engine discharged to the atmosphere (as in locomotives) rather than to a condenser, the thermodynamic effect was equivalent to rejecting heat to a condenser at a temperature of 100°C. If the boiler was at 130°C, the ideal efficiency was only 7%. Condensers have seldom been used on locomotives.

Carnot had views on the internal combustion engine, in which the heat is produced in the working substance (air) rather than having to be transmitted through metal plate. Some years before 1824, the brothers Niepce demonstrated an air engine fuelled by lycopodium powder (spores of a moss). Unfortunately this was far too expensive (its normal use was to produce sheets of flame at pantomimes). Carnot suggests using such fuel as powdered coal, but mentions the possible rapid abrasion of the piston and the cylinder which, so far as I know, has always inhibited its use in diesel engines. Strangely enough, Carnot barely touches on the use of liquid fuels (France has always been awash with alcohol) or gaseous fuels such as hydrogen or coal gas.

Another strange omission is any premonition of the mechanical refrigerator or heat pump. Carnot knew that a reversed heat engine could suck in heat at a low temperature t_2 and discharge it at a higher temperature t_1. However, it was not until the 1860s that Alexander Kirk introduced practical refrigerators. The working fluid was pressurised air and the cycle the reverse of that of the Stirling air engine. Carnot the engineer writes, in the last paragraph of his book, that fuel economy is only one of the conditions to be fulfilled in heat engines, and in many cases it is only secondary:

'It should often give precedence to safety, to strength, to the durability of the engine, to the small space which it must occupy, the small cost of installation, etc. To balance these factors properly against each other should be the leading characteristic of the man called to direct, to co-ordinate among themselves the labours of his comrades and to make them co-operate towards one useful end, whatever it may be.'

This is a definition of the good engineer (and a negation of the revolutionary idea that instructions merely form the basis of a meeting of a comrades' committee).

The Four Elements, Phlogiston, Caloric, the Luminous Ether—each of these may have proved an aid to thought at some time but long outlived its usefulness. What is leading us astray now? Quarks or the Big Bang, or possibly something in the biological sciences? No Nobel Prize is guaranteed to the winner of the guessing game.

Appendix 3
How to be
a consultant

By 1980 Peter Danckwerts had given up acting as a consultant but he found time to write down his thoughts drawn from experience in the role.

My career as a consultant is drawing peacefully to a close, after some 30 years of engagement, paid and unpaid, casual and long-serving. There are several different sorts of pitch a consultant can make. One is 'I have a first-class scientific education and can give a valuable opinion on any scientific topic '. Not an easy stance to sustain any more, now that industry is more riddled with science than it was 30 years ago, and one may always run up against the man who says 'I have a first-class classical education and can give a valuable opinion on any topic '. Or, for 'classical' read 'legal'—I have met several.

The second ploy is 'This is an infant subject. I have made significant theoretical contributions. Your employees were recently my students. You may expect me to open your eyes to possibilities you have overlooked '. But your students grow up, they keep in touch with theoretical advances and are engaged on specific research on their own problems which are unsuitable for university research. They come to regard you, possibly with affection, as an old fuddy-duddy.

The major requirement for a consultant is to keep the ball in the air. He may or may not have some information relevant to the topic under discussion, but at least he must not let the conversation fall into despondent silence. Lateral thinking is essential. For instance, when I was consulted about the absorption of carbon dioxide into the sea from power stations, it was useful to recall early experiments on the dispersion of radio-active effluent from Windscale off the coast of Cumberland. What, for instance, is the total volume of the world's oceans and the mean residence-time of the water in the Irish Sea as it flows from South to North? Or perhaps, when asked what instrumentation to use to ensure that emissions reach an acceptable standard, to say 'Use the same instruments as the Environmental Protection Agency '. You cannot be faulted.

Some consultants can exert such an ascendancy over their clients that they can maintain their reputations by Delphic proclamations such as 'The trees are tall but they do not reach the sky '(an apocryphal proverb mis-attributed to Nikita Khrushchev). Others can draw on vast practical experience from China to Peru. I remember a Cambridge don, the least convincing theoretician I have ever met,

who would frequently be called out to the Congo gold-mines when something
went wrong with the flotation process and the jungle became choked with froth.

One of my most rewarding consulting jobs was brought about by an article in
Private Eye. It concerned emissions from a processing plant which, it was claimed,
were causing great environmental damage—so much so that the plant should be
shut down. I learned a great deal about the process, and about similar plants
overseas with clean records. I was able to convince the Commission of Enquiry
that the emissions were not dangerous and could be reduced. But—and this is the
cardinal point for expert witnesses—I never had to appear before Counsel and be
torn to shreds by remarks starting out with 'Professor Danckwerts, I believe that
you claim to be an expert in so-and-so. Can you tell us of your previous experience
in the field?' This is the kind of mine-field only to be entered by the case-hardened
professional who can out-face an arrogant QC.

Suppose you are appointed as a consultant or government-sponsored adviser
to a company. It may be a multi-national with immense in-house resources at its
command. Why does it pay good money to an academic to come up and waste
its own experts' time several times a year? It may be that he is generally at the
forefront of his subject and that his university researches are directly relevant. It
may be that he is appointed a kind of Court Jester, given a good lunch and not
regarded as a particularly serious source of advice. A modest concern like the
Charcoal Burners' Research Association cannot afford to pay you but really stands
to benefit from some elementary scientific advice—if only they can be persuaded to
relate it to their centuries-old experience of the brown smoke and the blue flame.
As with Operational Research during the war, it is largely a matter of speaking
sympathetically and soothingly to the man who actually has to do the job.

The international dimension introduces an interesting extension to the whole bag
of tricks. I have advised or talked to, for instance, concerns in Japan, India, USSR,
Israel, USA and South America, apart from our cousins in the EEC. One method
of categorising various countries is to see whether they give you a drink or not.
Indians and Arabs don't drink for religious reasons. Jews have no religious
objection to drinking but are temperamentally uninterested. The UK and Europe
(particularly France and Germany) are sound. In the USSR lunch starts at noon and
carries on until 4pm, after which all hands put their feet up. In the USA one is on
uncertain ground. One company gave me a nasty canteen lunch with ice-cold
Coca Cola. I remarked how much more stimulating were the lunches I had enjoyed
in Europe. 'Oh my, ' they said, 'we should all go to sleep if we had a drink at
lunch-time. ' I had to disagree—in my experience a moderate dose of ethyl alcohol
reduces inhibitions and stimulates the imagination; after-lunch sessions may be the
more productive. I may add that the US company in question ran a country club for
its employees but refused to provide a swimming-pool in case one of them might
drown and prove a legal liability.

Appendix 4
Danckwerts memorials

1. MEMORIAL SLATE

Peter's widow Lavinia and his younger brother Michael devised a memorial slate set into the wall of the churchyard of St Andrew the Less—the Abbey Church—as close as possible to Abbey House. The epitaph is in Lavinia's own words:

> *Young and brave, warm and wise*
> *Lovingly remembered by his wife Lavinia*

2. DANCKWERTS-PERGAMON PRIZE

In 1986 Robert Maxwell made a gift of £2,000 on behalf of *Chemical Engineering Science* to the Cambridge chemical engineering department. Income from the money provides the basis for an annual prize in the department awarded to the PhD student presenting the best dissertation that year.

3.   THE DANCKWERTS MEMORIAL LECTURE

An annual lecture sponsored by Elsevier (who publish the text) and the
Institution of Chemical Engineers, the European Federation of Chemical
Engineering and the American Institute of Chemical Engineers.

Contributors so far are:

1986: Neal Amundson, University of Houston, USA

1987: Man Mohan Sharma, University of Mumbai, India

1988: Octave Levenspiel, Oregon State University, USA

1989: Mooson Kwauk, Institute of Chemical Metallurgy, Beijing, China

1990: Rutherford Amis, University of Minnesota, USA

1991: Sanjoy Banerjee, University of California, USA

1992: Harry Beckers, Group Research Coordinator
      Royal Dutch Shell, Netherlands

1993: Malcolm Lilly, University College London, UK

1994: Ramesh Mashelkar, National Chemical Laboratory, Pune, India

1995: John Davidson, University of Cambridge, UK

1996: Robert Brown, Massachusetts Institute of Technology, USA

1997: Edward Cussler, University of Minnesota, USA

1998: Klaus Wintermantel, BASF, Germany

1999: Julio Ottino, Northwestern University, Illinois, USA

2000: John Prausnitz, University of California, USA

2001: Roger Sargent, Imperial College London, UK

2002: Robin Batterham, Rio Tinto, Australia

2003: Jackie Ying, Institute of Bioengineering and Nanotechnology, Singapore

2005: Doros Theodorou, National Technical University of Athens, Greece

2007: Matthew Tirrell, University of California at Santa Barbara, USA

2008: Wolfgang Marquardt, Rheinisch-Westfälische Technische Hochschule,
      Aachen, Germany

2009: Jay Keasling, University of California, Berkeley, USA

2010: Roland Clift, Centre for Environmental Strategy,
      University of Surrey, UK

2011: Frances Arnold, California Institute of Technology, USA

Information on future lectures from Angela Welch at: a.welch@elsevier.com

Timeline

14 OCTOBER 1916: Peter Victor Danckwerts was born at 25 Junction Road now Wimbledon Park Road), Southsea, Hampshire.

1921: Family moved to Merton Lodge, 75 Havant Road, Emsworth, Hampshire.

SEPTEMBER 1926–JULY 1930: Attended Sir Montagu Foster's preparatory school at Stubbington House, Fareham.

SEPTEMBER 1930–JULY 1934: Attended Winchester College as a fee-payer in Du Boulay's (C) House.

JULY 1934–SEPTEMBER 1935: Gap year in Austria and Germany

OCTOBER 1935: Took up Jack Frazer scholarship at Balliol College, Oxford. Spent long vacations 1936 and 1937 as a travel courier.

JULY 1939: Started work as a chemist at Fullers' Earth Union, Redhill, Surrey.

28 JULY 1940: Joined the Royal Naval Volunteer Reserve (RNVR), special branch: war service began.

MID-AUGUST 1940: Posted to *HMS President*, Port of London: temporary sub-lieutenant, mine and bomb disposal.

29 DECEMBER 1940: GC award announced, investiture 27 July 1941.

14 OCTOBER 1941: Promoted temporary lieutenant.

OCTOBER 1941–MARCH 1942: Tyneside, Londonderry, naval bomb and mine disposal.

11 MARCH-08 APRIL 1942: *HMS Volcano*, Holmrook Hall, Ravenglass, Cumberland, training course.

09 APRIL-23 NOVEMBER 1942: *HMS Cormorant*, Gibraltar, bomb safety officer.

29 DECEMBER 1942: MBE award announced, investiture 16 March 1945.

23–31 DECEMBER 1942: *HMS Excellent*, Whale Island Portsmouth, training course.

01 JANUARY: Posted to *HMS Hannibal*, Algiers.

11 JULY 1943: Wounded by an Italian anti-personnel mine during the invasion of Sicily (D+1), Operation Husky.

NOVEMBER 1943: Repatriated from Alexandria to Liverpool.

29 NOVEMBER 1943–05 MARCH 1944: *HMS Victory*, Portsmouth. Further recuperation from injuries. Taken off bomb disposal work.

06 MARCH 1944: Combined Operations Headquarters, experiments and developments branch, Whitehall, London.

20 MAY 1946: 'Released to shore' from RNVR: war service ended.

1946–48: Commonwealth Fund fellow, Massachusetts Institute of Technology. Graduated as SM (Chemical Engineering Practice).

1948–54: Lecturer, Cambridge University, Shell department of chemical engineering.

1952: Joined editorial board of *Chemical Engineering Science* (Pergamon Press) for its second issue published in 1953.

01 OCTOBER 1954–SEPTEMBER 1956: Deputy Director of Research & Development, Industrial Group, UKAEA.

05 MAY 1955: Elected Member of Institution of Chemical Engineers, UK.

AUGUST 1955: Attended *Atoms for Peace* meeting, Geneva, Switzerland.

OCTOBER 1955: Argonne National Laboratory, Illinois, USA.

OCTOBER 1956: Became Pergamon professor of chemical engineering science, Imperial College London.

1957: Visits to Holland and Germany.

1958: Elected executive editor of *Chemical Engineering Science* (Pergamon Press) from 1959 by his fellow editors. Visit to Caracas, Venezuela.

AUGUST 1959: Visit to Belgrade, Yugoslavia.

SEPTEMBER 1959: Became Shell professor of chemical engineering and head of department, University of Cambridge.

10 JUNE 1960: Married Lavinia Anne Harrison (née Macfarlane).

1960: Travel to Italy: Venice, Varese, Padua, Milan; North America: Vancouver, Phoenix.

SEPTEMBER 1961: North American tour: Boulder, Colorado; Phoenix, Arizona; Los Angeles, Berkeley, California; Vancouver, Canada; Minneapolis, Minnesota; Princeton; New York.

NOVEMBER 1961: Tokyo, Japan.

APRIL 1962: Amsterdam, Netherlands.

DECEMBER 1962: Copenhagen, Denmark.

AUGUST–OCTOBER 1963: Round-the-world tour: Sydney, Brisbane, Australia; Wellington, Dunedin, Christchurch, Auckland, New Zealand; Fiji, Honolulu, San Francisco, Philadelphia, Wilmington, Baltimore, Ann Arbor, Syracuse, Montreal, New York.

22 OCTOBER 1963: Elected Member, Chemical Institute of Canada.

13 MAY 1964: Elected Foreign Honorary Member of the American Association for the Advancement of Science (AAAS).

SEPTEMBER 1964: Antibes, France; Amsterdam, Netherlands.

LATE 1964: Moved into Abbey House, Cambridge.

27 APRIL 1965–26 APRIL 1966: President of the Institution of Chemical Engineers.

JULY 1965: First visit to Soviet Union: Leningrad, Moscow, USSR.

AUGUST 1965: Stockholm, Sweden.

SEPTEMBER 1965: Brno, Prague, Czechoslovakia.

MARCH 1966: Toulouse, France.

MAY 1966: Boston, Massachusetts, USA.

MAY 1967: Amsterdam, Netherlands.

SEPTEMBER 1967: Kansas City, USA.

MARCH 1968: Amsterdam, Netherlands.

20 MARCH 1969: Elected Fellow of the Royal Society, London.

1969: Venice, Italy.

23 FEBRUARY 1970: Received the American Chemical Society's Murphree Award.

1970: Book *Gas-Liquid Reactions* published.

APRIL 1971: Tel Aviv, Jerusalem, Israel.

1972: Second visit to Soviet Union: Moscow, Leningrad, Novosibirsk, USSR.

1972: Germany.

1973: Lisbon, Coimbra, Portugal.

MARCH 1974: Mumbai, India.

JULY 1974: Stood down as head of chemical engineering department, University of Cambridge. Bowel cancer surgery.

SEPTEMBER 1975–AUGUST 1976: Sabbatical year at North Carolina State University, Raleigh, USA.

JULY 1977: Retired as Shell professor.

1978: Elected foreign associate of the US National Academy of Engineering.

OCTOBER 1978: Honorary Fellow of the Institution of Chemical Engineers, UK.

02 DECEMBER, 1978: Honorary doctor of technology, University of Bradford.

JULY 1981: DSc *honoris causa*, University of Loughborough.

DECEMBER, 1982: Stepped down as executive editor of *Chemical Engineering Science*.

29 JUNE 1983: DSc *honoris causa*, University of Bath.

24 OCTOBER 1984: Died in Addenbrooke's Hospital, Cambridge.

Sources

Almost all of the chapters draw on both published and unpublished material written by Peter Danckwerts, and many of them on his Autobiographical note, *Insights into Chemical Engineering*, Pergamon Press, 1982, ppix–xv.

CHAPTER 1: VOLUNTEERING FOR SPECIAL DUTIES *Payment and Victual Ledgers*, Royal Navy; Arthur Hogben, *Designed to Kill*, Patrick Stephens, 1987; John Frayn Turner, *Fight for the sea: naval adventures from WW2*, Naval Institute Press, 2001, pp 8–19; Rob Hoole, *World Naval Ships Forum*, www.worldnavalships.com/forums/showthread.php?t=4927; A B Hartley, *Unexploded bomb*, Cassell, 1958; Robert Lochner, A Backroom in Battledress—the fight against magnetic mines, *Blackwood's Magazine*, April 1947, pp348–360; *German bomb fuses*, Steve Venus aka The Fuzeman: http://bombfuzecollectorsnet.com/page5.htm; Departments of the US Army and Air Force, *German explosive ordnance*, TM901985-2, March 1953: www.lexpev.nl/manuals/unitedstates.html; AB Wood, From Board of Invention and Research to Royal Navy Scientific Service, *Journal of the Naval Scientific Service*, volume 20, no 4, July 1965, pp1–99; War Cabinet weekly resume no 57, p11, The National Archieves, catalogue reference: cab/66/12/35.

CHAPTER 2: MEETING WITH A PARACHUTE MINE Peter Danckwerts, Objects of enemy origin, *Blackwood's Magazine*, December 1945, pp361–368; BBC archives, *The earl and the secretary*, www.bbc.co.uk/ww2peopleswar/stories/43/a3924443.shtml; The Clacton-on-Sea magnetic mine explosion: www.youtube.com/watch?v=Kz34ZrHIWGM; Constantine Fitzgibbon, *The Blitz*, Macdonald & Co, 1957; Arthur Hogben, *Designed to kill*, Patrick Stephens, 1987; John Frayn Turner, *Service most silent*, White Lion Publishers, 1955; A B Hartley, *Unexploded bomb*, Cassell, 1958; Ivan Southall, *Softly tread the brave*, Angus and Robertson, 1960; M J Jappy, *Danger UXB*, Channel 4 Books, 2001; P E Jenkins, *Journal of Naval Science*, 1980, no 2 pp 104–113; no 3 pp 159–173; John Miller, *Saints and parachutes*, Constable, 1951; Instructions for rendering safe underwater weapons, German ground mines, CB3115 (3B), 1944; Danger unexploded bomb (1979 ITV series of 13 x 60 minute episodes): www.televisionheaven.co.uk/dangeruxb.htm www.youtube.com/ watch?v=Knw7WF8jWpg www.youtube.com/watch?v=bWOMVAA8Ar0

www.youtube.com/watch?v=w3ZcF8rQ_YY.

CHAPTER 3: OUT ON THE LOOSE Peter Danckwerts, Objects of enemy origin, *Blackwood's Magazine*, December 1945, pp361-368; Kenneth Denbigh, Peter Victor Danckwerts (1916–1984), *Oxford Dictionary of National Biography*, Oxford University Press, 2004; Arthur Hogben, *Designed to kill*, Patrick Stephens Ltd, 1987; J F Turner, *Service most silent*, White Lion Publishers, 1955, p97; M J Jappy, *Danger UXB*, Channel 4 Books, 2001; A B Hartley, *Unexploded bomb*, Cassell, 1958; Andrew Brown, *J D Bernal, the sage of science*, OUP, 2005, p180; William Dring, portrait of Peter Danckwerts, National Maritime Museum, ref: PT0799 www.nmmprints.com/ image.php?id=342617&idx=0&fromsearch=true; The teaching of chemical engineering, Profile of Professor P V Danckwerts, *The New Scientist*, 21 January 1960, pp146–47; Peter Danckwerts, In the land of the giants, *New Scientist*, 31 March 1983, pp904–5; Royal Engineers army bomb disposal data: www.bombdisposal club.org.uk/BD_history.htm; George Cross award: www.london-gazette.co.uk/ issues/35018/supplements/7107; Richard Ryan: www.artfact.com/auction-lot/ second-world-war-1939-45-george-cross-bomb-and-mi-1-c-wl6nvhjlry; *Navy News*, no 665, December 2009, p12: http://content.yudu.com/A1jcim/navynewsdec09/resources/12.htm; John Babington, Obituary of Captain Llewellyn Llewellyn, *The Times*, 23 February 1970; Llewellyn Llewellyn, Naval report S206, 7 August 1942.

CHAPTER 4: LIMPET MINES AND HUMAN TORPEDOES Peter Danckwerts, Gibraltar 1942, *Blackwood's Magazine*, May 1946, pp322–328; Peter Walker, BBC2 People's War, To and at Gibraltar 42/43: www.bbc.co.uk/ww2peopleswar/stories/89/a2934489.shtml; Winston Ramsey (ed), The Italian underwater attacks against Gibraltar, *After the battle*, no 21, Battle of Britain Prints International, 1978. www.afterthebattle.com; *Midget submarine and human torpedo attacks*, UK National Archives, document ADM 199/1812; *Italian limpet mine from Gibraltar*, UK National Archives, document ADM 253/265; *Effemeride*, 23 September 2010: www.museomarinaro.it/effemeride_di_autunno_2010.htm; William Schofield and PJ Carisella, *Frogmen—first battles*, Branden Books, 1987/2005, chapter 20; Paul Kemp, *Underwater warriors*, Arms & Armour Press, 1996, chapter 3; Frank Goldsworthy, *The Mail* (Adelaide, Australia), 18 February 1950, pp2, 8: http://trove.nla.gov.au/ndp/del/article/58181130; Frederick Galea, *Mines over Malta: the wartime exploits of commander Edward Woolley*, Wise Owl Publications, Malta, 2008; Peter's MBE: www.london-gazette.co.uk/issues/35838/supplements/5645; Andrew Bailey, A wartime tribute to Bill Bailey: www.mcdoa.org.uk/Bill%20Bailey%20Tribute.pdf; Rob Hoole, The Buster Crabb enigma, *Warship World*, Jan/Feb 2007: www.mcdoa.org.uk/Buster%20Crabb%20Enigma.htm; www.specialoperations.com/Foreign/United_Kingdom/Royal_Marines/Buster_Crabb.htm; William Fairchild (director), *The Silent Enemy* (feature film), UK, 1958.

Chapter 5: Chaos in North Africa Colin Smith, *England's last war against France: fighting Vichy 1940–42*, Weiderfeld & Nicholson, 2009; Cyril Ray, *Algiers to Austria*, Eyre & Spottiswoode, 1952; Richard Wilson, Beards, *New Scientist*, 21 January 1982, p179; Peter Danckwerts, Beards, *New Scientist*, 4 February 1982, p331; Peter Walker, BBC2 People's War, *HMS Minna in the Mediterranean*, www.bbc.co.uk/ww2peopleswar/stories/56/a2325656.shtml; Manifest of alien passengers, *SS President Coolidge*, Yokohama to the USA, 9 November 1934; Marie-Louise Perreux, Letters to Kedive Abbas Hilmi II, Archive HIL/315/1-21, 1940–43, Durham University Palace Green Library, Durham; Marie-Louise Perreux, *Croquis d'Asie*, V Heintz, Algiers, 1942 (Bibliothèque Nationale de France, Paris); Renée Pierre-Gosset, *Algiers 1941–1943*, Jonathan Cape, 1945; www.spartacus.schoolnet.co.uk/FRweygand.htm; http://en.wikipedia.org/wiki/Henri_Giraud; www.spartacus.schoolnet.co.uk/FRdarlan.htm

Chapter 6: Sicily and the aftermath Invasion of Sicily: www.history.army. mil/brochures/72-16/72-16.htm, www.movcon.org.uk/History/Anecdotes/Sicily %20Tour%20Notes.htm, www.kiltsrock.com/forum/index.php/topic/16460-a-history-of-the-51st-highland-regiment, and www.primopachino.it/bandi/pon2010/ microprocessori/libretto2.pdf; James Ladd, *Assault from the sea 1939-1945—the craft, the landings, the men*, David & Charles, 1976; Stephen Roskill, *The navy at war 1939–1945*, Wordsworth Editions, 1998, p290; US army and air force technical manuals, TM9-1985-6, TO39B-1A-8, *Italian and French explosive ordinance*, pp165–6: www.lexpev.nl/downloads/tm919856italianfrenchexpl. ord1953.pdf; Lex Peverelli, *Grenades, mines and boobytraps*: www.lexpev.nl/ minesandcharges/europe/italy/vaudagnab4.html; Minefields in desert terrain, *Intelligence Bulletin*, US Military, January 1943; The Admiralty, *Certificate for Wounds and Hurts*, 25 May 1944; Ted Brown, Operation Husky and *HS Talamba*, http://www.bbc.co.uk/ww2peopleswar/stories/80/a3136880.shtml; Alice Robertson, *Bangor Daily Commercial* (Maine, USA), Thursday 11 December 1947, p1; E N Poland, *The Torpedomen, HMS Vernon's story 1872-1986*, published privately, 1993; Jean Donnelly, *Wound healing—from poultices to maggots*, The Ulster Medical Society, 1998, pp47–51, www.ncbi.nlm.nih.gov/pmc/articles/ PMC2448900/pdf/ulstermedj00061%2D0049.pdf; Ronald A Sherman, Maggot therapy project, www.pathology.uci.edu/sherman/maggots.htm; John Plumridge, *Hospital ships and ambulance trains*, Seeley, Service & Co, 1975, p55; Geraldine Edge and Mary Johnston, *The ships of youth*, Hodder & Stoughton, 1945.

Chapter 7: Combined Operations Was mother a virgin? *Time*, 8 March 1976, Lord Ampthill, www.time.com/time/magazine/article/0,9171,879625,00.html; R V Jones, *Most Secret War*, Hamish Hamilton, 1978, chapter 27, Bruneval raid;

Imperial War Museum, *Great Panjandrum*, film number ADM 1038
www.youtube.com/watch?v=1cyBcGDzQzI&feature=related; BBC TV, *Round
and round went the great big wheel*, Dad's Army episode No 52, first broadcast
22 December, 1972; Peter Danckwerts, In the land of the giants, *New Scientist,*
31 March 1983, pp904–5; Andrew Brown, *J D Bernal, the Sage of Science*, Oxford
University Press, 2005, pp482–3; Anthony Michaelis, Tom Gaskell—Tale of a busy
scientific life, *BP Shield*, 1974, issue 1; Tom Gaskell, private diaries, 1946–61;
James Ladd, *Assault from the sea 1939–1945—the craft, the landings, the men,*
David & Charles, 1976, p186–9; BBC Written Archives Centre, *The world goes by*,
transmitted Sunday 12 May 1946 at 4.30pm; Peter Danckwerts, Objects of enemy
origin, *Blackwood's Magazine*, December 1945, pp361–368; *idem*, Gibraltar 1942,
Blackwood's Magazine, May 1946, pp322–328; *idem*, The story of King Red and
Co, *Blackwood's Magazine*, November 1946, pp289–297; The Blackwood archive,
National Library of Scotland, letters to and from Peter V Danckwerts 1945–46;
Operation Backfire: www.v2rocket.com/start/chapters/backfire.html; British
Intelligence Objectives Sub-Committee Final Report No 398, item no 21, trip
no 1680, 1 December 1945, *The German Activated Bleaching Earth Industry,*
HM Stationery Office.

CHAPTER 8: HOW REGULATION 18B WAS SIDE-STEPPED Cormac McGinley,
Sinking of the Arandora Star, www.bbc.co.uk/ww2peopleswar/stories/94/
a2618994.shtml; J B Rietstap, *Armorial Général*, 2nd edition, volume 1,
Gouda, 1884, p509; Stutterheim Historical Society, *British German Legion,*
www.border.co.za/stutt/legion.htm; Jane Diana McSporran (née Danckwerts),
Robert Templeton and Bedford village school, http://archiver.rootsweb.ancestry.
com/th/read/SOUTH-AFRICA-EASTERN-CAPE/2008-11/1227813729; Leslie
Blackwell, William Otto Danckwerts—from barnyard to London Bar, *Personality,*
26 February, 1970, pp131–133 (National Library of South Africa, Pretoria, ref:
P5776W05615); Hugh Owen, *The Lowther family: eight hundred years of 'a
family of ancient gentry and worship'*, Phillimore, 1990, pp148–152; *The
Peterhouse Annual Record*, 1914, pp11–12; AW Brian Simpson, *Cannibalism and
the common law: a Victorian yachting tragedy*, The Hambledon Press, 1994; Jehu
Junior, *Vanity Fair*, 23 June 1898; David Midwood, letter to Caithness Field Club,
www.caithness.org/geography/walksgordonwilson/strathnaver.htm; Sex literature:
an absurd prosecution, *Reynolds's Newspaper*, Sunday, June 19, 1898; Issue 2497;
County of Sutherland, Sutherland Vehicle Registrations, 1912, www.ambaile.org.
uk/en/item/item_writtenword.jsp?item_id=4956; Our London correspondence
(by private wire), *The Manchester Guardian*, 28 April 1914, p8; Obituary (William
Otto Danckwerts), *The Times*, 27 April 1914, p12.

CHAPTER 9: THE BRITISH DANCKWERTS A stoker's triumph: how the *Kent* caught and sank the *Nürnberg, Deeds that thrill the Empire*, volume 2, p305, www.battleships-cruisers.co.uk/monmouth_class.htm; Victor Danckwerts, Letter to Joyce Middleton, 12 December 1914; *The Daily Colonist*, Victoria BC, Canada, Sunday 6 June 1915, pp21–23; The Star Man's Diary, *Daily Star*, March 1938; Obituary (Vice-Admiral Victor H Danckwerts), *The Times*, 6 March 1944; Pendennis, Table Talk, *The Observer*, 16 March 1952; Editorial, Doctors' pay, *The Times*, 26 March 1952, p5; Ruth Levitt, *The reorganised National Health Service*, Croom Helm, 1976, p103; Celebrities at home, *The World*, 2 March 1898; R T Shannon, Richard William Evelyn Middleton (1846–1905), *Oxford Dictionary of National Biography*, Oxford University Press, 2004; Stephanie Pain, How to rule the waves, *New Scientist*, 19 June 1999; Sandwich and Middleton, www.henrycort. net/nisandmid.htm; Roger Morriss, Charles Middleton, first Baron Barham (1726–1813), Oxford *Dictionary of National Biography*, Oxford University Press, 2004; John E Talbott, *The Pen and Ink Sailor: Charles Middleton and the King's Navy, 1778-1813*, Frank Cass Publishers, 1998.

CHAPTER 10: EARLY LIFE AND PREP SCHOOL Peter Danckwerts, Cavity foam, *New Scientist*, 3 June 1982, p669; Peter Danckwerts, The perils of chemistry, *New Scientist*, 27 May 1982, p601; Peter Danckwerts, Static idea, *New Scientist*, 27 May 1982, p604; Donald P Leinster-Mackay, *The Rise of the English Prep School*, The Falmer Press, 1984; Massachussetts Institute of Technology, *The Wimshurst machine in action*, www.youtube.com/watch?v=Zilvl9tS0Og

CHAPTER 11: WINCHESTER MANNERS Patrick Shovelton, Obituary (Sir Laurence Pumphrey), *The Independent*, 04 February 2010, www.independent.co.uk/ news/obituaries/sir-laurence-pumphrey-diplomat-decorated-for-wartime-bravery- who-later-served-as-ambassador-to-pakistan-1888573.html; Leader, The shameless elite, *The Times*, 1982, 20 May; Letters, *The Times*, 1982, 22 May, 25 May and 28 May; David Avery, Thermite reaction, www.davidavery.co.uk/thermite

CHAPTER 12: TO BALLIOL VIA SALZBURG Heinrich Baltazzi-Scharschmid, *Die Familien Baltazzi-Vetsera im Kaiserlichen Wien*, Bohlau Verlag, Vienna 1980; Gillian Freeman, *The uncertain beyond*, Royal Opera House Covent Garden ballet programme for the ballet Mayerling, 2004; Katherine Arens, *Mayerling: women's lives, affairs of state*, Royal Opera House Covent Garden ballet programme, 2004; Peter Danckwerts, The perils of chemistry, *New Scientist*, 27 May 1982, p601; Kenneth Denbigh, *Biographical Memoirs of Fellows of the Royal Society (Peter Victor Danckwerts)*, 1986, volume 32, pp98–114; RP Bell and PV Danckwerts, *Journal of the Chemical Society*, 1939, p1774–75; Brian Cox and John Jones, *Biographical Memoirs of Fellows of the Royal Society (Ronald Percy Bell)*, 2001, volume 47, pp19–38; Malcolm Fluendy, *Ronald Percy Bell*, www.rse.org.uk/612_ ObituariesB.html; Tom Smith, The Balliol-Trinity Laboratories, *Balliol Studies*, ed J Prest, Oxford, 1982, pp185-224.

CHAPTER 13: THAT RESERVED OCCUPTION Jack Trevor Story, *New Scientist,* 21
July, 1983, p202; Wilfrid Raikes, unpublished letter, 29 April 1940, Surrey History
Centre, reference 7160/1/7: www.exploringsurreyspast.org.uk/GetRecord/SHCOL_
7160; Wilfrid Raikes, http://thepeerage.com/p18369.htm; Fullers' earth,
British Geological Survey Mineral Planning Factsheet, 2006; Fullers' earth:
www.fullersearth.com/index.html; Cockley Works: www.urbexforums.co.uk/
showthread.php? 8740-La-Porte-Fullers-Earth-Works-Redhill-June-2010;
Robert HS Robertson, *Fuller's earth: a history of calcium montmorillonite,*
Volturna Press, 1986, chapter 38; Robert HS Robertson (Obituary of Henry
Hathaway), *Nature,* volume 154, 30 September 1944, p422-3.

CHAPTER 14: PRACTICE IN MASSACHUSETTS The Commonwealth Fund:
www.commonwealthfund.org; John E Craig Jr, *History of the Harkness
Fellowships Programme of the Commonwealth Fund,* The Commonwealth
Fund, New York; National Library of Scotland, The Blackwood archive,
National Library of Scotland, letters to and from Peter V Danckwerts 1945–46;
Alice Robertson, *Bangor Daily Commercial,* Thursday 11 December 1947, p1, 12;
Peter Danckwerts, The Massachusetts Institute of Technology, *Discovery, the
Magazine of Scientific Progress,* Jarrold & Sons, Norwich, November 1952,
pp351–55; Peter Danckwerts, Chemical engineering comes to Cambridge,
The Cambridge Review, 26 February 1983, pp53-55; MIT on 'Doc' Lewis:
http://web.mit.edu/cheme/news/lewis/lewis_lecture_prev.html; Lewis by his former
students, *A dollar to a doughnut, the Lewis story,* privately printed at MIT, 1953;
Hoyt C Hottel, Memorial tribute to Warren Kendall Lewis, The National
Academies Press,http://books.nap.edu/openbook.php?record_id=578&page=177;
PL Thibaut Brian, Memorial tribute to Edwin Richard Gilliland, The National
Academies Press, http://books.nap.edu/openbook.php?record_id=578&
page=97; Hoyt C Hottel, Memorial tribute to Thomas Kilgore Sherwood,
The National Academies Press, http://books.nap.edu/openbook.php?record_id=
578&page=247; Dimensionless number, http://encyclopedia.kids.net.au/page/di/
Dimensionless_number; Peter Danckwerts and Stanley Sellers, *Food,* December
1952, pp459–461 and January 1953, pp23–25.

CHAPTER 15: CAMBRIDGE BOHEMIA George Porter: http://news.bbc.co.uk/2/hi/
science/nature/2230778.stm; Tom Gaskell, private diaries; John Prebble and Bruce
Weber, *Wandering in the Gardens of the Mind: Peter Mitchell and the Making of
Glynn;* Oxford University Press, 2003; *Jesus College Cambridge Annual Review,*
1992, p51; Peter Danckwerts, Chemical engineering comes to Cambridge, *The
Cambridge Review,* 28 February 1983, pp53–55; Peter Danckwerts, Famous men
remembered (Terence Fox), *The Chemical Engineer,* December 1984, pp44–5.

CHAPTER 16: THE SHELL ENDOWMENT Patrick Barrie, *Chemical Engineering's 60th Anniversary*, Cambridge, 16 July 2008; Kenneth Denbigh, *Biographical Memoirs of Fellows of the Royal Society (Peter Victor Danckwerts)*, 1 December 1986, volume 32, pp98–114; Peter Danckwerts, Chemical engineering comes to Cambridge, *The Cambridge Review*, 28 February 1983, pp53–55.

CHAPTER 17: LIFE ON TENNIS COURT ROAD Matthew Eisler, Francis Thomas Bacon and the fuel cell: www.todaysengineer.org/2005/Mar/history.asp; *Chlorella*: www.physorg.com/news204367138.html; Peter Danckwerts, Chemical engineering comes to Cambridge, *The Cambridge Review*, 28 February 1983, pp53–55.

CHAPTER 18: THE FRUITS OF ACADEMIC INDOLENCE Peter Danckwerts, *Gas-liquid reactions*, McGraw-Hill, 1970, pp1–5; Kenneth Denbigh, *Biographical Memoirs of Fellows of the Royal Society (Peter Victor Danckwerts)*, 1 December 1986, volume 32, pp98–114. http://rsbm.royalsocietypublishing.org/content/32/98.full. pdf+html; Walt Whitman, *Chemical and metallurgical engineering*, volume 29, p146, 1923; Ralph Higbie, The rate of absorption of a pure gas into a still liquid during a shorttime of exposure. *Transactions of the American Institute of Chemical Engineers*, volume 31, 1935, pp365–389; Michael Martin, *Chemical engineering at the University of Arkansas, a centennial history 1902-2002*, p55: http://books.google.com/books?id=VlXJIRMcSmAC&pg=PA55&lpg=PA55&dq=H igbie+chemical+engineering&source=bl&ots=q5aINurLrO&sig=G2uwatBgFeOjF Mr3ElXYtujgqa4&hl=en&ei=1W7VTNPPB4T78AainczsBg&sa=X&oi=book_resu lt&ct=result&resnum=7&sqi=2&ved=0CDEQ6AEwBg#v=onepage&q=Higbie%2 0chemical%20engineering&f=false; Peter Danckwerts, Significance of liquid-film coefficients in gas absorption, *Industrial & Engineering Chemistry*, volume 43, June 1951, p1460; John Cullen and John Davidson, *Transactions of the Institution of Chemical Engineers*, volume 35, p51, 1957 and volume 37, p122, 1959; Harry Kroto, Moments of Genius (Robert Hooke), BBC Radio 4, 13.30, Sunday, 12 December 2010 and http://news.bbc.co.uk/dna/mbradio4/alabaster/NF2766774? thread=7934775&skip=0; Peter Danckwerts, Continuous flow systems and distribution of residence-times, *Chemical Engineering Science*, 1953, vol 2, p1; Peter Danckwerts, A century of chemical engineering (review), *The Chemical Engineer*, November 1982, p406; Peter Danckwerts, The definition and measurement of some characteristics of mixtures, *Applied Science Research*, volume 3, 1952, p279; Julio Ottino, *The kinematics of mixing: stretching, chaos and transport*, Cambridge University Press, 1990; AHG Cents, FT de Bruijn, DWF Brilman and GF Versteeg, *Chemical Engineering Science*, volume 60, issue 21, November 2005, pp 5809-5818; Peter Danckwerts and others, *Chemical Engineering Science*, volume 18, 1963, pp63–72.

CHAPTER 19: ACOUSTICS, FOOD AND FIRE BALLOONS Peter Danckwerts, *Power & Works Engineering,* July 1950, pp233–239; Peter Danckwerts and Stan Sellers, *Food,* December 1952, pp459–461 and January 1953, pp23–25; Fire balloon experiments: www.overflite.com; Sticky tape experiments: www.nature.com/nature/videoarchive/x-rays, www.cosmosmagazine.com/news/2268/sticky-tape-gives-x-rays, Crepitation: en.wikipedia.org/wiki/nitrogen_triiodide, Jetex engines: http://jetex.org/index.asp

CHAPTER 20: ORIGINS OF A BESTSELLER Robert Cahn, The origins of Pergamon Press: Rosbaud and Maxwell, *European Review,* vol 2, no 1, John Wiley & Sons, 1994, pp37–42; RV Jones, *Reflections on Intelligence,* Heinemann, 1989; Sidney Kettle in *Robert Maxwell & Pergamon Press,* Elizabeth Maxwell (ed), 1988, pp490–492; Joe Haines, *Maxwell,* Macdonald and Co, 1988; Tom Bower, *Maxwell the outsider,* BCA/William Heinemann, 1992, p101; Arnold Kramish, *The Griffin,* Houghton Mifflin Co, 1986; Ben MacIntyre, Family at war with MI6 over secret files, *The Times,* 16 December 2006; Denis Shaw, *The Geochemical News,* no 114, January 2003, pp20–22; Ernest Volkman, *Espionage,* John Wiley & Sons, 1995, pp95–104; John Bridgwater in *One hundred years of chemical engineering,* Nikolaos Peppas (ed), Kluwer Academic Publishers, 1989, pp39–46; Neal Amundson, Inaugural Danckwerts Memorial Lecture, *Chemical Engineering Science,* vol 41, no 8, 1986, pp1947–1955; Peter Danckwerts, In the land of the giants, *New Scientist,* 31 March 1983, pp904–5.

CHAPTER 21: BRUSHES WITH ROMANCE Naomi Canning, private letters; Alex McLaren and JA Spink, Frank Philip Bowden, *Australian dictionary of biography*: http://adb.anu.edu.au/biography/bowden-frank-philip-9550; David Tabor, *Biographical memoirs of Fellows of the Royal Society (Philip Bowden),* volume 15, November 1969, pp1–38: http://rsbm.royalsocietypublishing.org/content/15/1.full.pdf+html?sid=03943d7f-bf2b-47c6-a108-4ee08163e539; Hugh Carey, Mansfield Forbes and his Cambridge, CUP 1984; P Gray and KJ Ivin, *Biographical memoirs of Fellows of the Royal Society (Fred Dainton),* volume 46, November 2000, pp85–124: http://rsbm.royalsocietypublishing.org/search?fulltext=Dainton&submit=yes&journalcode=roybiogmem%7Croyobits&x=0&y=0; Wilma Crowther: *The brown book,* Lady Margaret Hall, Oxford, December 1989, pp38–42; John Buckatzsch, An experimental study of certain tax assessments, *The Economic History Review,* volume 3, issue 2, December 1950, pp180–202; Ruth Scorr, *Once a Caian...,* Gonville & Caius College, Issue 7, Spring 2008, p4,(Gonville & Caius College archives, Cambridge); Leonard Miall, (Obituary of Peter Bicknell), *The Independent,* 7 June 1995.

CHAPTER 22: HALCYON DAYS IN ATOMIC ENERGY Tom Lehrer, *The Elements*, www.privatehand.com/flash/elements.html; Francis MacDonald Cornford, *Microcosmographia Academica, A Guide for the Young Academic Politician*, Bowes and Bowes, 8th edition, 1970; Peter Danckwerts, Visit to Argonne National Laboratory, *UKAEA internal report*, 7 November, 1955; Huw Dorkins, Flames at Windscale, *New Scientist*, 23/30 December 1982, p857; Obituary—Lord Hinton of Bankside, *The Times*, 23 June 1983, p16.

CHAPTER 23: GRAZING IN GENEVA John Krige, Atoms for peace, 1955, www9.georgetown.edu/faculty/khb3/Osiris/papers/Krige.pdf; Peter Danckwerts, Atoms for peace—the pilgrimage of 1955, *New Scientist*, 2 June 1983, pp647–8; Simon Sebag Montefiore, *Stalin: the court of the red tsar*, Weidenfeld & Nicolson, 2003, chapter 44; John Cockroft, The peaceful uses of atomic energy, *Nature*, volume 176, 10 September 1955, p482–4; Obituary—Richard Moore, *The Times*, 29 April 2003.

CHAPTER 24: DEVELOPING DOUNREAY Uranium: http://periodic.lanl.gov/elements/92.html; Plutonium: http://periodic.lanl.gov/elements/94.html

CHAPTER 25: LIVING THE LIFE IMPERIAL Peter Danckwerts, Famous men remembered (Dudley Newitt), *The Chemical Engineer*, October 1984, pp66-7; America's Cup: www.answers.com/topic/12-metre-class; Peter Danckwerts, Chemical engineering in a university, *Inaugural Lecture*, Imperial College of Science and Technology, London, vol 58, 1957, pp59-69; The teaching of chemical engineering, Profile of PV Danckwerts, *The New Scientist*, 21 January 1960, pp146–47; John Bridgwater, *One hundred years of chemical engineering*, Nikolaos Peppas (ed), Kluwer Academic Publishers, 1989, pp39–46; Graeme Jameson, *UniNews*, University of Newcastle, Australia, May 2005 p1; Joe Haines, *Maxwell*, Macdonald and Co, 1988; Tom Bower, *Maxwell the outsider*, William Heinemann, 1988.

CHAPTER 26: PLAYING WITH FIRE The Harrison family: www.decisionmodels.com/Blessigs/PDFs/PJ_Blessig2_Descendants.pdf; Veterans Aid: www.veterans-aid.net/history.html; Café de Paris: www.cafedeparis.com/club/history and www.westendatwar.org.uk/page_id__97_path__0p3p.aspx; The winter of the bombs, *BBC Home Service*, 29 September 1957 at 9.15pm and repeated 1970, BBC references LP23854-55(2) and T293W; Constantine FitzGibbon, *London's Burning*, Macdonald & Co, London, 1971, Chapter 7.

CHAPTER 28: THE PROFESSOR'S PROJECTS Peter Danckwerts, Chemical engineering comes to Cambridge, *The Cambridge Review*, 28 February 1983, pp53–55; The teaching of chemical engineering, *The New Scientist*, 21 January 1960, pp146–147; Peter Danckwerts and Man Mohan Sharma, The absorption of carbon dioxide into solutions of alkalis and amines, *The Chemical Engineer*, October 1966, pp244–280; Armando Tavares da Silva and Peter Danckwerts, The effects of diffusivity in the liquid on rates of gas absorption, *IChemE Symposium Series no 28*, 1968, pp48–55; Peter Danckwerts, Scotch and heavy water, *New Scientist*, 6 January 1983, p35.

CHAPTER 29: AT HOME IN ABBEY HOUSE Peter Danckwerts, The inheritors of Barnwell Priory, *Proceedings of the Cambridge Antiquarian Society*, volume LXX, 1980, pp211–234, available from the Society librarian, www.camantsoc.org/council.html; Tony Benis: www.associatepublisher.com/e/a/an/anthony_m._benis.htm; Gerard Hoffnung, *The Bricklayer's Lament*, www.youtube.com/watch?v=zZUJLO6lMhI; Clockwork Daleks: http://www.projectdalek.co.uk/files/showcase/showcase_john.html; Abbey House as haunted: Andrew MacKenzie, *Hauntings and Apparitions*, Heinemann for the Society of Psychical Research, 1982, pp186–195, www.zurichmansion.org/abbeys/abbey.html, Alan Kersey, *Cambridge Evening News*, 15 July 1982, pp12–13, James M Deem, The Nursery: the haunting of Abbey House, *The Haunted House*, www.jamesmdeem.com/hh.nursery.htm; Bernard Atkinson, (Obituary of George L Standart), *The Times*, 1 September 1978, p14; Jaromir Ulbrecht, From small acorns, large trees grow, *Chemical Engineering Progress*, November 2008 and private communication; Michael Haag, Stationary and non-stationary random processes, http://cnx.org/content/m10684/latest

CHAPTER 30: TRAVELS, TRAVAILS AND CANCER Professor's advice to schoolboys, *The Times*, 2 September 1963; Two-way nuclear power stations, *Queensland Courier-Mail*, 17 September 1963; Peter Danckwerts, In the land of the giants, *New Scientist*, 31 March 1983, pp904–5; Peter Danckwerts, To Russia with science, *New Scientist*, 22/29 December 1983, p943; Greville Janner, Tax on departing Jews, *The Times*, 13 September 1972; Nikolai Mikhailovitch Zhavoronkov, www.springerlink.com/content/x0rt603411747455/; Danckwerts and Sherwood in Moscow, *International Union of Pure and Applied Chemistry (IUPAC)*, www.iupac.org/publications/pac/10/4/; Peter Danckwerts, *Gas-liquid reactions*, McGraw-Hill, 1970; CO Bennett and J-C Charpentier, Review of *Gas-liquid reactions*, *Chem Tech*, November 1971, p655.

CHAPTER 31: A BREAK IN NORTH CAROLINA Letters and cards from Lavinia Danckwerts to her son Charles Harrison.

CHAPTER 32: THE CURTAIN FALLS Peter Danckwerts, The inheritors of Barnwell Priory, *Proceedings of the Cambridge Antiquarian Society,* 1981, pp211–234, available from the Society librarian: www.camantsoc.org/council.html; Peter Danckwerts, In the land of the giants, *New Scientist,* 31 March 1983, pp904–5; Donn Casey: population control expert, Obituary, *The Times,* 28 February 2009; WJ Hudson, Richard Gavin Gardiner Casey, *Australian Dictionary of Biography,* Volume 13, Melbourne University Press, 1993, pp 228–229: http://adb.anu.edu.au/biography/casey-richard-gavin-gardiner-9706; Dianne Langmore, Maie Casey, *Australian Dictionary of Biography,* Volume 17, MUP 2007; Jack Richardson, book review, *PhysicoChemical Hydrodynamics,* 1982, vol 3, no 1, p83; *Guinness Book of Records 1983;* section: Honours, Decorations and Awards, sub-section: Versatility; Royal Commission on the Historical Monuments of England, City of Cambridge, Part II , ISBN 0-11-300023-5, *Secular buildings,* pp366–368; South on Wednesday, *Cambridge Evening News,* 25 April 1984; Ronald Webber, *The peasants' revolt,* Terence Dalton, 1980, p53 and p96.

Acknowledgements

Peter Danckwerts didn't have much time for archiving, and the only substantial text he wrote about himself was a 2,500-word autobiographical sketch by way of introduction to a selection of his best scientific papers. Taken with his self-effacing nature and laconic style, oral history became important. In this and other ways a great many people helped me piece a story together; I am very grateful to them all. The Danckwerts family could not have been more supportive: Peter's brother Richard, his sister Diana, nieces and nephews Ingrid and Peter Michael Danckwerts, Ian and Liz Garson, Mike Gatehouse—they all gave me wholehearted help and I never had the feeling that I might be intruding. Each of Peter's step-sons contributed too; Mike got me off to a flying start, Brian's encouragement stemmed from a deep admiration for his step-father and Charles prepared a painstaking review of his mother's correspondence, not to mention a chronological list of family dogs. Anne Crole and Patsy Watkinson, daughter of Jane Plunkett, kindly shared with me their memories of the young Lavinia Macfarlane.

Miles Kennedy, as Peter's only doctoral student during his first stay in Cambridge, was a rich source of technical and social information on Cambridge immediately post-war in innumerable e-mail conversations. Sydney Andrew, Geoff Place and Bill Wilkinson also illuminated the early 1950s when they were undergraduates. Man Mohan Sharma and Tony Gillham helped enormously in my attempts to do justice to Peter's second period in Cambridge department (1959-74) when everyone who was anyone wanted to visit and most did.

Amongst Peter's friends and contemporaries, Ray Baddour summoned up the MIT practice course on which he and Peter were colleagues. Peter Gray recalled the early days of the Cambridge department, advised me of fruitful contacts, kept my nose to the grindstone and was not discouraged even when confronted with early proofs. John Davidson and David Harrison also shared their memories of department history. Roger Sargent and Peter's former secretary Emma Maccall conjured up life at Imperial College London in the late 1950s. Anthony Pearson explained to me some of Peter's insights into chemical engineering science, while Anthony Barber gave me sympathetic help in presenting them. Nikolai Kulov came to the rescue on Peter in the Soviet Union, and Hal Hopfenberg brought Peter's North Carolina sabbatical year to life.

I owe a special debt to several others. Anthony Gaskell gave me access to his

late father's five-year diaries with their laconic references to Peter in Combined
Operations days. Naomi Canning generously provided commentary on Peter's
social life in the early 1950s. Noel Cashford initiated me into the world of
parachute mines and Geoff Mason spent an inordinate amount of time researching
and sorting out the detail of WW2 mines and bombs while pretending he was doing
it for himself. Rod Boroughs, who kept an eye on Lavinia for more than a decade
after Peter's death, rescued vivid fragments of Danckwertsiana from the skip
outside Abbey House when she died in 2000 and gave me access to them. Dick
Fifield recalled his dealings with Peter in the 1980s from his desk at *New Scientist*
and lent me some contemporary files which he just happened to find upstairs in
his attic. My old friend Liesl Krammer da Silva deployed her Viennese contacts to
help with the mysteries of Peter's gap year in Austria, and Winchester College's
exemplary archivist Suzanne Foster shed light on Peter's teenage schooldays.
Late in the day Richard Darton alerted me to 'The Standart affair' and
Jaromir Ulbrecht gave me chapter and verse.

Peter wrote wonderfully well, and I was very keen to let him speak directly to
the reader, so I am particularly grateful to the current *New Scientist* editor-in-chief
Jeremy Webb who has allowed me to quote liberally from Peter's Indian summer in
the columns of that magazine. He considered this preferable to leaving the words
'hidden in yellowing copies of the magazine in libraries around the country'. And
at the last Nigel Hirst, Miles Kennedy and my college friend and favourite pedant
in matters touching on the use of English, Tony Porter, gave the text a final read-
through before the pages went to press.

Many others contributed and I thank them all. But the manner of presentation
and interpretation of it all, plus any errors introduced on the way, are entirely
down to me.

Peter Varey
Cambridge, April 2012

Index